About the editors

John-Andrew McNeish is associate professor at the Norwegian University of Life Sciences (NMBU) and senior researcher at Chr. Michelsen Institute (CMI). His research focuses on indigenous politics and participation in resource politics and governance. McNeish has authored and co-authored a number of publications including *Flammable Societies: Studies on the Socio-Economics of Oil and Gas* (Pluto Press, 2012), *Gender Justice and Legal Pluralities: Latin American and African Perspectives* (Routledge, 2013), and *Indigenous Peoples and Poverty: An International Perspective* (Zed Books, 2005). He is currently carrying out research for the Norwegian Research Council-funded *Extracting Justice* project.

Axel Borchgrevink is associate professor at Oslo and Akershus University College (HIOA). He is an anthropologist who has considerable international consultancy experience and has worked on a range of development issues in Africa, Asia and Latin America. He is a former co-editor of the journal *Forum for Development Studies*, and his book *Clean and Green: Knowledge and Morality in a Philippine Farming Community* was published in 2014 by Ateneo de Manila University Press.

Owen Logan is a photographer and research fellow at the University of Aberdeen, where he worked closely with the 'Lives in the Oil Industry' oral history project. Between 2007 and 2014 he was a contributing editor to Variant magazine and is co-editor with John Andrew McNeish of *Flammable Societies: Studies on the Socio-Economics of Oil and Gas* (Pluto Press, 2012). His work as a photographer has been widely exhibited and his images are in several public collections, including the Scottish Parliament. In connection with the *Contested Powers* project he co-curated, with Kirsten Lloyd, the exhibition *The King's Peace: Realism and War* at the Stills Gallery in Edinburgh in 2014.

CONTESTED POWERS

THE POLITICS OF ENERGY AND DEVELOPMENT
IN LATIN AMERICA

*edited by John-Andrew McNeish, Axel Borchgrevink
and Owen Logan*

Zed Books
LONDON

Contested Powers: The politics of energy and development in Latin America was first published in 2015 by Zed Books Ltd, Unit 2.8, The Foundry, 17 Oval Way, London SE11 5RR, UK

www.zedbooks.co.uk

Editorial copyright © John-Andrew McNeish, Axel Borchgrevink and Owen Logan 2015
Copyright in this collection © Zed Books 2015

The rights of John-Andrew McNeish, Axel Borchgrevink and Owen Logan to be identified as the editors of this work have been asserted by them in accordance with the Copyright, Designs and Patents Act, 1988

Set in Monotype Plantin and FFKievit by Ewan Smith, London NW5
Index: ed.emery@thefreeuniversity.net
Cover designed by Ed Marshall

All rights reserved. No part of this publication may be reproduced, stored in a retrieval system or transmitted in any form or by any means, electronic, mechanical, photocopying or otherwise, without the prior permission of Zed Books Ltd.

A catalogue record for this book is available from the British Library

ISBN 978-1-78360-093-9 hb
ISBN 978-1-78360-092-2 pb
ISBN 978-1-78360-094-6 pdf
ISBN 978-1-78360-095-3 epub
ISBN 978-1-78360-096-0 mobi

Printed and bound by CPI Group (UK) Ltd, Croydon, CR0 4YY

CONTENTS

Tables and figures | vii Map | viii

1 Introduction: recovering power from energy – reconsidering the linkages between energy and development 1
   John-Andrew McNeish and Axel Borchgrevink

2 Oil extraction and territorial disputes in the Maya Biosphere Reserve 40
   Virgilio Reyes

3 *Gracias a díos y al gobierno*: electric power struggles in Nicaraguan politics 66
   Axel Borchgrevink

4 Wind at the margins of the state: autonomy and renewable energy development in southern Mexico. 92
   Cymene Howe, Dominic Boyer and Edith Barrera

5 Oil and environmental injustice in Venezuela: an ethnographic study of Punta Cardón. .116
   María Victoria Canino and Iselin Åsedotter Strønen

6 'Everything moves with fuel': energy politics and the smuggling of energy resources 141
   Cecilie Vindal Ødegaard

7 The continuous negotiation of the authority of oil- and gas-dependent states: the case of Bolivia 167
   Fernanda Wanderley

8 Passive revolution? Social and political struggles surrounding Brazil's new-found oil reservoirs 195
   Einar Braathen

9 Doing well in the eyes of capital: cultural transformation from Venezuela to Scotland. 216
   Owen Logan

10 Latin America *trans*formed? 254
   John-Andrew McNeish

11 From the King's Peace to transition society. 291
   Owen Logan and John-Andrew McNeish

About the authors | 313
Index | 316

TABLES AND FIGURES

Tables
2.1 Protected wetlands of Guatemala. 47
2.2 Recoverable costs for oil extraction 57
8.1 Congress's decision on the distribution of royalties between the different territorial categories. 206

Figures
2.1 Ratio between royalties and recoverable costs of oil in Guatemala 58
4.1 Isthmus of Tehuantepec, Oaxaca, Mexico 93
9.1 Italian street musicians photographed by John Thomson . . . 220
9.2 Palma de Mallorca. The creativity of 'beggars' who perform in Spain is celebrated in an award-winning blog, *My Cruise Stories*, by Lois Beath 221
9.3 Photomontage by Karel Hajek (1934) 222
9.4 Onlookers at a street performance in 2013 about crime and sinfulness by an evangelical church in Caracas 233
9.5 Onlookers during the news digest on *Petare al Dia*, broadcast by Petare TV in 2013, in an open-air studio 239
9.6 Musicians and production crew on *Petare al Dia* 240

1 | INTRODUCTION: RECOVERING POWER FROM ENERGY – RECONSIDERING THE LINKAGES BETWEEN ENERGY AND DEVELOPMENT

John-Andrew McNeish and Axel Borchgrevink

A country without energy has no future. Simply put, a country without energy remains paralyzed; in a country where they fail to develop energy generation there are no investments ... Without energy they cannot publish the newspapers that are circulated in our country. Without energy they cannot transmit television programs, sports, culture, soaps, films, and politics ... nothing on television. Without energy there would be no radio in our country. The country would be completely silent ... (Daniel Ortega, president of Nicaragua, 20 March 2010)

Peru has enormous wealth in the mountain ranges because of the rainfall. It is estimated that 800 billion cubic meters of water fall on the mountains every year and flow down in the rivers towards the Pacific and Atlantic oceans. How can we make the most of it? Now that the price of oil has increased and will continue to increase, we should think of hydroelectric energy, which is renewable, almost unlimited, and clean ... it has to be done by large private or international capital that needs very long-term security to invest billions and to be able to recover the investment. But the dog in the manger says, 'Why should they make money off our waterfalls?' (Alan Garcia, former president of Peru, 28 October 2007)[1]

Extractivism is not a destiny, but may be the point of departure to conquer it. Certainly it can be found condensed in all of the world's territorial divisions – much of which is colonial. And to break with this colonial subordination it is not enough to fill the mouth with insults against this extractivism, to stop producing and sink the population into further misery, to return to rights without modification and partial satisfaction of the basic needs of the population. This is precisely the trap of inflexible critique in favor of opposing extractivism ... (Alvaro Garcia Linera, vice-president of Bolivia 2012: 106–7)

Introduction

This book examines how energy has been converted into political and economic power in recent times in Latin America, and the consequences this mode of acquiring power has for society, politics and economics in the region. As may be gleaned from the statements above, struggles over the extraction and exploitation of energy resources to a large extent define the controversial nature of contemporary Latin American politics. Reflecting on societal development, we are particularly conscious of the ways in which our environments and natural resources structure our societies and the way societies also change the environment. It is not merely technology, but the politics, social struggle and institutional formations that occur in the dynamic relationship between humans and the environment which are transformative.

In suggesting this, *Contested Powers* challenges the orthodox treatment of energy resources in development studies on two levels. This volume demonstrates that contests over natural resources refract the international, national and sub-national fault-lines of sovereignty. Indeed we see that in many cases these contests involve attempts to define the character of popular sovereignty, in Latin America and elsewhere. The case studies in subsequent chapters cast light on the historical meanings and socio-economic values which underpin and articulate what we call resource sovereignty. Secondly, we argue that it is vital to see the varied articulations of resource sovereignty in the larger context of a global division of labour and nature.

The model of an economy based on knowledge and creativity popularized by market economics is the polar opposite of resource sovereignty insofar as it disguises more complex dependencies on labour and natural resources at home and abroad. Given these insights, the book calls into question simplistic assumptions, i.e. that the exploitation of energy resources equals development and modernity, and the opposite idea, that natural resource wealth is a 'curse' on national development. We argue that these overly optimistic and overly pessimistic assumptions are two sides of the same technocratic coin. At a time when fossil fuel extraction and use are widely recognized as factors promoting global warming, it is crucial to arrive at a more realistic understanding of the dynamics of energy struggles.

Contested Powers is also intended as a necessary corrective to the slogan 'think globally act locally'. Our research in Latin America leads us to stress the constraints of local environments and cultural habitus (Wacquant 2005). Transformations at the local level need to be understood in relation to expressions of power at an international level. The book is structured to reflect this critical theoretical orientation. As such it moves progressively between the fractured politics of citizenship at the sub-national level and on to efforts to make states more socially and politically coherent. Historically, states and political systems have used energy infrastructures as a means to make nations more coherent (Mitchell 2011). We argue that a pivotal issue for the fate of nations in the twenty-first century will be the dynamic interaction between local and international politics surrounding the production and consumption of energy.

This introductory chapter presents the theoretical foundation of the book (and the joint research project on which it is based). It starts by outlining our position, in contrast to other more mainstream approaches. In order to ground the theoretical discussion, this is followed by a broad overview of energy struggles in Latin America. Thereafter, we review existing literature relevant to our project, with emphasis on the two bodies of literature we have found to be particularly inspiring: anthropological approaches to the study of energy and the recent development of critical institutionalism. The introductory chapter ends by describing the research project on which the book is based and giving a brief outline of the remaining chapters.

Resource sovereignties

There are several powerful discourses directly and indirectly applied to the field of energy development and resource management.[2] In the following we refer to three of them, i.e. the 'energy equals development' perspective, the 'resource curse' thesis, and lastly the 'creative economy' model. These three perspectives yield radically different conclusions about the importance of and potential for using energy resources for sustainable development. Yet we take issue with all three – for being technocratic, for having too simplistic a view of politics and for misrepresenting the relationship between the political and the material.

While policy documents and business statements often recognize the paradoxical risks that come with resource wealth, the overwhelming emphasis made by today's policy-makers and business leaders is that increased energy extraction should not be a matter for political discussion (Schaffer 2009). In assessing the merits of various pipeline and energy production projects, companies and governments consistently warn their counterparts to stick to 'commercial considerations' (ibid.: 1). Despite concerns about peak oil and the diminishing supply of other carbon resources such as coal, on the one hand, and the threats of global warming stemming from our carbon-dependent economy on the other, research centred on 'energy efficiency' and exploration sustains the notion that solutions lie in technical innovation. Indeed, in most policy discourses and industry statements there is a common tendency to conflate energy and development uncritically,[3] thereby reducing the challenge of development to a technological problem of developing the scientific and institutional capacity for effective energy production. In this sense the route to energy development is one of investment, of technical expertise and institutional establishment and stability. Insofar as this model of development also seeks to avoid, or to replace, fundamental political conflicts over the exploitation of natural resources it is technocratic – the goal of technocracy being to increase efficiency within the existing structures of power.

Yet underlying the production of energy a number of complex and incendiary political issues are evident. First among them is a variant of social territorialism expressed in what we call *resource sovereignty*, namely the controversial and uncertain ability of nation-states, or emergent polities within them, which seek sovereignty over territory to manage territorial resources in the public interest (McNeish and Logan 2012). Controversially the public interest in this context is not always identical to that of energy consumers either within or beyond such territorially based polities. More problematically still, the desire for resource sovereignty, whether seen in an increasingly dis-United Kingdom, a dangerously fragmented post-colonial Nigeria or a pluri-national Bolivia, is often at odds with the state's governing structures, and this means that state structures of power and accountability need to be devolved, rebargained or officially pluralized if national sovereignty is not to be called into fundamental question. In this sense resource sovereignty may often be regarded as a jagged proto-nationalism capable of threatening the officially sanctioned nationalism of states with significantly different formative histories, including of course those at the two ends of imperialism – colonizing and colonized powers (Blom Hansen and Stepputat 2001).

We argue that it is vital to see varied articulations of resource sovereignty in the larger context of a global division of labour and nature (Coronil 2000). As we suggested above, the model of an economy based on knowledge and creativity popularized by market economics since the 1990s is the polar opposite of our perspective on resource sovereignty. The commoditization of creativity and knowledge beneath the banner of a creative economy is proposed by governments and regeneration consultants alike as the post-industrial answer to dependency on labour and natural resources. It is argued that an economy increasingly focused on creativity, knowledge and intellectual property is also a more environmentally friendly economy. The core of this perspective is strongly expressed by the regeneration guru Richard Florida. He claims that 'Knowledge and creativity have replaced natural resources and the efficiency of physical labour as the sources of wealth creation and economic growth' (Florida 2005: 49). Looked at critically, however, Florida's diminishment of dependency on natural resources and manual labour, either at home or abroad, is extremely difficult to sustain (Peck 2005: 757). In *Contested Powers* we argue that the quick fixes offered by discourses of creativity and knowledge are bluffs, and in their own right will not lead to a greener economy. Rather, these discourses rearticulate the deeply inequitable global division of labour and nature which continues to hinder technological development in the global South by putting the capacity for important technological advancements – not least in energy production and consumption – in private hands.

Likewise we take issue with the technocratic and rational-choice foundations of another popular understanding, the so-called resource curse thesis. While research has provided considerable evidence that natural-resource abundance is often associated with various negative development effects, the relationship is nowhere near as conclusive, or as direct, as the terminology of a 'curse' suggests (Rosser 2006). Researchers have been too quick to posit a limited set of variables for understanding and responding to the dynamics surrounding resource governance. As a result an overly deterministic relationship has been characterized between natural resource abundance, *Homo oeconomicus* pathologies and various negative development outcomes. In responses to resource challenges, 'resource curse' theorists appear to ignore issues of ideology, history and political feasibility to account for the role of social forces and external political and economic environments in shaping development outcomes.

Recognizing the weaknesses of the resource curse thesis, and in particular the technocratic weakness of key writers in this field, e.g. Collier (2000, 2010), efforts are now being made to expand political ecology and political economy perspectives on resource governance so that they truly capture the political relationships involved in resource wealth governance, and acknowledge the social complexities and importance of historic grievances in shaping the present (Omeje 2008; Stevens and Dietsche 2008; Rosser 2006). An extension of theory and methodology away from rational-actor models and beyond quantitative research also heralds a return to a consideration of power and social class, and of imperialist relations and discourses. Importantly, parallels can also be made here between the 'non-renewable' literature focused on the operation of the resource curse, and the more 'renewables'-oriented research focused on the role of institutions in environmental governance. Indeed, the recent critique of the resource curse mirrors to a large degree the theoretical and policy-oriented moves that have been made in recent years from 'mainstream' to 'new' and more recently 'critical institutionalism' perspectives on environmental governance (Cleaver 2012). Below we give a brief review of this evolution in approaches. Here, it is sufficient to point out that the coincidence between developments in the resource curse and environmental governance literatures not only provides support for a critique of technocratic and rational-actor perspectives, but is suggestive of the need to extend the theory and practice surrounding resource and energy governance in a similar direction. They are mutually supportive of the importance not only of individual agency, but social agency over time and of a movement away from rational-actor theory to social constructivism. Moreover, critiques of the resource curse thesis and of institutionalism in governance are also indicative of possible directions needed to improve analysis and practice.

We stress that our critical focus on energy politics is crucial to wider debates about modes of development and sustainability. This is because energy resources are converted into political power in complex ways. Chapter 11 will return to this issue. Here it is worth pointing out that only some of the conversions of energy resources into political power are immediately obvious. For example the rise to power of the late Hugo Chávez, a political movement strongly underpinned by the radical redistribution of Venezuela's national hydrocarbon patrimony. Yet in Venezuela too, where a clear social contract was formed and its basis personified by President Chávez, it is evident that the possession of hydrocarbon reserves or other energy sources was not in itself a source of power. Other socio-economic forces are at work to compound their political value (Strønen 2014). Resource sovereignty is part of a political sequence, the ends of which will always be uncertain.

In his book *Carbon Democracy* (2011), Mitchell demonstrates that energy extraction and the development of governance structures are mutually constitutive in the course of history. He argues that no nation escapes the political consequences of our collective dependence on oil. It shapes the body politic and has created both the possibilities and limitations of modern democracy. Recognizing this does not mean deterministically tracing political outcomes back to forms of energy production, but investigating the material ways in which carbon must be transformed, beginning with the work done by those who bring it out of the ground. These vital transformations involve establishing connections and building alliances – which do not respect any divide between material and ideal, economic and political, natural and social, human and non-human, or violence and representation (ibid.: 7). By ignoring the apparatus of oil production, where human decision-making is vital, a particularly narrow conception of democracy is brought into being. Democracy is presumed, as in an oil-engineering project, to be a set of procedures and political forms that can be reproduced everywhere. As such, democracy has become an abstract idea, free from local histories, circumstances and material arrangements. In line with Mitchell's approach we study their histories and the ways in which these human transformations take place at different levels. Importantly this includes, in parallel to Mitchell, a critical concern with the machinations of imperial and global financial claims to power and sovereignty, but also a more specific concern with regional, national and local socio-economic transformations (prototypical ethical conflicts over certain human cultural values versus global market values). There is no doubt that energy resources are an incendiary factor in politics, and even when orchestrated from above, resource sovereignty highlights the shortcomings of public accountability and rights within supposedly democratic systems. Ultimately what becomes evident for everyone is the disconnection between democracy and the modern sense of civilization (based on human rights) which democracy is said to uphold. Yet, as we discuss in the concluding chapter, the production and distribution of energy also provide a crucial means of cohesion in such fractured polities.

From the case studies in this volume and our research on the social politics of energy more broadly we see a complex series of interdependencies and social contracts at play. This reality demands a more pluralized discussion of 'sovereignties', in line with the understandings developed by Blom Hansen and Stepputat (2001). A pluralized and multilayered understanding of sovereignty encourages us to question rigid accounts of the relationship between carbon and democracy, energy and development, or energy and political expression. If anything the social struggle for *sovereignty* is even more multifaceted than the economic struggle for competitive advantage and leverage over labour and nature.

Thus we explore the meaning of resource sovereignties as a concept and its connection to social territorialism in a diverse array of empirical cases. Energy resources, a wider material and at times more esoteric grouping of human interests than natural resources owing to their physical and non-physical properties, are equally if not more prone to be sites upon which complex claims and struggles for use, belonging, ownership and sovereignty are made. In summary, efforts to understand energy need to capture what is fundamentally a social relationship. A question we need to ask, hinted at in the quotations at the outset of this chapter, is whether we value the energy we can take from our surroundings more or less than the means of extraction. For example, should oil be as 'cheap' as an extractive industry can make it, or should it be as 'expensive' as market controls can make it? And what are the long-term consequences of such moves? With such questions in mind energy politics ought to be seen in relation to people's ways of life which are socially developed over time, in other words against our habitus. In many ways we can see that energy politics articulate the tension between expressions of habitus and those larger settings and power structures that remove crucial economic decision-making to the transnational level.

Contested powers in Latin America

While generally considered a middle-income region with abundant energy alternatives, and where over the last two decades great advances have been made in securing economic and political stability, Latin America remains a region where there are clear challenges in ensuring that the benefits of wealth reach the poorest sectors of the population.[4] Given this uneasy balance we propose in this book that study of the region's energetic inner workings provides a relevant and promising perspective on what are as yet unclear and paradoxical global processes. As well as being a region where the boom in natural resource commodity prices has led to a dramatic expansion in energy resource extraction, it is also a region that stands out from others in terms of the number and diversity of proposed and actual energy development projects, and its experimentation and innovation in energy generation, including extensive use of renewable energy alternatives. Controversy and contestation can be seen to follow energy development projects wherever they are established, and they are often responsible for uncovering and renewing long-standing social cleavages and conflicts. Moreover, issues of access to, distribution and pricing of energy – whether electric or combustible – have also been highly contested throughout the region. Indeed, energy issues play a major role in political processes in Latin America, and are clearly at the centre of many of the current political debates and conflicts in the region (see McNeish this volume).

Oil has been produced in several Latin American countries since the early twentieth century, with Brazil, Mexico and Venezuela being the current largest producers. Colombia and Argentina have significant petroleum production, and the sector is of special importance for Ecuador, accounting for about 50 per cent of the country's export earnings. Likewise, hydrocarbons have a similar share in the value of Bolivia's exports, with natural gas making up the largest part. The real 'petro-state' of the continent, however, remains Venezuela, where oil and gas make up 90 per cent of exports. In *The Magical State: Nature, Money and Modernity in Venezuela,* Coronil (1997) described how the promise of oil wealth shaped public fantasies, myths of progress and the Venezuelan state itself. As Gledhill (2008) has shown, similar effects can be seen also in countries where oil accounts for a smaller fraction of national production. Thus, he argues, oil is at the centre of 'popular imaginaries' also in countries such as Brazil and

Mexico, for instance closely tied to particular forms of nationalism and claims to national sovereignty against perceived US imperialism (see Gledhill 2011: 166). A striking case is Gledhill's account of poor Mexicans queuing up to donate their jewellery and other valuables to help the Mexican state pay compensation to the foreign companies affected by the 1938 nationalization of the country's petroleum resources (ibid.: 174).

Such popular imaginaries of hydrocarbon resources shape the way that conflicts over petroleum policies and projects develop, and the way that different actors and groups are aligned in alliances or confrontations. The special position of the petroleum industry in the public mind was undoubtedly the main reason why it was shielded to a great extent from the wave of privatization of state companies that rolled over the continent throughout the 1980s and 1990s. Although this sector experienced some liberalization, Latin American state petroleum companies such as Mexico's Pemex, Venezuela's PDVSA and Brazil's Petrobras maintained their central role.[5] Widely shared notions of oil and gas as national patrimony are obviously at the heart of the current attempts to use hydrocarbon resources for the benefit of the whole nation – including the poorest – in Chávez's Venezuela (Logan and Canino & Åsedotter Strønen in this volume) and Morales' Bolivia (Wanderley this volume). They similarly fuel Brazilian dreams of becoming an energy superpower on the basis of its recent offshore discoveries (Braathen this volume).

However, while there are widespread expectations of the benefits that hydrocarbon resources can bring to nations, there are also observations and strong feelings that in actual practice too much resource wealth is siphoned off by foreign and regional elite interests. The discrepancy between state-sanctioned imaginings of what could be done with sovereign energy resources and the perceived reality create resentments and spur resistance and political action throughout the region. This is evidenced, for instance, by the strong feelings evoked in debates either over nationalization or privatization of hydrocarbon resources. Popular protests such as the '*gasolinazo*' (Wanderley this volume) and the Territorio Indígena y Parque Nacional Isiboro Secure (TIPNIS) controversy (McNeish this volume) in Bolivia were motivated by such contrasting claims of ownership and legality. Fuel smugglers in Peru also justify their actions with reference to similar statements and arguments (Ødegaard this volume).

The so-called left turn in Latin American politics in the last decade owes a great deal to this dissatisfaction and to increasing public perceptions that dominant neoliberal policies have not been able to spread the benefits from the exploitation of the region's resources, and that fairer policy alternatives are called for. However, as the chapters on Venezuela and Bolivia in this book amply demonstrate, even if there is agreement on rejecting the neoliberal prescriptions, there is no easy consensus on alternative economic models or development alternatives. While many of the new platforms for energy policy in the region are influenced by anti-imperialist and socialist ideas of state regulation and the improved redistribution of wealth and services, there are as many levels of disagreement within these contexts of change as there are within those countries remaining faithful to neoliberalism. Indeed, tensions and radical disagreements over energy policy in the region appear to be rising with increasing frequency in Latin American politics and society despite the expressed contrasting political persuasions of their governments.[6]

Friction commonly arises between those living in the areas where the energy resources exist and those wishing to extract or harness them, whether they be national governments, multinational corporations or other external actors. Extraction raises the thorny issue of who owns or should benefit from the resources. In Bolivia this question resulted in calls for autonomy and secession as lowland gas-producing regions clashed with the central government over the distribution of hydrocarbon rents. Elsewhere, local populations have protested against the environmental costs of hydrocarbon production. Indigenous groups in the Ecuadorean Amazon have fought a landmark lawsuit against Chevron, claiming compensation for contamination of the environment and consequent health problems caused by the company's oil extraction (Sawyer 2007). However, such struggles are not usually successful. Reyes (present volume) describes, for example, how Guatemalan oil production in a supposedly protected area has caused serious negative repercussions both for local peasants and for the environment. In Venezuela, the small-scale fishermen described by Canino & Åsedotter Strønen (present volume) have been largely powerless to halt the damage to their livelihood that the activity of the nearby petroleum refinery implied. The TIPNIS case, described by McNeish (present volume), demonstrates how competing claims for protection of territory and environment as well as for development have not only challenged state attempts to expand the construction of infrastructure and hydrocarbons exploration, but have revealed a series of serious and persistent social cleavages in the country.

In many cases, areas of potential for hydrocarbon extraction or hydropower development overlap with territories traditionally belonging to indigenous peoples. Thus, claims of indigenous rights to territory and self-determination contrast directly with the powerful economic and development interests of accessing these resources. While most Latin American states have ratified the International Labour Organization (ILO) convention 169 on the rights of indigenous and tribal peoples, and thus in principle recognize indigenous rights to traditional territories, these rights are seldom respected in practice, particularly when there is significant potential for profitable oil and gas exploitation.[7] For example, around three-quarters of the Peruvian Amazon – home to numerous indigenous peoples, including what is probably the largest number in the world of isolated Indians – is under contract for oil and gas exploration to international companies (Finer et al. 2008). On the one hand, this has led to protests, demonstrations and legal challenges from national and indigenous organizations.[8] Parallel and in response to the destabilization resulting from these events, there have, however, also been efforts in recent years to find ways to make operational the principle of free and prior informed consent (FPIC). Prior consultations with affected population groups, in order to seek consensus on the development of extractive activities, including oil and gas exploration, have now become part of the standard practice of resource governance in the region (Schilling-Vacaflor 2013).

In Latin America the slogans of new policy directions in energy management can also be seen to struggle in the regional context of complex 'legal pluralism', i.e. the coexistence and overlap of a mixture of legal orders including national, religious and 'customary' or 'traditional' law.[9] At a certain level market-led ideas of individual material well-being, just like more socially formulated radical ideologies as well as indigenous cosmologies which closely identify with the natural environment, all make claim to be holistic and to ultimately encompass the common good. Indeed, the political discourses surrounding energy often appear as short cuts to this greater good, but the exploitation of natural resources for the production of energy is contingent on immediate goals and vulnerabilities across all levels of individual activity and social organization.

A politics of renewables

Latin American energy governance is of course not limited to hydrocarbons.[10] Renewable sources of energy have long been important on the continent, primarily in the form of hydropower, while in Brazil the production of ethanol fuel has been an important strategy aimed at ensuring energy sufficiency. As oil prices rose over the last decade and international concern over global warming spread, renewable energy production increased throughout the region, and expanded to include new forms of bio-fuels and wind power, as well as to solar and geothermal production. However, it is clear that renewable energy in itself is no panacea for ending the social conflicts over energy issues. Current means of creating renewable energy are not necessarily sustainable, particularly now in a moment in which as a result of geo-political maneuvering oil prices have fallen.

Many of the same problems and cleavages described for oil and gas extraction surround hydropower production. Reflecting the ample water resources of the region, Latin America is the world region where the highest percentage of electricity production comes from hydropower. While the world average is around 16 per cent, Latin America generates roughly 65 per cent of its electricity from hydropower. Moreover, the potential for further hydropower development remains huge, and new mega-projects are under way in Brazil, Chile and Venezuela, among other countries. Yet these developments have carried considerable environmental and social costs. Large-scale dams result in the displacement of people and the destruction of resources and ecosystems they have depended on. The World Commission on Dams (WCD) documents a consistent history of displaced people not being properly compensated for their losses.[11] The WCD has also developed a set of recommendations for ensuring that affected populations are consulted and compensated, and that environmental consequences are properly assessed, which have made governments, international funding institutions and investors more careful. Yet new hydropower projects are invariably controversial. The Belo Monte project in Brazil, which if completed will be the world's third-largest hydropower complex, aptly illustrates this. It has pitted local populations – indigenous and non-indigenous – and national and international environmental and rights organizations against power companies and a government focused on growth and economic development. While the project has been slowed by a series of mass mobilizations, and experts are highly critical of the environmental and social impact assessments for the project, several court cases and workers' strikes, the Brazilian federal government has been determined to move ahead with the project. The Belo Monte scheme, and many others like it, demonstrates that even if the energy is renewable, there is the same potential for conflict as found in non-renewable cases. This should perhaps not surprise us given the same context of economic modernism and a renewed global tendency towards capitalist accumulation by dispossession (Harvey 2007).

The expansion of biofuels production – in Brazil as well as throughout the continent – is also hotly contested. While hailed as an important step against the emission of greenhouse gases, its environmental and social consequences are vehemently questioned. Critics claim that the cultivation of biofuels comes at the expense of the production of staple foods, thus reducing local and global food security. It is noted that the expansion of biofuels production is accompanied by 'land grabbing' or land acquisitions, resulting in the marginalization of small farmers. In the Amazon, biofuels cultivation is indirectly also a major cause of the extension of the agricultural frontier, leading to deforestation and biodiversity loss. Wind power is another 'green' form of energy production encountering local opposition. The largest expansion of wind power in the world is presently taking place on the Tehuantepec peninsula in Oaxaca, Mexico. The peninsula has been identified as one of the world's most ideal locations for this kind of energy production owing to the natural current of air that blows across the area from the Gulf of Mexico to the Pacific. While foreign companies have a key role in this development, the process is strongly backed by the Mexican federal government. However, it is seen by the local population as the forced and little-compensated takeover of common and agricultural land, and resistance to the wind-park project is growing. Protests by the local population have resulted in death threats being received by community leaders and physical confrontations with the federal police (see Howe, Boyer & Barrera this volume).

Beyond national and regional borders

Energy distribution in the region is also an issue that links national policy debates to the regional level. Within Latin American countries such as Nicaragua (see Borchgrevink this volume) the formation of the national energy grid and its physical distribution and delivery have prompted heated discussions. A national electricity grid is a powerful symbol of the internal socio-political coherency of the modern nation-state. The fact that large population groups are not connected only underlines the partial reality of modernity. Efforts to expand the grid are at the same time attempts at integrating the nation and at extending the state's reach. Imbalances in coverage and delivery raise questions about the linkages between energy poverty and expressions of political clientelism and corruption in different national contexts. The question of pricing of energy services – whether of electricity or fuel – pits market logics against fundamental claims of the rights to energy, and has led to massive protests in many Latin American countries, including Bolivia, Peru and Nicaragua, as described in this volume.

Energy contestation takes place not only within national borders. There are important international dimensions. Exploration for new hydrocarbon resources and plans for the construction of large-scale dams have on several occasions over the last decades led to border disputes and militarized action between Latin American countries.[12] Over the last decade there have been tense diplomatic relations between Bolivia and Chile rooted in Bolivia's historic grievance about their loss of access to the sea in the War of the Pacific, and also in Bolivia's contemporary aspirations to build a pipeline and be able to ship gas from the port of Arica. The smuggling of fuel from Bolivia to Peru, as documented by Ødegaard (present volume), is another instance of the way effects of national energy policies may flow across borders.

Clearly one country's energy policies have repercussions on others. Governmental decisions to encourage the distribution and sale of energy resources across borders, decisions to nationalize natural resources or new levels of taxation on energy companies affect trading partners and the countries' position in the global system. While policies of this sort in Bolivia and Venezuela were considered controversial when first carried out, it is of some note that they have not led to significant disruptions in terms of private contracts and investments – reflecting perhaps the power of oil- and gas-producing countries in a period of high international fuel prices and dwindling world reserves. Venezuela under Chávez used this power for an active foreign policy that has not only demonstrated the country's autonomy vis-à-vis the traditional influence of the USA, but has used its energy resources to significantly diminish US influence over the continent as a whole. While the regional Bolivarian Alliance for the Peoples of Our America (Alianza

Bolivariana para los Pueblos de Nuestra América – ALBA) may not have had the impact Chávez hoped for, it has been of fundamental importance for the left-leaning governments of Cuba and Nicaragua and an important symbolic counterweight to US political and economic intervention and influence. The Banco del Sur, a regional bank planned by member countries of ALBA, established an economic alternative to that of the Bretton Woods institutions. In competition with Venezuela, Brazil has also used the sale and distribution of its energy resources to strengthen its political power in the region. Indeed, Brazil's expansion of the trade in its energy resources, investments in energy production and granting of economic and technical assistance throughout Latin America and farther abroad to lusophone countries in Africa have helped establish Brazil as a leading emerging power. The exploitation of the recently discovered *pre-sal* oil deposits off its coast is also likely to provide Brazil with further influence as an economic powerhouse in coming years.

In sum, then, energy governance and politics are not only keys to understanding Latin American national economies, development strategies, social movements, political conflicts and environmental problems. Energy governance and politics are also crucial forces in international relations within the continent as well as the rest of the world.

Development, modernity and power

In recognizing the interplay between resource sovereignties and expressions of social territorialism in Latin America and more broadly, we demonstrate that development is not a given, whether through the development of energy resources or otherwise. At the same time our aim here is not to argue that there is no relationship between energy and development. Clearly, improved access to energy may lead to increased productivity, economic growth, higher incomes, better health, education and transport services, as well as other benefits associated with development. Rather, we recognize that the connection between energy and development, as with democracy, must be nuanced to the point where competing ideas of development come to the fore. At the level of social theory this competition highlights differing notions of both *stateness* (Blom Hansen and Stepputat 2006) and *modernity* (Latour 1993).

By revealing the social nature of material cultures, writers such as

Pieterse (1998) and McMichael (2011) offer a deeper understanding of economics and development whereby development is not only anchored in formal institutions and structures, but also in the lives of its subjects. Our work draws attention also to how subjects of energy projects legitimize or contest them; and thus attempts to demonstrate how energy development is accomplished, the terms through which it is challenged, and how new possibilities are formulated. In *Contested Powers*, as in *Flammable Societies* (McNeish and Logan 2012), this means that we draw on a theory of value which questions the universality of dominant European theories of accumulation, and we argue that the ultimate stakes of politics or social order are the struggle, not to appropriate value, but to establish what value is (Graeber 2005). Arguably shifts from the struggle to appropriate value from labour and nature to the collective struggle to interrogate valuation, and vice versa, is the movement of history.

Notwithstanding such an emphasis, it is also important to recognize that special claims for the existence of a 'good faith economy' which respects honour, virtue and true worth are likely to be ethnocentric (Bourdieu 2007: 159–97). Building on different strands of economic anthropology and sociology, we see economic or monetary value as intrinsically linked to – although not reducible to – other types of values: ideas of what is good and bad, right and wrong, desirable and undesirable. Value is constituted through social interaction, in the same large-scale and diachronic processes that have produced the social groups, differences, relationships and institutions that make up societies and states.

Modern secular and liberal democratic society is often described as a stage in history shaped by the development of science. In Enlightenment thinking it was claimed that scientific thought and discovery enabled us to distinguish clearly for the first time between nature and society, to organize collective life in stable and transparent systems of relations. The perfection of what remains for many an unquestioned ideal is, however, under increasing attack as it collides with other systems of knowledge and belief as a result of globalization, and it becomes more and more obvious that technical change does not remove uncertainty. Bruno Latour (1993) suggests that instead of the certainty of modernity, we should recognize that we have always inhabited a mixed world, made up of imbroglios of the technical, the natural and the human. As such he suggests: 'we have never been modern'. Any technical apparatus combines different kinds of materials and forces, involving various combinations of human cognition, mechanical power, chance, stored memory, self-acting mechanisms, organic matter and more. In this book we highlight that to talk about energy and development means dealing with socio-technical systems in which the social, natural and technical are intertwined. As such we combine an 'anti-essentialist' (Escobar 1999) approach to political ecology with political economy's concerns with the operation over time of economic institutions and structures of power at different levels.

Social scientists are by now only too well aware that knowledge is not evenly spread out, but is distributed in socially complex ways among positioned subjects. Obviously the same goes for our worldviews, i.e. the ways in which we make sense of and know the world. However, to the extent that they spring out of the same processes that generate groups and differences within society, their distribution will not be random, but will to some extent conform to the divisions of society. As will appear from our case studies, the competing parties in energy conflicts relate to and seek to legitimize their positions with reference to competing epistemologies. And the intractability of many of these conflicts is in part due to the breakdown of communication because of the incompatibility of, or friction between, the parties' epistemologies, in what Anne Tsing (2004) has termed 'zones of awkward engagement'.

Of course, understanding these zones of awkward engagement – as with any other form of conflict – requires an understanding of power and the structures through which it works. Despite globalization and changes to the state system we see that states remain the pre-eminent system for constituting power and shaping its exercise in modern society. Therefore, to understand energy conflicts in Latin America we need to understand how they are conditioned by the state. However, given the fact that there are particular characteristics to states and to processes of state formation in Latin America that may differ from other parts of the world, there are elements here that cannot be automatically generalized.

High levels of inequality are a common feature in most Latin American states, with large excluded sectors of the population receiving little if any of the benefits of economic growth, development and democracy over the last twenty years. This has been spoken about by the political left in different countries and described as a form of embedded *exclusion* with long historic roots, with both economic and political dimensions. The highly skewed structure of landholdings is one dimension, and results in a Latin American tenancy structure where in general the best lands are concentrated in relatively few but extremely large holdings, while remaining properties are too small or with too poor soils to easily support a family. Historically, this pattern has been implemented through the dispossession of indigenous peoples or other marginalized populations and displacement to more remote and less productive lands. While there is great variance between Latin American states in terms of the numerical importance of indigenous peoples and their economic and political conditions, they can nevertheless be seen as emblematic of the excluded: i.e. commonly scoring highest on most poverty indicators, such as income, health or education; until recently having had little participation in the formal political system of the state; experiencing persistent and systematic cultural and social discrimination.

In such a situation of great inequality, it is of course not surprising that the extraction of resources of great economic value generate conflicts, in particular when they are tapped or harnessed in the same remote and marginal ecological areas to which the excluded groups have been displaced. This needs also to be understood in the context of the changing face of violence and power. The intimate link between states and violence is far from simple. Theoretically it can be deduced from as diverse approaches as Weber's definition of the state[13] to Foucault's (1991) understanding of 'governmentality' as the conduct of conduct, and Agamben's (2005) concept of *Homo sacer*, where in 'states of exception' some sectors of the population are in the interest of security reduced to 'bare life', and thus deprived of any rights. There is also a significant body of literature that argues empirically that violence is much more pervasive and closer to the surface in Latin America than elsewhere, and that throughout the continent state power has been underpinned by the undisguised use of or threat of violence, and the consequent spread of fear, most evident in earlier years of military dictatorship and counter-insurgency. The many brutal dictatorships, the proclivity of the military to assume political roles, the pervasiveness of death squads, the brutal civil wars, and the willingness of political opposition to use violent methods are all indications of the violent nature of states in Latin America. Recent manifestations of violence linked to drugs and to criminal gangs, but also paramilitaries, private security companies and vigilantism, suggest, however, that while states and militaries continue to attempt to monopolize the use of violence, they are now joined and challenged in this endeavour by an increasing array of non-state or semi-state actors. Recognition of the different articulations of power, and the way they interact, is vital if we are to understand the changing identity and possibilities of states in the region and the complex dynamics and interests articulated by energy conflicts. We return to this issue later on in our discussion of energetic states in the concluding chapter.

An anthropology of energy?

As you read through the book it will increasingly be evident that *Contested Powers* is deeply influenced by anthropological and sociological approaches to the complex issues of development, modernity and power. This is due to the disciplinary background and persuasion of the majority of our project members. However, even for those who come from other disciplines such as political science, law, or the arts and humanities, there has been recognition of the strength of anthropology in attempting to disentangle and disaggregate the operation of politics. In particular an anthropological lens reveals the importance of moving towards and understanding of resource sovereignties and expressions of social territorialism. This is particularly true in relation to energy politics, where we recognize the head start made by anthropology in considering informal and non-institutionalized aspects of the struggle over energy resources. This does not mean, however, that we consider this book to fit neatly within the field of anthropology. Indeed, while existing anthropological ideas on energy and its politics are useful as inspiration and as a foil for sparring with, *Contested Powers* proposes an interdisciplinary approach that tests and in some ways defies disciplinary limits.

The existing anthropology of energy questions the assumptions of a neat separation between institutional, technical, scientific and social elements of energy observed in other disciplinary approaches. Earlier anthropological writing acknowledged the manner in which the wealth and conflict produced by resource and energy extraction enter processes of both national and local identity (Coronil 1997), and the way in which global and local culture collide and generate new cultural forms and expressions that attempt to make sense of, celebrate and counter the destructive nature of commodification (Taussig 1983;

Nash 1993 [1979]). Indeed, rather than treat energy as something apart from human life, earlier anthropologies of energy importantly acknowledge its value as an explanation of processes that occur both within our bodies and without (White 1959; Nader 1981, 2010). In doing so a bio-politics is combined with a realpolitik of energy where simplistic claims of development and modernity from energy cover the ulterior motives of imperial power and capitalist expansion. Boyer (2011, 2014) calls for a renewed examination of this pervasive and possibly purposeful opacity to energy politics through his coining of the term 'energo-politics', i.e. power over (and through) energy.

Boyer proposes that Mitchell's (2009) work offers an entrance into another way of viewing the genealogy of modern power and modern statecraft; different from if parallel to the much-analysed Foucauldian phenomenon of 'biopolitics' – power over life and population. He argues that power over energy has been the *companion* and the *collaborator* of modern power over life and population from the beginning. And from this perspective we may see that the proximity to power in the field of energy production invites the accumulation of political capital. Recently a new wave of anthropological texts have married Marxist political economic and political ecological perspectives to study a rising number of socio-environmental conflicts, related court cases and the impacts of extractive neoliberalism (Sawyer 2007; Behrends et al. 2011). They also study close up the contradictions of a system of capital based on fossil fuels and the cruel paradoxes of the local cultural phantasmagoria produced by exposure to the energy economy (ibid.).

For Wilhite (2005) and Strauss et al. (2013) anthropological work on energy should not only be concerned, then, with macroeconomy and politics, but also the economics and politics of everyday individual and household consumption. In so suggesting, they are critical of earlier work by economists and natural scientists that in their application of rational choice theory have reduced the world to an environment where individuals operate 'at arm's length'. Wilhite (2005) notes that the importance of social relations, culturally determined practices and changing material culture have been largely absent from the analysis, despite the results of many contextualized studies which point to the fallacies in economic models. He proposes that anthropology, with its concern for these kinds of details, is equipped to reframe energy demand as taking place in the interaction between consumers and producers of energy choices, both of which are socially constructed.

In *Contested Powers* we are admittedly inspired by this nascent anthropology of energy. However, we also think it important to highlight that in its current form it has some significant limits. The anthropology of energy forms of course part of a much larger 'tradition' within anthropology in relation to the study of human relationships to the natural environment, nature and the management of natural resources (Dove and Carpenter 2007; Haenn and Wilk 2005). This linkage to some extent relieves anthropology of its superficiality in relation to a subject that clearly requires much deeper and more detailed empirical and theoretical study, but not entirely so. Both the anthropology of energy and that of the environment, while making important strides in terms of describing complexity, and recognizing linkages between materiality and performance, fail to explicitly demonstrate the connections between theory and practice and more importantly how political power is to be recovered from energy. Despite various attempts by major figures within anthropology (Barth;[14] Hylland-Eriksen 2006; Hale 2001; Fals Borda and Anisur Rahman 1991) and fashions for applied, public and activist anthropology, the core of the discipline remains largely analytical with little willingness or sense of purpose to engage and transform public debate and politics.

We argue that this is unacceptable at a time when economic and climatic crisis threaten to push national and global politics not in the direction of sustainability, but farther in the direction of polluting practices of ever quickening energy production through carbon resource exploitation and more and more extreme efforts to secure capital in the face of social protest and militancy. Faced with the task of not only studying the status quo of energy politics, but aiming to identify collective alternatives, a wider array of social theory and methodological approaches is brought into tension with anthropology in this volume. In line with Abramsky (2010), a wider class analysis of energy is also used to situate the contemporary evolution of the energy sector in general, and the expanding renewable sector, within wider global systemic dynamics and processes.

It is claimed in statements by politicians and energy companies alike that the age of carbon is far from over. Although the 2012 Rio+20 summit concluded that the future of the economy is green, it also underlined that the basis of energy production will remain brown for some time to come. We are told that to avoid the recognized environmental and social problems associated with the current energy economy it is not a drastic change to the model and type of energy production which is needed but the increased application of a series of already recognized and patented technological, economic and institutional solutions. And yet a growing number of analysts and activists recognize the potential that twin energy/climate crises could have in terms of substantially increasing the already brutal inequalities that exist today, and hitting the world's most vulnerable people hardest (ibid.: 8). We agree with Abramsky that a 'worldwide energy revolution' requires the clearer definition of a politics of transition to a new sustainable model of energy production. Indeed, people everywhere, relying on energy to meet their basic subsistence needs, are beginning to question the 'inevitability' of dramatic instability in prices, claiming that access to energy is a human right and not a privilege (ibid.: 9.

In *Flammable Societies* (McNeish and Logan 2012) we discussed the complex divisions of labour and nature underpinning the global market and the likelihood that relatively low oil and gas prices (relative to the long-term value of these resources) cannot be politically sustained. We have argued that there are an increasing number of expressions of resource sovereignty focusing on the right to control the exploitation of these natural resources. Contrary to the interpretations that describe a 'resource curse', we also argue that resource-based conflicts, even when sometimes violent, are in many cases also important workings out of the terms of popular sovereignty. This is true from Norway to Nigeria, from Bolivia to Venezuela and beyond.

However, in envisaging such positive developments stemming from the assertion of resource sovereignties, it is important to see that there is no historical guarantee that such a scenario is synonymous with egalitarian anti-capitalist politics. The complex issue of popular sovereignty and economic self-determination (only partly expressed by varied demands of sovereignty over energy resources) is an increasing focal point also of far right politics. Moreover, given our prognosis of the dangerous yet potentially positive struggles linked to resource sovereignty there is a need to reflect on the concept of political power somewhat more generally, and give some thought to the factors that influence both the centralization and decentralization of decision-making and power – issues which have a direct bearing on the technical production and diversification of sources of energy as well as some of the ideological dilemmas.

Critical institutionalism

In his book *Seeing Like a State*, James Scott (1999) focuses on state visions of social organization associated with the modern improvement of life and what he calls high modernism. At the level of urban planning Brasilia is an urban archetype in this utopian craze for order and standardization – a city built from scratch which also eradicates or denies much of the everyday life of cities. Scott picks out a range of similar large-scale programmes, such as compulsory *ujamaa* 'villagization' in Tanzania, to show how such strategic development plans damage people's lives and their environment. For example, a peasant farmer who is pressured into resettling and trying to cultivate different land may well be effectively deskilled because the farmer's knowledge is not standardized and is instead part of a physical habitus. Similarly, when it comes to issues of social agency, the conception of proletarian political power seen from the position of a Lenin sitting on the Communist Party's central committee is very different from a Rosa Luxemburg considering the political development of workers' sense of agency and the scope of their concerns at a grassroots level. In such cases Scott puts a premium on local knowledge and asks his readers to appreciate the importance of thinking and acting critically from the bottom up rather than from the top down. Scott argues that planners have made the world more legible to the state but they have failed to make the state more legible to its citizens. This is why, Scott claims, these high modernistic undertakings fail.

We see the pertinence of Scott's arguments to the politics of energy because both state and private interests work to centralize and maximize systems of production. Notwithstanding socialist policy rhetoric and the authentic struggles taking place in Latin America, the differences between the approaches of national and private energy corporations are outweighed by their similarities. Scott's work is also important to us because it clearly grounds a wider critique of technocratic thought and practice in resource governance – relevant to governments on both the right and left of the political spectrum. Here we are reminded of insights that were of importance in our earlier work on the politics of oil and gas. As argued above, there are serious problems with the technocratic and rational choice foundations on which the influential 'resource curse' thesis is founded. Recognizing these weaknesses, efforts are now being made within political ecology and political economy to move beyond a mainstream perspective on environmental governance so that they capture the political relationships involved in resource wealth governance, and acknowledge the social complexities and importance of historic grievances in shaping the present.

As a field of study, environmental governance grew out of the acknowledgement within political science and economics of the role of institutions in grouping different actors together, and in steering their action (Evans 2012: 46). As Rydin (2010: 96–7) writes: 'Institutions bind actors together into arrangements and patterns of behaviour that exhibit strong path dependencies ... actors learn to behave in accordance with institutional norms and this reinforces certain behaviour.' A review of the literature on environmental governance finds a great deal of agreement on three core principles: a commitment to collective action to enhance legitimacy and effectiveness; a recognition of the importance of rules to guide interaction; and acknowledgement that new ways of doing things are required to go beyond the state (Kooiman 2000).

Mainstream institutional governance emphasized that collective action is possible if it makes rational sense. It was broadly optimistic about the possibility of identifying basic principles underlying effective institutions, and assisting people to use these principles to 'design' institutional arrangements through a conscious and rational process. From this perspective, epitomized by the Nobel-prize-winning work of Ostrom (1990, 2005), the role of institutional governance is to provide information and assurance about the behaviour of others, to offer incentives to behave in accordance with the collective good and to monitor opportunistic behaviour. Although still seated in a rational-actor perspective, Ostrom's sensitivity to complex institutional settings is widely recognized for its role in dismantling earlier assumptions about the 'tragedy of the commons' (Hardin 1968).

Aiming to demonstrate that government and wider institutional decision-making are not only matters of individual behaviour, March and Olsen (1984) coined the term 'new institutionalism' to emphasize that decisions are shaped to a large degree by pre-established rules and procedures through which institutions respond to real-world issues. However, new institutionalism also emphasizes that institutions are not static things, but dynamic entities that require constant maintenance and reproduction through sets of procedures and rules that become habitual (Lowndes 1996).

The overall strength of the mainstream and new institutionalist school has been their theoretical and empirical demonstration that the management of common property through collective action is possible; that there are certain conditions that facilitate this and that people govern resources through a range of formal and institutional forms (Cleaver 2012). However, critique is increasingly made of persisting rational-actor assumptions regarding the direct relationship between well-designed community-level institutions on the one hand and well-managed forest and improved livelihoods on the other. Citing Scott (1999), Cleaver highlights that in its focus on planning and design, legibility and codification and the engineering of 'good governance' arrangements, mainstream institutionalism incorporates features of high modernism. As such, she also claims in parallel to Scott that its modernist project of designing institutions (and systems) for natural resource management is partially doomed to failure (Cleaver 2012: 172).

Of interest to our search for a more accurate frame for understanding energy politics and governance is a more recent shift from an initial rational-actor position to social constructivism, and calls for models that seek a balance between structure and agency. *Critical institutionalist approaches*, the most recent development in the field of environmental governance, claim to differ from earlier schools because of their broad focus on the interactions between the natural and social worlds rather than a narrower concern with predicting and improving the outcomes of particular institutional processes (ibid.: 13). In contrast to earlier approaches, critical institutionalism suggests that institutions managing natural resources are only rarely explicitly designed for such purposes and that their multifunctionalism renders them ambiguous, dynamic and only partially amenable to deliberate crafting (ibid.: 13). In addition to exploring complexity, these perspectives explore the uneven costs and benefits of public participation and the ways in which power works through local institutions. Cooperation in institutions is not so much about direct exchange and anticipation of benefit, but about the generalized concept of the need for accommodation and reciprocity between neighbours.

Critical institutionalists adopt a 'thicker' model of human agency (ibid.: 15). For them, strategic livelihood choices (about the use of resources) are critically influenced by social concerns, by psychological preferences and by culturally and historically shaped ideas about 'the right way of doing things'. Drawing principally on the work of Giddens (1984) and Bourdieu (1987), Cleaver (2012) argues for a critical realist approach to environmental governance where the interrelationship between structure and nature is recognized and studied. In building her case for critical institutionalism, Cleaver (ibid.) also draws on the work of Mary Douglas (1987) and her use of the earlier French anthropologist Lévi-Strauss's coining of the concept of 'intellectual *bricolage*'.[15]

From this theoretical perspective, rather than being designed, or even crafted, institutions are patched together, consciously and non-consciously, from the social, cultural and political resources available to people based on the logic of dynamic adaptation. *Bricolage* is furthermore acknowledged by Cleaver to be an 'authoritative process, shaped by relations of power' (Cleaver 2012: 49). Here the configuration of societal resources shapes the 'institutional stock' from which institutions can be assembled, and the choice of instruments and mechanisms that can be applied. In emphasizing that invented institutions are shaped by past arrangements and relationships of authority, Cleaver recognizes that she repeats a perspective already well captured in earlier political economy (ibid.: 194).

It is evident from the above that critical institutionalism arrives at very similar insights with regards to the governance of renewable resources to those we and others propose in relation to non-renewables and a critique of the resource curse. Common to both is a grounded critique of technocracy and rational-actor perspectives in environmental governance. Common to both is also the recognition of the importance of history, power and the close relationship between biophysical context, social institutions and moral economy. Given that energy resources cross the analytical boundaries of standard understandings of renewable and non-renewable, acknowledgement of this commonality is of vital importance. Indeed, it appears to provide clear theoretical and empirical backing for our emphasis on the study of resource sovereignties, an idea that aims to bracket together and highlight the linkages between historic claims for territory and resources and the formation and expression of current interpersonal relationships. However, parallels between the critical literature on the resource curse and environmental governance also help to highlight where each is stronger and weaker with regard to an overall understanding of resource and energy governance.

Critical perspectives on the energetic state

In earlier work (McNeish and Logan 2012) we argued that one of the major shortcomings of technocratic thought in development and resource governance has been its unwillingness to address the full extent and influence of values, such as entrepreneurial and competitive ethos, or virtues of cooperation and solidarity in the formation of political and ideological praxis. Although aiming for equilibrium, the combination of neoclassical economic thought and liberal practical reason puts an artificial distance between economics and society. Scott and other anthropological writers such as Graeber (2001) importantly remind us that the relationships between different models of knowledge and value open up the possibility of better defining the sociobiological nature of commodities without losing sight of the violent interplay of positions and structures at work at all levels up to that of global capitalism. By setting Marx and Mauss together, and with clear echoes of Polanyi's classic work *The Great Transformation*, Graeber (2005) demonstrates that the construction of meaning (material and social value) involves imagining totalities (systems of meaning). According to Graeber, recognition of these processes has political utility insofar as the ultimate freedom is not the freedom to create or accumulate value, but the freedom to decide (collectively or individually) what it is that makes life worth living.

One of the defining characteristics of a state is the attempt to hold a monopoly of violence. Not only is this monopoly frequently controversial, in many instances it is also far from extensive, apart from the obvious example of civil wars. If the state is an altogether less coherent entity than it may first appear, energy infrastructures and the control of energy flows are an important means through which states attempt, with varying degrees of success, to *make* nations and make political systems more consensual (i.e. to establish shared value standards). It is in this regard that we refer to energetic states in this volume.

While critical institutionalism makes a significant contribution to the study of renewable environmental resources through its recognition of the need to understand the messy power-laden micro-politics of the interactions within and between the social and natural worlds, its emphasis on *bricolage* retains some of the diagnostic rationalism of earlier approaches by failing to seat these dynamics within a wider context that includes the formation of states and global economic and political linkages. In Cleaver's recent book she writes in conclusion that: 'in addressing analysis we need to place institutions in a wider governance framework that links scales and domains of action' (Cleaver 2012: 13). In many ways this is what *Contested Powers* seeks to do. While Cleaver emphasizes the need to have a multi-scalar understanding of power dynamics in institutional formation, her own work remains largely focused on meso-level processes, i.e. the interaction between state and community. Although there is expressed concern with realistically 'remapping' the institutional landscape, there is little concern with the seating of this within an international political economy. As a result, the critical institutional perspective she and others have developed looks stunted in terms of its analysis of power. Indeed, placing again the environmental governance literature in relief with the debates on the resource curse, there is also evidently a very similar lack of reflection about the way in which global economic and political forces impact on institutional and state formation. There also appears to be a lack of explicit reflection over the way in which the constitution of environmental governance links to these processes and to particular ideological expressions and interests.

While a set of basic principles can be stated, as above, as the foundations of environmental governance, it is important to acknowledge the linkage between the study of environmental governance and the growing recognition in the 1990s and 2000s of the dangers of a looming international crisis caused by climate change, agreements on cooperation for sustainable development and their linkage to international commitments on 'good governance'. As such it is coincident with the formation and operation of a range of international organizations such as the UN Environmental Programme and the Global Environment Facility in addressing global and national environmental challenges. In this conjuncture it was posited by a range of social scientists that institutional environmental governance provided 'a third way between the two poles of market and state, incorporating both into a broader process of steering in order to achieve common goals' (Evans 2012: 4). As with the resource curse, a particular set of liberal ideological assumptions have therefore been fundamental to the formation of the environmental governance field. For critical institutionalism, recognition of the role of these ideological assumptions and the interplay with knowledge formation should be an essential part of the self-reflexive development that a critical realist approach demands.

Moreover, recognition of the historical conjuncture in which the field came into existence should remind us of the need to return to some older principles and insights from classical political economy. In particular there is an evident need, in a search for a more comprehensive understanding of their operation across scale, of bringing in the way in which institutions, states and wider governance regimes have come into being in the great sweep of global history and capital.

The story of this book

The research for this book was conducted with funding from the Latin America Programme at the National Research Council of Norway. As has been indicated above, the project was to a large degree an offshoot of concerns and insights formed in the course of the earlier research for the *Flammable Societies* book. In the course of this work it became evident to us that much of the theory and argument used to analyse and comment on the socio-economics of oil and gas would also be a valid framework for the study of wider energy and resource questions. At the same time we also acknowledged that by widening our focus to consider energy politics in general there would be a need to both sharpen and deepen our use of key analytical terms such as power and resource sovereignties. We think that this is reflected well in the content and structure of the current volume.

Given previous experience and specific regional characteristics, Latin America also appeared to be an appropriate environment in which to carry out this new work. Moreover, by extending our gaze to energy resources in general, we are also aware that there would now be the possibility – if not requirement – to go beyond the development of a critical position within the academic debate on resource governance and actively respond to the unavoidable linkages that exist between this field and the larger pressing questions of climatic and environmental change. A focus on energy reminds us not only of the connections between us and the environment, but of the non-abstract or material nature of these relationships. These are not distant forces, but forces that are transformed through politics, social struggle and institutional formation.

The extended case study method used in the volume brings to ethnography some of the reflexive insights gained from historical sociology. Our approach does not focus on stable orders; rather we emphasize historically changing social practices and conflicts, and the competition of social actors over resources in the context of plural, or even contradictory, norms and rules. The case study method utilized here furthermore highlights the interplay of structure and agency. As such, we draw attention to the basic observation that 'social theory can avoid many difficulties with concepts of structure and agency when it makes an effort to regard social life less in terms of individual agency "reproducing" structure in any mechanical sense and more in terms of on-going dynamic interrelations between actors [...] interacting with each other by reference to shared understanding' (King 2005: 230). This basic premise helps to focus attention on micro–macro connections, which are the 'genesis' of changing social structures. But just as importantly, and without conflating structure and agency, the interplay between (unreflexive) systemic structures and (reflexive) human agency means that ultimately we are concerned with the transformation of normative structures, not their reproduction or discursive validation.[16]

Our project financed several periods of fieldwork for its team members between 2010 and 2012. However, rather than rely on extended fieldwork periods alone, *Contested Powers* has built on the contributors' knowledge of the research contexts, on the mutual trust between the researchers and informants, and more generally on our concern with both oral and secondary written sources. This approach to multi-sited ethnographic research has also been refined in recent years by a series of global ethnographers seeking to link study of large-scale processes with fine-grained observations of everyday life.[17]

We have argued above that national energy systems and state governing structures are mutually constitutive through history (Mitchell 2011); that there are intrinsic links between state formation and violence in the Latin American region; and that the simultaneous existence of multiple claims to sovereignty gives rise to what we have call jagged resource sovereignties. Chapter 2, by Virgilio Reyes, takes up these themes through the description of oil exploitation in the Maya Biosphere, where threats of violence underlie the tenuous social order in Guatemala. The chapter studies the conflicts over land raised by state development policies, the activities of oil companies, and the recent ingress into the zone by migrant peasants, ranchers and drug traffickers. It demonstrates how the state actually promotes territorial conflict by backing a development model based on extraction, in the face of social needs for livelihoods and inclusion. Chapter 3, by Axel Borchgrevink, also illustrates the close link between energy choices and contests over state power. In this analysis of the Nicaraguan electricity sector, attention is shifted to the distribution and consumption of energy. It shows that the privatization of electricity distribution is a complex story that can be understood only within a broad context. This includes the historical development of Nicaragua's energy sector; the country's recent political history of revolution and counter-revolution; the legacies of these turbulent times in the form of political structures and divisions; and an appreciation of the Nicaraguan state and the shifting relations it has had to society and to the market. Furthermore, the ideas, ideologies and worldviews that shape people's perceptions of these issues must be taken into account.

In Chapter 4, Cymene Howe, Dominic Boyer and Edith Barrera study attempts to build the world's largest wind power project on the Tehuantepec peninsula in Oaxaca, Mexico. While foreign companies have a key role in this development, the process has been strongly backed by the Mexican federal government. It was seen by the local population as the forced and unfairly compensated takeover of common and agricultural land, and resistance to the wind-park project grew. Among other things, the study shows that renewable energy projects can involve the same kind of conflicts with local affected populations commonly witnessed in hydrocarbon extractive industries. In Chapter 5, María Victoria Canino and Iselin Åsedotter Strønen provide such a case. Their work investigates the conflict between small-scale fishermen and the petroleum refinery complex of the Coro Gulf in Venezuela. In this conflict, the fishermen are doubly disempowered, both because their empirical-popular knowledge of reduced marine resources is discredited in the face of environmental standards based on scientific knowledge, and because any questioning of petroleum activities confronts the hegemonic national development discourse based on oil production.

In Chapter 6 Cecilie Ødegaard expands the notions of sovereignty by looking at the interstices of states. The smuggling of energy resources from Bolivia to Peru is the result of price differences between the countries, partly because of state subsidies in Bolivia. Her chapter sees smuggling in terms of the existence of parallel, or overlapping, sovereignties in these highland areas. *Contrabandista* (smuggling) activity not only negates state legitimacy and territoriality, but also gives expression to an understanding of development and progress that goes against the equation in public discourse between energy and development. State attempts to reinforce control, on the other hand, demonstrate how various claims to sovereignty are vested in commodity circuits and their regulation.

The subsequent three chapters deal with ways in which the legitimacy of left-leaning governments in Latin America is underpinned by hegemonic discourses on national development through the redistribution of rents from hydrocarbon production – and the contradictions and conflicts to which this gives rise. In Chapter 7 Fernanda Wanderley discusses the growing social conflicts resulting from the '*gasolinazo*' in Bolivia in 2010, when the government sought unsuccessfully to eliminate fuel subsidies. The articulation between the social-economic and the political-symbolic in hydrocarbon governance opens for questioning the sovereignty of the state, its relation to society and the pluri-national state project initiated in 2006. Einar Braathen, in Chapter 8, studies the social and political struggles surrounding the Brazilian discovery of large offshore oilfields. The discussion is framed against the background of the political context of 'Lulismo', which the author presents in terms of the Gramscian concept of *passive revolution*. His chapter examines the social origins, mobilization capacity and political impacts of three different initiatives to contest the dominant views and public policies regarding the exploitation of these oil reservoirs. The relative success of these competing expressions of resource sovereignty at different levels depends on local popular mobilization as well as on the implied challenges to the federal state. In Chapter 9 Owen Logan takes the analysis of petroleum-dependent leftist governments farther by examining the cultural policies of Venezuela's Bolivarian state. Focusing on the *el Sistema* youth orchestras, Logan presents the paradox of a self-professed revolutionary government promoting the conformity of a bourgeois cultural form. In this new discourse, the accumulation of social-symbolic capital is regarded as a civilizing influence and a remedy for exclusion and crime. The chapter makes comparisons with the import of the idea of *el Sistema* to Scotland, and argues that resource revenue underpins a shift away from theories of exploitation in both countries. It furthermore suggests that when determined by contests over the exploitation of natural resources the politics of popular sovereignty are empowering and disabling at the same time.

John-Andrew McNeish places previous chapters into the wider

context of energy contestation in Latin America in Chapter 10. Starting from an appreciation of optimistic expectations regarding the region's economic growth and democratization, the chapter moves to a sobering discussion of political economic and epistemic limitations to further energy-fuelled development. Drawing and extrapolating on McNeish's qualitative research on the Bolivian TIPNIS case, the chapter explores the rising tendencies for socio-environmental conflict linked to energy and natural resource extraction. Using a discussion of energy politics in Bolivia, the chapter moves to an exploration of wider economic and governance models and consideration of contrasting regional projects to use energy, and related infrastructure construction, as a means to cultivate stability and political power. As such, the chapter moves through multiple scales to analyse the linkages between energy and regionalism, and beyond to the place of Latin America in global capitalism. The chapter highlights the existence of opposing logics in the region, and argues that they evidence to a large degree the fiction of Latin America as a coherent idea.

Together, these chapters demonstrate how an anthropologically inspired approach, building on detailed, grounded case studies, can reveal how energy is not a recipe for development in its own right. Indeed, they reveal how the energy-driven models for development in the region are shaped by and highly reliant on competing expressions of resource sovereignty. They also reveal that because of political economic and epistemic limitations, there is a tendency for progressive reforms to be toned down, if not abandoned altogether. As a result the contests and expressions of resource sovereignty underpinning left-leaning 'extractivism' can be seen less as a left turn and more as the reconstruction of a civil peace by and mainly for the middle classes. The shaping of that 'peace' can also be traced by the disruptions to it, which are examined in our case studies. Logan and McNeish pick up the same thread in the concluding chapter, entitled 'From the King's Peace to transition society'. Taking up the discussion of energy and the fault-lines of sovereignty we have introduced here, they question the political discourses and cultural rhetoric that obscure socio-economic class and ethnic cleavages on the continent. The conclusion argues that the current model for the conversion of energy resources into the political and economic power that underpins a tenuous civil peace in Latin America reproduces inequality in the region and obscures the obstacles to meaningful energy transition. If the latter is to be achieved, critical and less ideologically rose-tinted understandings of political power are required, such as those discussed in the final chapter. We associate a much-desired energy transition with environmentally conscious action groups and social movements that already exist, but more importantly with the urgent need for a truly internationalist strategy to confront the global division of labour and nature, and the related environmental crisis to which our patterns of energy use are leading us.

Notes

1 See: www.peruviantimes. com/30/president-alan-garcias-policy-doctrinethe-dog-in-the-manger-syndrome/2860/. The original title of President Garcia's article is 'El Perro del Hortelano', a commonly used phrase to describe someone who begrudges others what they are not enjoying themselves. It is taken from the play with the same title (*The Dog in the Market Garden*) written by Spanish poet and playwright Lope de Vega (1562–1635).

2 We refer to 'discourse' as a set of organizing principles that form a language in action, not merely to textual, oral or visual themes and rhetoric.

3 See, for instance, the webpage of the Sustainable Energy for All Initiative created by UN General Secretary Ban Ki-moon: 'Through innovation in energy products and investment in deployment, businesses can create jobs and supply millions of people with the tools they need to make a better life. Policy makers can do their part to remove legal and regulatory barriers that stand in the way of business innovation and investments. Civil society groups can encourage governments to make more sustainable choices and provide community-based models of energy innovation' (www.sustainableenergyforall.org).

4 Although improvements are beginning to occur in various countries in the region, World Bank figures continue to highlight that inequality statistics in many countries of Latin America are among the highest in the world, with the richest 10 per cent of the population receiving 41 per cent of total income and the poorest just 1 per cent. Approximately 47 million people in the region are mired in extreme poverty.

5 The Mexican National Oil Company was only very recently privatized. See www.forbes.com/sites/doliaestevez/2013/12/11/mexico-reverses-history-and-allows-private-capital-into-lucrative-oil-industry/.

6 Recent cases in point include the violent confrontations that took place in Bagua, Peru, in 2009, surrounding the TIPNIS march in 2011 and repeated clashes over the building of the Belo Monte Dam in 2010–12.

7 This is the theme of a new project now being worked on by one of the book's editors. See www.umb.no/noragric/article/extracting-justice.

8 In 2009 these protests left more than thirty people dead after the Peruvian military police attacked peaceful protesters at Bagua.

9 Slogans such as resource solidarity, resource nationalism, participation and co-management should be examined against the complex practices of social reciprocity, regionalism, co-option and corporatism.

10 We recognize that when discussing hydrocarbons we should also consider coal resources, which are found

throughout the region and are a focus of economic activity in Colombia, Peru and Chile. Given the tremendous work currently being done by other reserchers, including some significant recent volumes focused on mining (Bebbington 2014), we have chosen not to focus on these resources.

11 www.internationalrivers.org/files/attached-files/world_commission_on_dams_final_report.pdf.

12 The most notable of these being the Peru–Ecuador confrontation focused on the 'Cenepa' river area from 1995 to 1998.

13 www.sscnet.ucla.edu/polisci/ethos/Weber-vocation.pdf.

14 www.publicanthropology.org/interview-with-fredrik-barth/.

15 See Douglas (1987). *Bricolage* is a French word: 'to make creative and resourceful use of whatever materials are at hand, regardless of their original purpose'. Lévi-Strauss (2004) originally used the term to characterize how people in 'primitive' societies think, especially in myth-making.

16 Margaret Archer argues that the distinction made between structure, as *unreflexive*, and agency, as *reflexive*, is required if one is not to conflate structure and agency, or take the position of methodological individualism. Archer emphasizes the need to reground 'collective reflexivity'. See 'Margaret Archer on reflexivity', www.youtube.com/watch?v=bMpJ5wnuB64.

17 The ethnographies of Buroway (2000), Ferguson (2006), Scott (1999), Tsing (2004) and Ongh (2006) all demonstrate the possibilities of detailing what is happening in local settings without losing sight of the fall and rise of ideas, processes and positions, or shifts in the organization and reach of capitalism. Importantly for a project aiming to take a 'cross-section' through the socio-economics of a global industry, these approaches to ethnography present the local and the global as mutually constitutive, and propose means to avoid common pitfalls of other analytic tools that 'dominate', 'silence', 'objectify' and 'normalize' the experience and knowledge of others.

References

Abramsky, K. (2010) *Sparking a Worldwide Energy Revolution: Social Struggles in the Transition to a Post-Petrol World*, Oakland, CA: AK Press.

Acheson, J. (2006) 'Institutional failure in resource management', *Annual Review of Anthropology*, 35: 117–34.

Agamben, G. (2005) *State of Exception*, Chicago, IL: University of Chicago Press.

Apter, A. (2005) *The Pan-African Nation: Oil and the Spectacle of Culture in Nigeria*, Chicago, IL: University of Chicago Press.

Bebbington, A. (2014) *Subterranean Struggles: New Dynamics of Mining, Oil, and Gas in Latin America*, Austin: University of Texas Press.

Behrends, A., S. Reyna and G. Schlee (2011) *Crude Domination: An Anthropology of Oil*, New York: Berghahn Books.

Berkes, F. (2007) 'Community based conservation in a globalized world', *Proceedings of the National Academy of Sciences of the United States of America*, 104(39): 15188–93.

Blom Hansen, T. and F. Stepputat (eds) (2001) *States of Imagination: Ethnographic Explorations of the Postcolonial State*, Durham, NC: Duke University Press.

— (2006) *Sovereign Bodies: Citizens, Migrants and States in the Postcolonial World*, Princeton, NJ: Princeton University Press.

Bourdieu, P. (1987) 'The force of law: toward a sociology of the juridical field', *Hastings Law Journal*, 38: 805–53.

— (2007) *Outline of a Theory of Practice*, Cambridge: Cambridge University Press.

Boyer, D. (2011) 'Energo-politics and the anthropology of energy', *Anthropology Today*, 52(5): 5–7.

— (2014) 'Energopower: an introduction', *Anthropology Quarterly*, 87(2): 309–33.

Buroway, M. (2000) *Global Ethnography: Forces, Connections and Imaginations in a Postmodern World*, Berkeley: University of California Press.

Chotray, V. (2007) 'The anti-politics machine in India: depoliticisation through local institution building for participatory watershed development', *Journal of Development Studies*, 43(6): 1037–56.

Cleaver, F. (2012) *Development through Bricolage: Rethinking Institutions for Natural Resource Management*, London: Earthscan/Routledge.

Collier, P. (2000) 'Doing well out of war: an economic perspective', in M. Berdal and D. Malone (eds), *Greed and Grievance: Economic Agendas in Civil Wars*, Boulder, CO: Lynne Rienner.

— (2010) *Plundered Planet: Why We Must – and How We Can – Manage Nature for Global Prosperity*, Oxford: Oxford University Press.

Coronil, F. (1997) *The Magical State: Nature, Money and Modernity in Venezuela*, Chicago, IL: University of Chicago Press.

— (2000) 'Towards a critique of globalcentrism: speculations on capitalism's nature', *Public Culture*, 12(2): 351–74.

Douglas, M. (1987) *How Institutions Think*, London: Routledge & Kegan Paul.

Dove, M. and C. Carpenter (2007) *Environmental Anthropology: A Historical Reader*, Hoboken, NJ: Wiley Blackwell.

Escobar, A. (1999) 'After nature: steps to an anti-essentialist political ecology', *Current Anthropology*, 40(1).

Evans, P. (2012) *Environmental Governance*, London: Routledge.

Fals Borda, O. and M. Anisur Rahman (1991) *Action and Knowledge: Breaking the Monopoly with Participatory Action Research*, New York: Apex Press.

Ferguson, J. (2006) *Global Shadows: Africa in the Neoliberal World*, Durham, NC: Duke University Press.

Finer, M., C. Jenkins, S. Pimm, B. Keane and C. Ross (2008) 'Oil and gas projects in the western Amazon: threats to wilderness, biodiversity and indigenous peoples', *PLOS One 3*, 8, Florida, R. (2005) *Cities and the Creative Class*, Psychology Press.www.plosone.org/article/info%3Adoi%2F10.1371%2Fjournal.pone.0002932.

Foucault, M. (1991) 'Governmentality', in G. Burchell, C. Gordon and P. Miller (eds), *The Foucault Effect: Studies in Governmentality*, Chicago, IL: University of Chicago Press.

Garcia Linera, A. (2012) *La Geopolitica Amazonica: poder hacendal-patrimonial y acumulación capitalista*, La Paz: Vice-presidencia de Bolivia.

Giddens, A. (1984) *The Constitution of Society: Outline of the Theory of Structuration*, Cambridge: Polity Press.

Gledhill, J. (2008) 'The people's oil: nationalism, globalization and the possibility of another country in Brazil, Mexico and Venezuela', *Focaal*, 52: 57–74.

— (2011) 'The persistent imaginary of "the people's oil": nationalism, globalisation and the possibility of another country in Brazil, Mexico and Venezuela', in A. Behrends, S. Reyna and G. Schlee (eds), *Crude Domination: An Anthropology of Oil*, Oxford and New York: Berghahn Books.

Graeber, D. (2001) *Towards an Anthro-*

pological Theory of Value: The False Coin of Our Own Dreams, London: Palgrave Macmillan.
— (2005) 'Value as the importance of action', *The Commoner*, 10 (Spring/Summer), www.commoner.org.uk/the_commoner_10.pdf.
Haenn, N. and R. Wilk (2005) *The Environment in Anthropology: A Reader in Ecology, Culture and Sustainable Living*, New York: NYU Press.
Hale, C. (2001) 'What is activist research', *Social Science Research Council*, 2(1/2): 13–15.
Hardin, G. (1968) 'The tragedy of the commons', *Science*, 162(3859): 1243–8.
Harvey, D. (2007) 'Neoliberalism as creative destruction', *Annals of American Academy of Political Science*, 610(21): 22–44.
Hylland-Eriksen, T. (2006) *Engaging Anthropology: A Case for a Public Presence*, London: Bloomsbury Academic.
King, A. (2005) 'Structure and agency', in A. Harrington (ed.), *Modern Social Theory*, Oxford: Oxford University Press.
Kooiman, J. (2000) 'Societal governance: levels, models and orders of social political interaction', in J. Pierre (ed.), *Debating Governance: Authority, Steering and Democracy*, Oxford: Oxford University Press.
Latour, B. (1993) *We Have Never been Modern*, Cambridge, MA: Harvard University Press.
Lévi-Strauss, C. (2004) *The Savage Mind: Nature of Human Society*, Oxford: Oxford University Press.
Lowndes, V. (1996) 'Varieties of new institutionalism: a critical appraisal', *Public Administration*, 74(2): 181–97.
March, J. G. and J. P. Olsen (1984) 'The new institutionalism: organizational factors in political life', *American Political Science Review*, 78(3): 734–49.
McGinnis, M. (2010) 'Building a programme for institutional analysis of socio-ecological systems: a review of revisions to the SES framework', Working paper, Workshop in Political Theory and Policy Analysis, Indiana University.
McMichael, P. (2011) *Development and Social Change: A Global Perspective*, Thousand Oaks, CA: Sage.
McNeish, J.-A. and O. Logan (2012) *Flammable Societies: Studies on the Socio-Economics of Oil and Gas*, London: Pluto Press.
Mitchell, T. (2009) 'Carbon democracy', *Economy and Society*, 38(3): 399–432.
— (2011) *Carbon Democracy: Political Power in the Age of Oil*, London: Verso.
Nader, L. (1981) *Barriers to Thinking New about Energy. Physics Today*, www.aip.org/tip/INPHFA/vol-8/iss-4/p24.pdf.
— (2010) *The Energy Reader*, Hoboken, NJ: Wiley Blackwell.
Nash, J. (1993 [1979]) *We Eat the Mines and the Mines Eat Us: Dependency and Exploitation in Bolivian Tin Mines*, New York: Columbia University Press.
Omeje, K. (ed.) (2008) *Extractive Economies and Conflicts in the Global South: Multi-Regional Perspectives on Rentier Politics*, Aldershot: Ashgate.
Ongh, A. (2006) *Neoliberalism as Exception: Mutations in Citizenship and Sovereignty*, Durham, NC: Duke University Press.
Ostrom, E. (1990) *Governing the Commons: The Evolution of Institutions for Collective Action*, Cambridge: Cambridge University Press.
— (2005) *Understanding Institutional Diversity*, Princeton, NJ: Princeton University Press.
Ostrom, E., R. Gardner and J. Walker (1994) *Rules, Games and Common Pool Resources*, Ann Arbour: University of Michigan Press.
Peck, J. (2005) 'Struggling with the

creative class', *International Journal of Urban and Regional Research*, 29(4): 740-70.

Pieterse, J. N. (1998) 'My paradigm or yours? Alternative development, post development, reflexive development', *Development and Change*, 29(2): 343-73.

Poteete, A. (2009) 'Defining political community and rights to natural resources in Botswana', *Development and Change*, 40(2): 281-305.

Robbins, P. (2004) *Political Ecology: A Critical Introduction*, Oxford: Blackwell.

Rosser, A. (2006) 'The political economy of the resource curse: a literature survey', IDS Working Papers 268.

Rydin, Y. (2010) *Governing for Sustainable Urban Development*, London: Earthscan.

Sawyer, S. (2007) *Crude Chronicles: Indigenous Politics, Multinational Oil and Neoliberalism in Ecuador*, Durham, NC: Duke University Press.

Schaffer, B. (2009) *Energy Politics*, Philadelphia: University of Pennsylvania Press.

Schilling-Vacaflor, A. (2013) 'Prior consultations in plurinational Bolivia: democracy, rights and real life experiences', *Latin American and Caribbean Ethnic Studies*, 8(2): 202-20.

Scott, J. (1999) *Seeing Like a State: How Certain Schemes to Improve the Human Condition Have Failed*, New Haven, CT: Yale University Press.

Stevens, P. and E. Dietsche (2008) 'Resource curse: an analysis of causes, experiences and possible ways forward', *Energy Policy*, 36(8): 943-68.

Strauss, S., S. Rupp and T. Love (eds) (2013) *Cultures of Energy: Power, Practices, Technologies*, Walnut Creek, CA: Left Coast Press.

Strønen, I. Å. (2014) 'The revolutionary petro-state: change, continuity and popular power in Venezuela', Unpublished PhD thesis, University of Bergen.

Taussig, M. (1983) *The Devil and Commodity Fetishism in South America*, Chapel Hill: University of North Carolina Press.

Tsing, A. (2004) *Friction: An Ethnography of Global Connection*, Princeton, NJ: Princeton University Press.

Wacquant, L. (2005) 'Carnal connections: on embodiment, membership and apprenticeship', *Qualitative Sociology*, 28(4): 445-71.

White, L. (1959) 'Energy and tools', in A. White (ed.), *The Evolution of Culture: The Development of Civilization to the Fall of Rome*, New York: McGraw Hill.

Wiersum, K. F. (1997) 'Indigenous exploitation and management of tropical forest resources: an evolutionary continuum in forest–people interactions', *Agriculture, Ecosystems and Environment*, 63(1): 1-16.

Wilhite, H. L. (2005) 'Why energy needs anthropology', *Anthropology Today*, 21(3): s1-3.

Young, O. R. (ed.) (1997) *Global Governance: Drawing Insights from the Environmental Experience*, Cambridge, MA: MIT Press.

2 | OIL EXTRACTION AND TERRITORIAL DISPUTES IN THE MAYA BIOSPHERE RESERVE

Virgilio Reyes

In referring to the interaction between society and energy, McNeish and Borchgrevink (see volume introduction) signal that there have been three dominant perspectives, i.e. technocratic, in which energy is equal to development, the resource curse and the creative economy. All of these perspectives are viewed as having an over-simplistic understanding of the interaction between politics and resources. As a counter-proposal McNeish and Borchgrevink propose the need to consider, as captured in the concept of resource sovereignty, the division of labour and the appropriation of nature. From this perspective resource sovereignty can be seen as a space of dispute between capital, the state and new forms of counter-hegemonic forces.

From this perspective, this chapter examines the way extraction activities generate focal points of social conflict and territorial dispute. I do this by charting the political economy surrounding the fifteen-year span of the oil contract awarded to the Perenco company by the Guatemalan state in 2011. The circumstances of this case stand out owing to the irregularities and political manipulations used to renew the contract. The contract was widely questioned and rejected by civil society and part of the academic community (IARNA 2010) owing to the effects oil extraction would have on the local area. The oilfield, known as the Xan oil well, is one of the largest sites for the production of crude oil in Guatemala and is located in the Maya Biosphere Reserve and in one part of the Laguna del Tigre-Río Escondido Biotope, where economic activity of this type is formally forbidden by law.

Oil extraction brings with it diverse effects on local ecosystems and human populations. In the Maya Biosphere there are clear indications of growing social conflict and competition for access to natural assets targeted for different purposes. These involve different sectors of both legal and illegal civil society. On the one hand, a rural population has migrated to the area in search of land for subsistence farming; on the other, there are the landowners and drug traffickers who seek to expand (sometimes working together) their activities in the region. On top of all this, the national government has intervened in order to maintain control of the area through an increasing police and military presence, and efforts to evict families occupying the area.

The presence of different stakeholders, with varying and clashing interests in the same territory, reflects different sources of territorial authority, and gives rise to what Agnew and Oslender (2010: 191) call overlapping territorialities. The competition for space from non-state stakeholders creates processes of reterritorialization that mark out boundaries within nation-state borders, thus constituting a context of disputed sovereignty. As the chapter makes clear, there is also a direct connection between these features and the now concluded thirty-six-year civil war.

Post-war Guatemala

Guatemala currently has a population of 14 million inhabitants, of which 43 per cent are considered an indigenous population of Mayan descent. The ethnic character of the make-up of the country's population is important given that public policies in general, and those related to energy in particular, must formally take the cultural factor into account when promoting private and public investment in this sector and when assessing the impact that the projects may have at local level.

Another aspect that characterizes the country is the incidence of poverty at national level. According to the National Living Conditions Survey (ENCOVI in Spanish) (NSI 2011: 4), it is estimated that in 2011 13.33 per cent of the population were living in extreme poverty and another 40.38 per cent were living in non-extreme poverty. In other words, the general poverty level in Guatemala is around 53.71 per cent and affects approximately 7,519,400 inhabitants. This section of the population is highly vulnerable to structural changes linked to the country's development model. By way of illustration, a report from the Central American Institute for Fiscal Studies (ICEFI in Spanish) and UNICEF on the cost of eradicating hunger in Guatemala for the period 2012–21 asserts that the country is one of the most unequal in the world. It has one of the highest indices of malnutrition in Latin America and the Caribbean, and globally it occupies sixth place on this issue. According to this report, such a situation is produced by relationships of social inequality, made evident by the national average (49.8 per cent) for chronic malnutrition. Disaggregation of this national figure reveals a level of 65.9 per cent for malnourished indigenous boys and girls. This level is higher than the 59 per cent for Afghanistan, the country with the highest level of malnutrition in the world (ICEFI and UNICEF 2011: 7).

The current socio-economic situation in Guatemala can be further characterized by a multiplicity of effects and reactions derived from a development model reliant historically on mining activities for export and energy extraction more recently. Agriculture is still important but, given the energy crisis reflected in the volatility of oil prices, activities such as dams for the generation of electricity, open-cast mining, the introduction of agro-fuels and the extension of oil contracts are viewed by the government as productive activities which attract high levels of private investment.

This model of economic development is especially problematic, owing not simply to the location of natural resources in territories set aside for conservation but also because these sites are often populated by indigenous and rural populations living in poverty. In the face of the pressure exercised by those who represent the extraction-based development model, opposition from the local population and the existence of protected areas are obstacles to the alliance of transnational capital investment and national economic groups. Here we must also emphasize the presence of an underground economy that also competes for control of the same spaces in order to carry out unlawful activities, such as fuel smuggling and drug trafficking. The combination of these factors at the core and periphery has unleashed a particular type of territorial conflict. It is underpinned by disputes over the rationales and purposes of economic production.

Although it has its own nuances, oil extraction in Guatemala is not exempt from the influence of macro-social factors in the mining industry. Mining constitutes one of the conflict fronts in the country owing to the direct and collateral effects derived from exploration, exploitation, transport and marketing. The levels of tension that exist between the stakeholders involved, such as the government, businessmen, organized civil society and communities, exist because of the way in which social, political and legal contexts have been formed historically. The national oil industry has been enshrined and privileged above all other economic activities under the pretext of generating economic benefits for society as a whole.

Taking current oil exploitation in protected areas of Petén as their reference, critics have expressed doubt about whether, given the political manner in which it has been handled, the exploitation of natural resources is a sustainable and feasible strategy for the country's development (IARNA 2010: 8). Guatemala is still in a period of recovery in the aftermath of the post-war era,[1] a fact that has implications for the nature of territorial disputes over natural assets defined as strategic on account of the value stakeholders assign to them. The loss of credibility in public institutions is shown by the cases of corruption reported frequently in the media, as well as the failure of the state to understand, deal with and seek viable solutions to social conflict. Added to these are acts of intimidation against defenders of human rights and leaders of social struggles for the defence of the territory, cases of which rarely pass from criminal investigation into final resolution in court.

As a result, the 'temperature' of social problems can be seen to be rising. Complaints and de facto measures exercised by civil society through the media and the courts reject the way in which the Guatemalan state negotiates, promotes and controls the sale of the national natural heritage to transnational companies. The response of the hegemonic groups in power has been to criminalize social protest, characterizing non-conformists as terrorist organizations opposed to development and seeking to profit from social conflict. International non-governmental organizations are accused of financing campaigns that falsely discredit the public benefits of mining activities.[2]

The political history of mining

The most important link between oil exploitation and the development model dates back to the structural changes that were promoted in Guatemala with the so-called October Revolution (1944–54). This ended the military dictatorship regime led by General Jorge Ubico, who was followed temporarily by General Federico Ponce Vaides. In this period, and specifically in 1945, a new Constitution was enacted which stated in its Article 95 that contracts to exploit minerals or mineral and hydrocarbon deposits could be agreed for a period of up to fifty years following approval from Congress. The bill also indicated that the deposits and their derivatives could be exploited only by the state, by Guatemalans or by domestic Guatemalan companies. In 1947, the sector was further regulated by a new law on oil (Decree 468), in which the National Petroleum Institute was created under the purely nationalist vision of that period. However, in 1949 the policy changed and Decree 649 was issued, declaring that foreign companies could explore and transform the transport of oil.

In this period, interference from the United States began, opposing the changes created within Guatemalan policy, which were subsequently branded as communist oriented, thereby creating the short-term conditions for armed intervention by external forces and the mobilization of the political right in the country. In 1952, during the second Government of the Revolution led by Coronel Jacobo Arbenz Guzmán, Decree 900 (the Agrarian Reform Law) was issued, introducing the idea of modernizing the Guatemalan countryside from what had been a system of quasi-feudal relationships in which the peasant population had been dispossessed by landlords and dictatorial governments. Decree 900 placed peasants at the centre of the transformation of the countryside in a vision of capitalist-style development with a nationalist content. Within the framework of the Cold War and with the expropriations carried out against the United Fruit Company (the US-based fruit production and export company) the regime was judged to be communist and was finally defeated.

The counter-revolution promoted anti-communist policies aimed at avoiding peasant-based social movements by promoting agrarian redistribution mechanisms. However, these policies failed to deliver owing to the corruption and political instability which Guatemala experienced until the signing of the Peace Agreements in 1996. Leaving behind a model of agrarian reform through expropriation, other policies were tested. The first was called *agrarian transformation*, in which the state distributed national land to the rural population under the private property regime, and the second was a policy of colonization managed by the national government in the lowland tropical department of Petén. These policies were tied to the framework of a military strategy aimed at bringing the department under national control.

Oil exploitation in the anti-communist context acquired other nuances when the counter-revolutionary government of Castillo Armas declared oil exploitation and extraction a national priority, with the advice of the US government. Thus, in 1955, the Petroleum Code was issued and, with it, exploration began again in the Petén department and at Amatique Bay in the department of Izabal. In Izabal ten wells were opened between 1958 and 1962. Significant political changes also occurred between 1984 and 1986. During this period Guatemala established a new framework for peaceful coexistence after years of *coups d'état*, election fraud and military governments. In 1985, a new Constitution recognized the importance of environmental protection and the need for it to be regulated by the state. In January 1986, a civilian government was elected and promoted a number of environmental laws which, among other things, would regulate productive activities through regulations for environmental impact assessment[3] and a new system for territorial planning.[4] Consequently, in 1989, a series of protected areas were created by Decree 4-89, with the intention of safeguarding the country's strategic natural assets and providing a legal barrier to both land invasions and industrialists interested in the exploitation of wood and energy. In the period from 1986 to 1989, a new framework of opportunities to settle in the Petén was presented to the migrant population. Although military actions between the guerrilla fighters and the army continued, the intensity of combat had notably decreased. Furthermore, negotiations were being carried out to cease hostilities within the framework of what were to become the Esquipulas Agreements,[5] resulting in a short-term improvement in political conditions.

For national conservationists, many of whom were involved in public environmental management, and foreign institutions interested in biodiversity, these reforms created the conditions not only for new migratory flows motivated by land, but also for government and business to make new investments in capitalizing on the resources of the Petén, including archaeological and ecological tourism. Conservationists feared that agrarian migration, encouraged by the improvement in roads and communications, would result in the increasing illegal occupation of lands that they had argued should be allocated to biodiversity conservation.

With these expectations and the existence of a favourable political atmosphere, conservationists lobbied and pushed in the courts to have the remaining forests in Petén considered as protected areas. When Law 4-98 was approved, the National Council of Protected Areas (CONAP in Spanish) was constituted as the lead state agency for conservation and, in 1990, the Maya Biosphere Reserve (MBR) was also created by Decree 5-90.

Given that the agrarian question was one of the key structural problems in the country, the legal constitution of the MBR and the different types of zoning it ushered in were seen by some as a threat. Indeed, the zoning of the new territorial model gave rise to two phenomena. The first was an acceleration of land occupancies via the strategy of '*agarradas*'.[6] These degenerated into cycles of huge forest fires. The second was efforts by the communities to stay within areas of restricted human settlement, although over time population growth required them to push for the legal regularization of settlements and lands.

The signing in 1996 of the Peace Agreements between guerrilla fighters and the government changed the conditions for how social conflict was approached in the democratic order. Paradoxically, however, the process of dismantling state institutions started after the signing resulted in the removal of units that had provided assistance to the population.[7] Access to land in this new context was to be made through market mechanisms that would define incentive packages for national and transnational private investment in extractive activities such as mining and oil.

Laguna del Tigre: a strategic territory

Laguna del Tigre is listed as the most important wetland in Central America and is even internationally recognized by the Ramsar Convention as the second-most important in Latin America.[8] In the list of protected wetlands, Guatemala appears with 628,592 hectares distributed over the sites shown in Table 2.1.

As observed in the Table 2.1, Laguna del Tigre is without doubt the most significant wetland in Guatemala with 53 per cent of the total species recorded in the Ramsar list of wetlands of international importance. Another way of looking at the importance of the wetland is that there are 680 bodies of water which, compared with the estimated total for the rest of the MBR, is equivalent to 70 per cent. In themselves, these indicators emphasize why the territory cannot be overlooked in the management of protected areas in the north of the country.

This part of the MBR is situated in San Andrés, one of the twelve municipalities within the department of Petén, and is regulated by two forms of management, the origins of which respond to two different historical moments in which the state assumed responsibility

TABLE 2.1 Protected wetlands of Guatemala

Site	Department	Area (ha)	Registration date	Percentage of total registered
Lachuá Eco-region	Alta Verapaz	53,523	05/06	10
Manchón-Guamuchal	San Marcos	13,500	25/04/95	2
Laguna del Tigre National Park	Petén	335,080	26/06/90	53
Yaxhá-Nakum-Naranjo National Park	Petén	37,160	02/02/06	6
Punta de Manabique	Izabal	132,900	28/01/00	21
Refugio de Vida Silvestre Bocas de Polochic	Izabal	21,227	20/03/96	3
Reserva de Usos Múltiples Río Sarstún	Izabal	35,202	22/03/07	5
TOTAL		628,592		100

Source: Produced using Ramsar Convention data

for defining natural conservation areas. Since the mid-1970s, one of the types of regulation, from the perspective of wildlife protection, was the creation of biotopes. In the mid-1980s the Investment and Development Company of the Petén[9] (FYDEP) granted the University of San Carlos various pieces of land to be administered under this category. The Laguna del Tigre-Río Escondido Biotope was created in 1986 among other areas similarly constituted in Petén. The second significant step was the creation of the Laguna del Tigre National Park, created in 1990 and administered by CONAP. This constituted one of the central areas of the MBR[10] with an area of 385,080 hectares of which 289,912 (slightly more than 75 per cent) are for the park and the remaining 25 per cent for the biotope.

The framework set out in Decree 5-90 determined the modes of access to and the use of different areas. This makes it clear that no form of human intervention is permitted, with the aim of benefiting scientific research on wildlife and cultural heritage. To protect the biodiversity of the area permanent human settlements, and farming activities, are also not allowed. Only those economic activities that owing to their nature are considered low impact and may appear in the form of controlled ecotourism are recognized by law. This is due in part to the composition of the soils, which are karst in origin, low in density and have a fragile structure that tends towards a physical profile that is more forest than agricultural (CONAP 1999).

The technical information listed in the Laguna del Tigre *Master Plan 2007–2011* recognizes the existence of 44 species of mammals, 188 species of birds, 17 species of amphibians and 55 species of fish. It reports that the concentration of crocodiles (*Crocodylus moreletii*) is the highest in Guatemala; furthermore, the howler monkey (*Alouatta pigra*) and the river turtle (*Dematemys mawii*), some of which are regional endemic species, are present. Of great importance are the three species of felines identified in the park: the jaguar (*Pantheraonca*), the puma (*Felis concolor*) and the lynx (*Leoparduswiedii*), all species in danger of extinction (CONAP 2006: 31). It is also known that the area is a critical habitat for the scarlet macaw (*Ara macao*), which nests here, and the San Pedro river is an important conservation area for bats (ibid.: 32). Briefly, it is to be noted that one of the features of the protected area is the heterogeneity of its biodiversity, a feature that stands out when it is reported that of the seventeen natural ecosystems that exist within the MBR, at least thirteen can be found inside the Laguna del Tigre (ibid.: 33).

It is not just the issue of biodiversity that makes Laguna del Tigre in Petén important. It is also an area of concentrated cultural heritage. It is well known that Petén is one of the most important areas in which Mayan culture flourished, and across the length and breadth of the whole department it is possible to find numerous archaeological remains from this pre-Columbian civilization. In the area along the banks of the San Pedro river there is a site administered by the General Directorate of Cultural Heritage called El Perú or Waka which has more than six hundred structures and is one of the most important archaeological centres to be found in the country.

According to CONAP, before the park and biotope were created, there had been no evidence of permanent or mobile human settlements; the only visible activity at that time was oil extraction. Nonetheless, several communities that are settled in the park claim that there were villages dating back decades before the creation of the MBR.

In the context of territorial planning in the north of Petén in the middle of the 1980s, the ecological importance of Laguna del Tigre meant that it emerged as a strategic area within the Maya Biosphere

Reserve. Beyond constituting an area of high ecological importance, Laguna del Tigre acquired significance geopolitically owing to its proximity to the border shared with Mexico and the discovery of oil, plus opportunities for commercial logging. Recognizing these commercial opportunities, a variety of stakeholders tried to establish both legal and illegal appropriation mechanisms. In some cases these interventions were assisted by the political weakness of existing institutional regulations.

This dynamic caused Laguna del Tigre to become a prized site for extractive and settlement activity. However, its appeal was heightened owing to the communication infrastructure already installed in the territory in order to provide access to an oilfield. This further facilitated the presence of the agrarian population and other interested groups linked to land speculation and organized crime.

The Xan oil well

As a result of the history of internal armed conflict, the military governments in Guatemala encouraged private mining activities with the goal of benefiting the country through revenue collected via taxes and royalty payments. In 1980, during the government of General Romeo Lucas García, the Xan-I oil well was established in what is today Laguna del Tigre. The well was initially operated by Texaco Exploration Guatemala Inc. (MEM 2007). In 1985, during the government of General Humberto Mejía Víctores, the Guatemalan state, through the Ministry of Energy and Mines, signed a contract (2-85) with the Hispanoil and Basic Petroleum International Limited consortium in which a concession for twenty-five years[11] was granted over an area of 9,953 hectares. When Hispanoil left the country in 1986, Basic Resources operated the contract until Perenco acquired the oil contracts in 2002 (Solano 2010: 3).

When the oilfield opened no environmental impact assessment was made because such legislation and instruments were not created until years after the start of operations at the Xan well. Those in favour of extending contract 2-85 used this as an argument for the non-retroactivity of the law when legislation was applied to protected areas. The Xan well is currently in the exploitation phase and, according to existing information,[12] the results from field drilling between 1985 and 2010 amounted to: 10,200 barrels of crude oil per day, 125,500 barrels of water per day and 2.02 million cubic feet of natural gas per day.[13]

In general terms, the Xan well is responsible for 94 per cent of national oil production and in nine years it has generated approximately US$607 million. The oilfield has an infrastructure that facilitates Perenco's operations in processing and transporting the product to the Piedras Negras. There is a refinery called La Libertad in Petén, situated in the municipality of the same name, which has a capacity to process 5,000 barrels of oil per day (bpd) and to produce 3,000 bpd of asphalt. This installation is also intended to produce fuel for the company's own use.

The infrastructure comprises a network of pipes and pumping stations that transport the product along 475 kilometres, beginning in the Xan field and reaching the terminal in Piedras Negras, located in the municipality of Puerto Barrios on the Atlantic Ocean.

Population and land occupation

When the MBR was established in the area corresponding to the Xan oilfield, there were no significant signs of human settlements. When the company set up the necessary water and land routes to facilitate the transfer of materials and employees, it failed to manage the migratory flows into Laguna del Tigre. Initially, transport to the territory started with one ferry on the San Pedro river. Subsequently a ballast road was opened that connected Las Flores municipality to the communities in Sacpuy and Sacluc, and access routes connecting the municipalities of San Andrés and La Colorada with the Paxban forest concession. Finally the company built a landing strip for the air transport of supplies.

This infrastructure is still fully functional, and the population benefits not just from this but also from the institutional vacuum on the part of the state in terms of protecting the area.

Illegal trafficking in land has also placed the territory in the spotlight. Between 1995 and 1998, an agrarian movement for land occupation grew in Laguna del Tigre. In order to stop the advance on the agricultural border and control the extensive growth of the human settlements, CONAP signed seven agreements with the communities settled at that time (CONAP 2006: 37).[14] Nonetheless, a Peace Brigades International (PBI) report highlights that nine communities, which together represented 650 households, equivalent to 3,250 inhabitants (Peace Brigades International 2010: 34), were settled in this period. Seven Community Management Units were created which, with the support of a non-governmental organization (NGO), would be tasked with managing the area. However, as the PBI report notes, the agreements do not consider the levels of possession, rent or concession taking place in other areas of Petén. As a result, families sold their land and went to occupy other areas within the MBR. According to CONAP, the failure to comply with the Agreements of Intention and lack of financing for follow-up technical support caused an explosion in population growth and the beginning of extensive cattle-raising (CONAP 2006: 38). Based on studies from Propetén, the same source reveals that in 2000 there were already twenty-four settlements with an estimated population of 1,205 inhabitants, although CONAP's Monitoring and Evaluation Centre (CEMEC in Spanish) puts the figure for 2001 at twenty-five human settlements and 4,900 inhabitants. Currently, thirty-seven communities[15] are settled throughout the nucleus and biotope areas and the population varies from 15,000 to 20,000 inhabitants.

Interest groups and social conflict

The private sector, principally through the Guatemalan Chamber of Commerce, has played a belligerent role with regard to the functions and actions of CONAP. Proof of this can be found in the 2004 Gremial Forestal request to the Guatemalan Congress to suspend CONAP's administration of new protected areas. CONAP was accused of technical and financial mismanagement and the private sector claimed that it would be able to administer the forests more efficiently according to market principles

In 2006, seventeen years after the introduction of the Protected Areas Law and four years after contract 2-85 with Perenco had expired, the Guatemalan Chamber of Industry filed an appeal of unconstitutionality in the national court. It claimed that there was a duplication of roles between CONAP and the Ministry of the Environment, and that there were areas of sensitive biodiversity in the country lacking sufficient protection from state bodies. However, following a series of protests and campaigning by civil society focused on the false basis for legal action, the Constitutional Court denied the appeal.[16]

Farmers originating from Petén and others from the east of Guatemala also took advantage of the situation in Laguna del Tigre. Indeed, several of them were accused by the country's authorities of being drug traffickers who, taking advantage of the proximity of the border with Mexico, had built infrastructure to facilitate illegal activities, such as small airstrips. Other known mechanisms included the anonymous appropriation of land, levelling of land and even, when the land was within a protected area, seeking to register it with false documentation in the Property Register of Guatemala. In 2006, the case of ten illegally held farms in the protected areas of Petén was made known, of which two, with an area of 4,590 hectares located in Laguna del Tigre,[17] were acquired by Waldemar Lorenzana, a notorious drug trafficker.[18] He had apparently acquired them from the inhabitants of the thirty-seven communities settled there. Maps of the area also show the issue of governance is made difficult by the de facto distribution of settled territory.

The government has attempted to use irregular meansto secure evidence of drug trafficking in the area. The status of human settlements in protected areas and the problems of drug trafficking have given rise to evictions of several communities, and some of them violently. Indeed, irregular settlements in the area have been called narco-communities, although it has been impossible to prove this as a relevant characterization of the entire population of the existing communities.

In January 2009, during the government of Álvaro Colom, residents from the El Vergelito community were detained by two CONAP park rangers and pressured to give them property deeds for the lands they occupied and to contribute to development strategies for the area, as had been negotiated in the permanency agreements of 1996.[19] The National Civil Police and the Guatemalan army were sent to the site and as a result of the action two people from the community died. In an investigative report produced by the apostolic Vicariate of Petén criticism was levelled at the government's lack of political will to seek a solution through dialogue. The Vicariate and the Human Rights Ombudsman had previously attempted to facilitate round-table talks between the authorities and the communities settled in the protected areas in order to seek a solution to the problem. However, according to the report state institutions failed to take part in anything but the first meeting.

Similarly, in August 2011, almost a year after the oil contract with Perenco was approved, the same argument of the alleged collaboration of the population with drug traffickers was used by the government to justify the eviction of the Nueva Esperanza community. During the eviction the houses, health centre and school were burned.[20] Owing to the proximity of the border, the inhabitants of this community, approximately four hundred people, fled into exile in Tenosique in Tabasco, Mexico. They returned to the country at the beginning of 2013.

This social conflict reveals not only the collateral effects of oil exploitation, but also the failure of the Guatemalan state to assume responsibility for the development and governance of the human population in the region. In other words, it reveals the state's institutional and financial shortcomings in addition to ideological partiality in terms of the social demands for land, education, health and housing, and the violation of residents' human rights.

The profoundly chaotic situation of informal access arrangements to natural assets in Laguna del Tigre has resulted in consecutive years of forest fires and deforestation in the areas. These fires graphically demonstrate the problems of governance in the territory and highlight the discourse on territorial sovereignty of other stakeholders. The fires geographically indicate the alliances these stakeholders manage to establish, or the economic and political resources they manage to control.

Even when the consequences of the impacts of the mining activities on the site are known, the state's ability to react to recover national heritage has been slow. However, in some cases clientelism has been resorted to in order to win support from the population. The cases of clientelism included support to political candidates during electoral campaigns or to local community leadership. They also cushioned the effects of conflict caused by efforts to establish settlement in the protected area.

Creating favourable public opinion

The attitude of the government significantly changes when addressing investment in private productive projects on national lands. To further create the political conditions necessary for the approval of the extension of the contract with Perenco by Congress, an Oil Fund was created. One of the fundamental problems with oil regulation in Guatemala is the calculation of royalties from extractive activities. Within the framework of the oil law of 1983, the royalties and the percentage of the hydrocarbons given to the state should constitute the economic basis of a Fund for the Economic Development of the Nation. According to its regulations, the Fund is exclusively aimed at developing the interior of the country, as well as research into new and renewable sources of energy for which further law would have to be promulgated. It was only at the time of extending the contract with Perenco that discussion of the activities of the Fund was resumed.

In 2007, Congress passed the Oil Tax Fund Law, which established that the funds obtained from royalties would be invested in Rural Development Councils (CODERUR in Spanish) (Solano 2005: 1). For the fund to be sustainable, the law ordered an increase in oil operations throughout the Republic, and the further development of activities related to exploration, exploitation, management and transport (ibid.). In other words, a way was envisaged to help Perenco to expand despite the drop in production of several of its operating oil wells; but the difficulty was expanding into protected areas.

The government's claim that increased activities in the oil sector would lead to increased local investment generated widespread manifestations of support by mayors in the Petén. They were mobilized by the idea that oil would benefit the communities in the geographical areas where the operations were being carried out. Very few local government officials openly questioned the hidden intentions of the law, i.e. to benefit Perenco through the provision of local investment incentives and the generation of public opinion favourable to extension of the contract. In 2008, the law was relaunched as the Fund for the Economic Development of the National Law, FONPETROL, Decree 71-2008. The aim of the reform was to regulate the collection and administration of funds that the national government obtains from royalties and from its share in hydrocarbon quotas, in addition to other revenue from oil activity. This fund was assigned to the Ministry of Public Finance. According to Decree 71-2008, the funds would continue to benefit the interior of the country and research into and development of new and renewable sources of energy. The Fund would be administered inter-institutionally by the Ministry of Energy and Mines and the Department for the Executive Coordination of the Presidency of the Republic.

The funds would be distributed as follows: of the total collected, 5 per cent would be earmarked for Departmental Development Councils, (CODEDEs[21] in Spanish), distributed proportionately in accordance with the number of inhabitants; 20 per cent would be distributed to departments where oil operations were being carried out; 3 per cent would be for public entities responsible for the validity and recuperation of the protected areas established by law; and 72 per cent would go to make up part of the Government's Common Fund. In spite of the protests from environmental and peasant organizations, the law was finally approved in November 2008 with the support of 109 deputies out of a total of 158. The National Council for Protected Areas attempted to oppose the new law and the establishment of oil extraction in the Maya Biosphere Reserve Area. The public management plan for Laguna del Tigre had already noted the collateral consequences of establishment of the oilfield, which had created breaches and facilitated access to the area, enabling the mobility of the population interested in land and other activities. This argument did not, however, suffice as grounds to suspend the extension of the contract. Shortly afterwards the Executive Secretary of CONAP resigned.

Oil contracts to the detriment of national interests

Once public interest had been created for oil extraction in Laguna del Tigre, the process for extending the contract with Perenco started. In 2010, then deputy Anibal García based his opposition to this process on how contract 2-85 had been constituted. Initially, the money the state collects through income tax is derived as a percentage of the oil shared between the company and the state. According to the ranges established by the Hydrocarbons Law 109-83, Article 73(g), and the terms established in the contract, the company receives approximately 46.5 per cent and the state 53.5 per cent of crude oil.[22] The total of the royalty is in accordance with point (f) of the aforementioned law which notes that: 'Twenty percent (20%) is set for crude oil with a weight of thirty degrees (30°) API, plus the difference between the participation included in the original contract, applicable to a production of one thousand (1,000) barrels a day and the minimum participation of fifty five per cent (55%) provided for in the Law on the Oil Regime for the Nation. The percentage indicated above will increase or decrease by one per cent (1%) for each API degree greater or lesser than thirty API degrees (30°), respectively; and in no case will the royalty be less than five per cent (5%).' According to the terms of contract 2-85, the royalty is provided for in law, and taking into account the fact that the Xan well contains a lot of sulphur and is classified with an API of 15.9 degrees, the royalties would be around 6 per cent. According to data from the Ministry of Energy and Mines for 2009, the royalties paid for the 2-85 contract to the national exchequer were US$9,957,330.56 while for 2010 they were around US$12,807,037.63; for 2011, an income of US$20,183,659.82 was reported. For 2011 alone, the total income reported from oil activity in Guatemala was US$132,233,939.99.[23]

One of the controversies connected to oil contracts in Guatemala is the issue of the so-called recovery costs. In 1983 legislation, a series of items was outlined by the state allowing private businessmen to recover as incentives for investment, therefore not only reducing their production and operation costs, but also cushioning the risks. Complaints about this model focused on the fact that the amount received by the state in royalties was far less than what it reimbursed for operating costs (García Hernández et al. 2010: 5).

Initially, Article 222 of Decree 1034-83 established recoverable and non-recoverable costs. The former consisted of exploration investments and operational expenditure, which previously had to be approved in accordance with exploration and/or extraction programmes. Non-recoverable costs included, among others, fees and wages paid to foreigners; payments with receipts or accounts records; parent company administration and management expenditure; as well as any cost, expense and investment not covered by the contract, etc. However, according to García (ibid.: 12), during the government of Oscar Berger, on 29 December 2004, the Minister of Energy and Mines processed a record from the Directorate General of Hydrocarbons that repealed Article 222, opening the way for non-recoverable costs to be transformed into recoverable costs. This measure, despite its damage to the interests of the country, was specified in Government Order 165-2005, which justified the reforms to Order 1034-83 with the argument that international energy market conditions had changed. It was further argued by the government that improvement had to be made to the conditions signed in 1983, whereby private investment in the country would be encouraged. The recoverable costs were specified in Article 57 of the aforementioned agreement.

As seen in the table above, the conditions offered by the Guatemalan state for oil investments are generous towards private business. Indeed, they noticeably change the principles that govern-

TABLE 2.2 Recoverable costs for oil extraction according to government agreement 165-2005

Wages and salaries	Geology, geochemicals and geophysics	Maintenance of installations, access, equipment and machinery covered by the contract	Equipment and accessories for industrial safety
Temporary and contract salaries	Roads	Insurance and finance premiums in favour of the state	Medicinal and pharmaceutical products covered by the contract
Overtime	Logistics and transport	Legal consultation that is not for settling proceedings against the state or labour issues	Fuel and lubricant costs for machinery, equipment and vehicles for operations covered by the contract, such as capital costs, excluding VAT
Employment benefits dictated by current laws on the matter	Drilling, civil works, drilling fluids, well registrations, well cementation, well testing	Internal consulting and auditing	Purchase and rent of buildings
Medical fees for care in clinics covered by the contract	Well completion	Food expenditure in the camps covered by the contract	Clothing for field workers
Life and medical insurance	Well reconditioning	Capital costs for field and duly registered headquarters	Administrative expenditure within the Republic to carry out planning functions

Source: Based on Government Agreement 165-2005

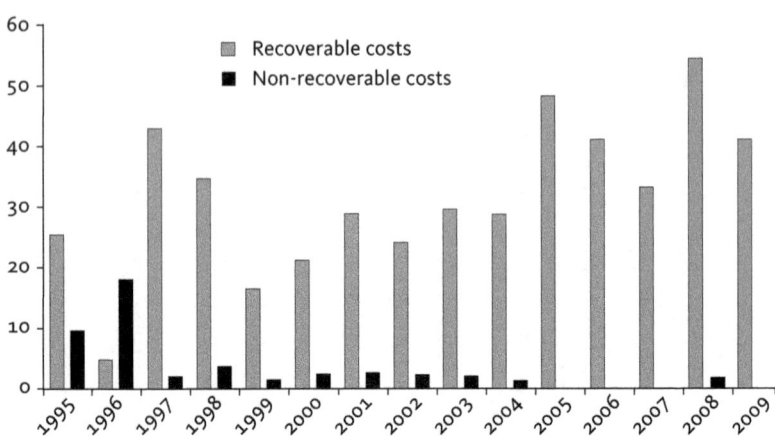

2.1 Ratio between royalties and recoverable costs of oil in Guatemala (*source:* Taken from García Hernández et al. 2010: Extension of Contract 2-85)

ments have successively displayed since 1954 with regard to the nature of public assets in the defence of national interests. García (ibid.: 17) points out that from 1995 to 2009, the Guatemalan state paid Perenco approximately US$53,000,000.00 in recoverable costs. Based on reports from the Ministry of Energy and Mines in 2008, they note that at that time the highest income from royalties in the previous fifteen years was recorded, rising to US$251,291.14; however, recoverable costs exceeded income by almost 255 per cent. As can be seen in Figure 1.1 above, royalties and recoverable costs are out of balance.

The controversial role of the state

The controversial role played by CONAP at the time contract 2-85 was approved stands out owing to the irregularities committed by then Executive Secretary Sergio Véliz Rizzo when he intentionally altered the *Master Plan 2007–2011* for the Laguna del Tigre National Park. Part of CONAP's duties as governing body for the protected areas in Guatemala is the registration, approval and supervision of the master plans that regulate the territory and the management and development of each protected area included in the Guatemalan System of Protected Areas (SIGAP in Spanish). The importance of this instrument lies in the fact that it expresses the policies and directives that need to be followed to ensure the appropriate use of resources.

All productive activity carried out within the different management areas must be authorized by CONAP.

Changes to the original master plan were premised on the effects of illegal activities – forest fires, wood and wild fauna trafficking – but oil activity, while mentioned in the original version, was completely removed from the final document. The area where the oil exploitation field is situated was included in the master plan as a recovery area specifically because of the consequences of human presence. As a result, prohibition of new roads and new oil exploitation was also considered (CONAP 2006: 115). Sergio Véliz was relieved of his duties and the new secretary of CONAP reported him to the Public Ministry for crimes of misrepresentation, breach of duties and resolutions that violated the Constitution. Finally, on 27 July 2010, Government Agreement 214-2010 approved the modification, expansion and extension of contract 2-85 for oil exploitation operations. The agreement granted CONAP responsibility for monitoring oil activities together with Perenco. In other words, the contract was approved on the basis of false information.

It is evident that beyond the illegal activities of local functionaries the central idea transmitted to the public was that the development of the petroleum industry would mean the development of the country. Again, this reflects the problems of a technocratic perspective, as critiqued by McNeish and Borchgrevink at the start of this volume.

The consequences

One of the consequences of approval of the oil contract extension was the immediate militarization of the area. On 13 September 2010, Government Order 260-2010 was enacted, by which the Guatemalan state signed a contract allocating to Perenco the support of the Jungle Brigades. They would assist the company with security for its installations, as well as reforestation of the Laguna del Tigre Park and implementation of environmental sanitation projects. From each barrel extracted from the Xan well, 30 cents per dollar would be earmarked for the aforementioned battalion, 10 cents per dollar for reforestation of the park and 15 cents per dollar for environmental sanitation to be distributed in the CODEDEs, related to oil activities in Petén. In November 2010, the Green Battalions were created. Their mission was to protect Laguna del Tigre and the border with Mexico; this military contingent would have a force of 250 men divided into six detachments. Beyond the official version of protection of the MBR, the measure can be considered as a mechanism to protect oil investments and to discourage other stakeholders who were attracted by the opening of the area.

Final reflections

In the debate on the political ecology of extractive activities, some authors have argued for the existence of a new current of neo-extractivism in Latin America. This idea attempts to capture the contradiction between the redistributive policies of countries in the region that have progressive governments and the continued deployment of economic models still largely based on conventional resource extraction.[24]

It might be true that in some cases a reformist kind of extraction exists. However, other less socially conscious models also still persist. There is a kind of extraction we could call stratified, i.e. where the population, according to the social position it occupies locally or regionally, has differing levels of access to natural assets. Even when it manages to exploit them, it does not manage to establish a process of significant accumulation.[25] Clearly a plundering form of extractivism still exists in peripheral countries such as Guatemala. This *accumulation by dispossession* may be exemplified by the case dealt with in this work. The industry extracts without contributing significantly to the development of the country and is a clear focal point for social conflict. However, despite the state's social failures, the case above shows its coercive power to impose a stratified modus operandi geared to the exploitation of natural resources. As can be observed above, the development of the oil industry in Guatemala was premised, despite legal rhetoric, on meeting the joint interests of foreign and national capital. Moreover, this was motivated by anti-communist policies that arose from the counter-revolution of 1954. Given that oil is an important source of revenue, the dispute over territory acquires a complex connotation when international capital in both its legal and illegal forms, indigenous and peasant populations and other stakeholders with specific local interests coexist in the same territories.

Disputes over territory caused by oil extraction result in a series of territorial juxtapositions of scale. This reveals the importance of the issue of sovereignty in the sense understood by Agnew and Oslender

(2010). As the example of Laguna del Tigre reveals, resource extraction was started in the territory by oil companies that arrived in the 1970s and 1980s. With the change of regime in Guatemala in 1986, and the national move to democratization, territory was officially reconceived according to the paradigm of sustainable development, leading to the emergence of protected areas. This had a strong impact on established interests in Petén because of the constraints that were then put in place and the way that the Maya Biosphere Reserve was divided into zones. There was no longer room for human settlements, oil extraction and traditional agricultural crops.

However, the post-war state in Guatemala demonstrated serious shortcomings in terms of its capacity to operate public policies and comply with the law. The public institutional structure was fundamentally weak. Scant or zero follow-up intervention processes in the territory resulted in the appropriation of space in different ways, legal or otherwise, which brought with it greater consequences for ecosystems and the population.

It is clear that establishment of the oilfield in the biotope and the nucleus area of the MBR, including Laguna del Tigre, has opened the way for the continuous invasion and occupation of land by peasants, farmers and drug traffickers. By using access routes into the interior of the reserve and taking advantage of the scant capacity of CONAP to react, they have managed to turn this ecologically important area into a cattle ranch and a corridor for drug traffic.

On top of all this, the extension of contract 2-85 revealed how far governments would go to strengthen the institutional structure in favour of private capital. It can be broadly said that oil in Guatemala is not exploited in the national interest despite legal foundations, given the fact that the royalties are subject to reimbursement of operational costs. Rather than providing an incentive, the state practically subsidizes the production of crude oil. The impunity and corruption that have benefited Perenco and guaranteed its operational security for fifteen years leave no doubt as to the government's lack of capacity and will to exercise its sovereignty and authority over natural assets that should have been more carefully protected and sustainably exploited in the interests of national energy security.

To conclude, it is perhaps pertinent to ask how we should understand the Guatemalan state in the current condition of social conflict and penetration by organized crime. These conditions affect both access to and distribution of benefits derived from energy exploitation. There is no single answer to this question, but it is possible to observe that the post-war Guatemalan state shows signs of deep institutional fragility and is therefore unable to effectively channel or pacify discontent caused by the dispute over natural resources. There is a clear need for institutional reforms to regulate the activities of the oil, mining and hydroelectric industries. This needs to be done in a way that promotes social investment and protection, sustainable management and a new relationship with nature.

Notes

1 The internal armed conflict between the government and guerrilla fighters from the Guatemalan National Revolutionary Unity lasted thirty-six years and left approximately 200,000 dead in its wake.

2 Recently, the Human Rights Ombudsman of Guatemala issued a resolution denouncing the threat to human rights defenders by an organization called Foundation Against Terrorism, which had been active in issuing false or distorted information to discredit the activities of organizations with links to human rights. For more information, see: www.pdh.org.gt/documentos/resoluciones/viewdownload/8-resoluciones/791-resolucion-por-denuncia-de-amenazas-a-defensores-de-derechos-humanos.html. http://www.pdh.org.gt/documentos/resoluciones/viewdownload/8-resoluciones/791-resolucion-por-denuncia-de-amenazas-a-defensores-de-derechos-humanos.html.

3 The Law on Protection and Improvement of the Environment, Decree 68-86, was issued, through which the National Environmental Council (CONAMA in Spanish) was created.

4 On the one hand, the Law on Protected Areas and its regulation was issued through Decree 4-89, which gave birth to the Guatemalan system of protected areas, and the National Council of Protected Areas was created. Then in 1990, through Decree 5-90, the Maya Biosphere Reserve was declared. Within this framework, inside the MBR, the Laguna del Tigre-Río Escondido Biotope was established.

5 The Esquipulas Agreements, I and II, were part of the preamble of democratic governments in Central America to negotiating peace in the region.

6 *Agarradas* refers to the mechanism whereby peasants demarcate an area of national territory to occupy. They clear the land, burn it and sow on it.

7 The service organizations that disappeared dealt with crops and livestock. This aggravated the situation in the field owing to the lack of technical assistance, which could not be provided by NGOs and the business community.

8 The Ramsar Convention, also known as the Convention on Wetlands of International Importance, especially habitats for water birds, was signed in 1971 and has been in force since 1975. Its distinctiveness lies in the fact that its aims are focused on the conservation of wetlands and their ecosystem functions, economic values and biological and cultural attributes.

9 FYDEP (Fomento y Desarrollo del Petén) was a public company administered by the army to develop Petén and

thereby integrate it with the rest of the country. It disappeared when Guatemala entered its democratic period.

10 The nucleus areas of the MBR are made up of biotopes and national parks; by law, the national parks comprise the Laguna del Tigre National Park, Sierra del Lacandón National Park, Mirador-Río Azul National Park and Yaxhá-Nakúm-Naranjo National Park. The Biotopes are Laguna del Tigre-Río Escondido Biotope, El Tzotz Biotope, Dos Lagunas Biotope and Cerro Cahuí Biotope.

11 The signed contract expired on 12 August 2010.

12 The information comes from the environmental impact assessment on well drilling, submitted by Perenco to the Ministry of Environment and Natural Resources in order to extend the contract in 2010.

13 The oil drilled from this site is classified as heavy; it has a gravity of 15.3 degrees API, a density of 0.9639 and a sulphur content of 6 per cent. In Guatemala, crude oil with such characteristics is used as raw material to make asphalt.

14 The most important communities at that time were Paso Caballos, Buen Samaritano, Mirador Chocop, Rancho Sucely, Cruce Santa Amelia, Los Reyes and Los Tubos.

15 Laguna del Tigre is populated by the communities of Estrella del Sur, El Sacrificio, San Martín Nuevo Paraíso, La Paz, La Gloria, La Lámpara, Buenos Aires, La Fisga, San José La Cumbre, San Luis Frontera, El Fracaso, Lagunitas Xan, Sucely, El Pacífico, Laguna Larga, La Florida, Los Tubos, Los Reyes, El Reloj, Santa Amelia, Nuevo Amanecer, Río Escondido, Bella Vista, La Mestiza, Mirador, La Bronca, El Buen Samaritano, Los Almendros, Seis Islas, Santa Rosita, Santa Marta, Paso Caballos, La Profundidad, Los Cerritos, La Tubería, Santa Clara and El Petén.

16 For some environmental organizations, the motive for the legal action taken by the Chamber of Industry was to open up the possibility of exploiting natural assets, currently prohibited by the constitution of Protected Areas.

17 www.prensalibre.com/noticias/Detectan-fincas-inscritas-forma-ilegal_0_131388643.html.

18 Waldemar Lorenzana was wanted by US justice for drug trafficking. He was captured by the Guatemalan authorities and the courts authorized his extradition.

19 The permanence agreements of 1996 were a measure that attempted to regulate the presence of the population within the protected area, given the scant resources CONAP had to deal with this situation and the political context of the signing of the Peace Agreements.

20 The situation occurred when Petén was under a state of siege from May 2011 on account of the massacre of twenty-seven peasants on a farm in Petén which was attributed to the Zetas.

21 CODEDEs (Consejos de Desarrollo Departamental, que corresponden al sistema de descentralización oficial de Guatemala).

22 In accordance with the Hydrocarbons Law, Decree 109-83, the hydrocarbons are shared when the state obtains at least 50 per cent of net production, which occurs when this production exceeds 10,000 barrels a day and does not exceed 20,000 barrels. For each production range, there is a percentage allocated to the state and to the company.

23 See: http://www.mem.gob.gt/wp-content/uploads/2012/05/1._INGRESOS_A_CAJA_FISCAL_ANO_2011.pdfwww.mem.gob.gt/wp-content/uploads/2012/05/1._INGRESOS_A_CAJA_FISCAL_ANO_2011.pdf.

24 Eduardo Gudynas is undoubtedly one of the representatives of the discourse on neo-extractivism in

South America: www.extractivismo.com/documentos/capitulos/Gudynas ExtractivismoSociedadDesarrollo09.pdf.

25 Here we are referring to the fact that at local and/or regional level, there are natural-asset mining activities that satisfy the needs of households or rural companies, which, owing to their scale, cannot be compared with the exploitations of transnational companies.

References

Agnew, J. and U. Oslender (2010) 'Territorialidades superpuestas, soberanía en disputa: lecciones empíricas desde américa latina', *Tabula Rasa*, July–December, pp. 191–213.

Albacete, C. (2003) *El Parque Nacional Laguna del Tigre y Biotopo Laguna del Tigre-Río Escondido en el contexto de la reserva de la Biosfera Maya*, Guatemala: Trópico Verde – Parks Watch.

COLLECTIF Guatemala (2011) *PERENCO: explotar petróleo cueste lo que cueste. Informe sobre las consecuencias sociales y ambientales de las actividades de la empresa PERENCO de Guatemala Limited*, France.

CONAP (1999) *Plan Maestro 1999–2003 del Parque Nacional Laguna del Tigre y Biotopo Laguna del Tigre-Río Escondido*, Guatemala: National Council of Protected Areas.

— (2001) *Plan Maestro de la Reserva de la Biosfera Maya 2001–2006*, Technical Co-editions Series no. 30, Guatemala: CONAP.

— (2006) *Plan Maestro 2007–2011. Parque Nacional Laguna del Tigre y Biotopo Laguna del Tigre-Río Escondido*, Guatemala: CONAP.

Congress of the Republic of Guatemala (1945) *Constitución Política de la República de Guatemala*, Guatemala: Congress of the Republic of Guatemala.

— (1949) *Decree Number 6-49. Ley de petróleos*, Guatemala: Congress of the Republic of Guatemala.

— (1968) *Decree Number 4-68. Ley de petróleos*, Guatemala: Congress of the Republic of Guatemala.

— (1983) *Decree Number 109-83. Ley de hidrocarburos y su reglamento general*, Guatemala: Congress of the Republic of Guatemala.

— (1985) *Constitución Política de la República de Guatemala*, Guatemala: Congress of the Republic of Guatemala.

— (1989) *Decree Number 4-89. Ley de áreas protegidas, reformas a la ley decretos 18-89, 110-96 y 117-97, y reglamento*, Guatemala: Congress of the Republic of Guatemala.

— (1990) *Decree Number 5-90. Declaratoria de la Reserva de la Biosfera Maya*, Guatemala: Congress of the Republic of Guatemala.

— (1996) *Decree Number 10-96. Ley de emergencia para la protección de la Reserva de la Biosfera Maya*, Guatemala: Congress of the Republic of Guatemala.

— (2004) *Decree Number 16-2004. Ley de emergencia para la defensa, la restauración y la conservación del Parque Nacional Laguna del Tigre*, Guatemala: Congress of the Republic of Guatemala.

— (2008) *Decree Number 71-2008. Ley del fondo para el desarrollo económico de la nación*, Guatemala: Congress of the Republic of Guatemala.

Environmental Observatory (2012) *De traiciones a la patria y corrupción en Guatemala: el caso del petróleo en el Parque Nacional Laguna del Tigre*, Guatemala: URL, USAC and FLACSO.

García Hernández, R. et al. (2010) *Prórroga del contrato 2-85: una historia de traición a Guatemala*, Guatemala: Calas.

Geoambiente (2011) *Estudio de evaluación del impacto ambiental del programa de perforación de pozos de desarrollo. Período 2010–2025 del contrato 2-85 de PERENCO Guatemala Limited, en el municipio de San Andrés. Departamento de el Petén, Guatemala.*

Governmental Agreement Number 1034-83 (1983) *Reglamento General de la Ley de hidrocarburos*, Guatemala: National Palace.

Governmental Agreement Number 165-2005 (2005) *Reformas al Acuerdo Gubernativo Número 1034-83, de fecha 15 de diciembre de 1983, que contiene el Reglamento General de la Ley de hidrocarburos*, Guatemala: National Palace.

Gudynas, E. (2010) *Diez tesis urgentes sobre el nuevo extractivismo: contextos y demandas bajo el progresismo sudamericano*, www.extractivismo.com/documentos/capitulos/GudynasExtractivismoSociedadDesarrollo09.pdf, accessed 2 July 2013.

IARNA (2010) *Laguna del Tigre: la necesidad de respetar y fortalecer su condición de Parque Nacional*, Guatemala: URL.

ICEFI and UNICEF (2011) *Protegiendo la nueva cosecha: un análisis del costo para erradicar el hambre en Guatemala, 2012–2021*, Guatemala: ICEFI and UNICEF.

MEM (Ministry of Energy and Mines) (2007) *Historia de la exploración y explotación petrolera*, www.iciaad.com/Historia%20de%20la%20extraccion%20petrolera%20en%20Guatemala-1.pdf, accessed 2 July 2013.

NSI (National Statistics Institute) (2011) *Encuesta nacional de condiciones de vida*, Guatemala: National Statistics Institute.

Oil Watch (2010) *Enfoque: Petroleras en acción, selva en destrucción. Análisis de situación*, Year II (no. 6), 30 April.

Parks Watch (2002) *Perfil del Parque Nacional Laguna del Tigre y Biotopo Laguna del Tigre-Río Escondido Tikal*, Guatemala: Parks Watch Guatemala.

— (2003) *Perfil del Parque Nacional Laguna del Tigre y Biotopo Laguna del Tigre-Río Escondido*, Guatemala: Parks Watch Guatemala.

— (2005) *Diagnóstico del Parque Nacional Laguna del Tigre y Biotopo Laguna del Tigre-Río Escondido*, Guatemala: Parks Watch Guatemala.

Peace Brigades International (2010) *Petén: las comunidades invisibles*, Special Bulletin, Guatemala: Guatemala Project.

Ramsar (2010) *Laguna del Tigre*, Switzerland: Ramsar Convention Office.

Solano, L. (2005) *Guatemala: petróleo y minería, en las entrañas del poder*, Guatemala: Inforpress Centroamericana.

— (2010) 'Petroleras en acción, selva en destrucción', *Enfoque. Análisis de situación*, 6.

Webpages

www.prensalibre.com/noticias/Detectan-fincas-inscritas-forma-ilegal_0_131388643.html

3 | *GRACIAS A DÍOS Y AL GOBIERNO*: ELECTRIC POWER STRUGGLES IN NICARAGUAN POLITICS

Axel Borchgrevink

'Thanks to God and to the government for bringing us this energy project, since we have been waiting for it for such a long time!' This quote from David Olivas, a small farmer in the remote Nicaraguan region of Madriz, opens a news item published on the webpage of Nicaragua's Ministry of Energy and Mines (MEM). The project in question extends power lines to the community of Las Vegas, and provides the 450 inhabitants with electricity for the first time. Even though this is traditionally a liberal village, the Sandinista government treats everybody equally, and provides services also to those who voted for the other party, the article goes on to tell us. However, this alleged impartiality is somewhat compromised by the next quote reproduced by the ministry's information department, from another villager, Julio César Alaniz: 'We have seen progress that no other government has provided. I used to belong to the liberal party but now I have joined the Sandinistas, for the projects and benefits they have brought to the community' (MEM 2011).

This communication from the Ministry of Energy and Mines illustrates a main theme of this chapter: how energy development in Nicaragua is deeply enmeshed with politics. The chapter deals with the electricity sector, and the privatization of electricity distribution is the main case. This is a complex story, and it can be understood only within a broad context, including the historical development of Nicaragua's energy sector, the country's recent political history of revolution and counter-revolution, the legacies of these turbulent times in the form of political structures and divisions, and an appreciation of the Nicaraguan state and the shifting relations it has had to society and to the market. Furthermore, the ideas, ideologies and worldviews that shape people's perceptions of these issues must be taken into account. Throughout, the aim is to show how energy issues set ideological ideas in play, mobilize existing and new political actors, illuminate linkages between different levels of contestation, and serve to reconfigure state–society–market relations. In short, *how energy issues are fundamentally and inescapably social processes* that can be understood only if the political and cultural dimensions are taken into account.

The first part of the chapter presents a brief background on Nicaraguan political history and the country's energy sector. Thereafter, the privatization of electricity distribution and the conflicts this has generated are described. The aim is to show the processes at various levels and from contrasting perspectives, but with an emphasis on how the changes are experienced by ordinary electricity users. While the relatively uncontroversial conclusion is that privatization failed, the chapter provides material on how this failure is popularly experienced. The last section describes how the configurations of Nicaragua's civil society and its relations to the state have changed over the years, and relates these changes to the dynamics of electricity politics. Briefly, it is argued that Sandinista organizational dominance over grassroots activists contributed to massive protests against electricity policies under the Bolaños government, but undercut the potential for mobilization against basically similar policies after Ortega became president. In the conclusion, I seek to draw out how different positions in the debates surrounding energy politics are formed by the horizon of ideas, beliefs and experiences which give them meaning.

The chapter is mostly based on written sources as well as on almost thirty years of research and contact with Nicaragua. However, shorter periods of fieldwork in October/November 2011 and April 2012 have been crucial for gathering first-hand information from electricity users. Interviews with inhabitants and activists of the Managua barrio of Batahola Sur supply the grounding for this analysis of electricity systems.

Background: Nicaragua, state, society and electricity

Political history An understanding of recent political history is fundamental for understanding the current configuration of Nicaraguan society – its political divisions, its organizational structure and the ideological frames through which people make sense of political conflicts and development issues. The revolution in 1979, which ended the more than forty years of dictatorship of the Somoza dynasty, installed the Sandinista National Liberation Front (FSLN) in power. The revolutionary government sought to transform the country according to socialist and welfare-oriented principles, and initially had considerable success in terms of improving health and educational services. However, economic blockade from the USA, the civil war against the US-backed *contra* forces and the steadily deteriorating economy all took their toll, and in the 1990 elections, Sandinista president Daniel Ortega lost by a wide margin. The eleven years of FSLN rule left the country significantly poorer, and most of the institutional and economic reforms of the period have since been revoked. Yet the revolution in many ways still shapes Nicaraguan political culture. For those who lived through the period, the experiences, aspirations and disillusions of these years have left deep traces that continue to shape their political perspective – whether on the left or on the right. Nicaragua remains a politicized and polarized society, and the main political divisions have their roots in the eighties.

Between 1991 and 2007, Nicaragua had three liberal presidents. The first, Violeta Chamorro, oversaw a transition period, when privatization, structural adjustment, reduction of the state apparatus and liberalization of the economy undid a lot of the changes of the FSLN years. There was considerable political turbulence as the broad alliance of fourteen political parties that brought Chamorro to power broke apart. Still, Nicaragua's fledgling democratic institutions were to some extent consolidated. The same cannot be said of the following periods. Arnoldo Alemán was president between 1997 and 2002. While he continued the liberalization policies of Chamorro, he also made a strategic deal with the FSLN and Daniel Ortega – the infamous *el pacto* – which effectively divided up power and control over the Nicaraguan state between the Partido Liberal Independiente (PLI), and the FSLN for more than a decade. Among other things, the pact gave the two parties joint control over the Supreme Court, the office of the Auditor General, and the Supreme Electoral Council. Still, Alemán's successor, Enrique Bolaños – supported in the 2001 elections by Alemán himself – proved uncontrollable, and had Alemán prosecuted and eventually convicted of corruption on a huge scale. But Alemán continued as a power-broker and had a large and loyal group of *diputados* in the National Assembly. Together with the Sandinistas, the opposition held the majority and in periods made it virtually impossible for the president to govern.

During the first years of opposition, the FSLN experienced internal conflict. There was a split in 1995, and several from the old FSLN leadership formed a new party, the Movimiento de Renovación Sandinista. Still FSLN remained Nicaragua's best-organized party, with structures reaching down to the local level throughout the country. In 2006, Ortega was elected president once again. While some of the old rhetoric of the left is intact, especially when condemning global capitalism and US foreign policy, he has softened his national politics considerably, in particular appeasing business and religious sectors. Throughout the term, the FSLN's hold on power was strengthened, by establishing new and party-controlled neighbourhood organizations, implementing extensive social programmes made possible through generous support from Venezuela, and maintaining an ever-tighter hold over electoral institutions. While charges of fraud abounded, Ortega won an impressive 62 per cent of the vote in the 2011 elections, allowing him to enter his third term as president in January 2012.

Electricity in Nicaragua The electrification of Nicaragua was slow compared to the other Central American countries. In 2004, 52 per cent of the population had access to electricity, compared to 66 per cent in Honduras, 81 per cent in El Salvador, 83 per cent in Guatemala and 98 per cent in Costa Rica (Acevedo 2009: 14). Electricity consumption per capita is just over half the regional average (Lecaros et al. 2010: 43). Furthermore, Nicaragua is the country with the highest dependence on hydrocarbons for electricity generation: in 2002, only 23 per cent of the country's electricity came from renewable resources, against a regional average of 59 per cent (Mostert 2007: 42), and Nicaragua is at the bottom when it comes to utilizing its potential for developing renewable energy: only 6 per cent (against a regional average of 16 per cent) of the hydroelectric potential had been developed by 2007 (Lecaros et al. 2010: 43). Finally, Nicaragua has by far the highest rate of losses in electricity distribution, and the prices of electricity remain among the highest in Central America (Herrera Montoya 2005a: 32).

While the history of electricity in Nicaragua goes back to 1902, when the first electric street lights came on in Managua, development was limited to a few urban centres until the mid-1950s. From that time and throughout the 1960s, the sector expanded considerably. At the end of the Somoza reign, there was an installed capacity of 300 megawatts (MW), with around 85 per cent of installed capacity being state owned (ibid.: 8–12, 23–5).

The main change during the revolutionary period (1979–90) was the development of the Momotombo geothermic plants, which yielded another 70 MW of renewable energy. Total capacity when the Chamorro government took over was 381 MW. Peak load during the first half of the 1990s was around 330 MW, leaving a very small safety margin. Thus, there were frequent shortages and power cuts (Acevedo 2009: 1).

Liberal energy policies During the presidency of Violeta Chamorro, a radical structural adjustment programme was carried out. As a legacy of the Sandinista government, Nicaragua had a huge public sector; 351 state corporations were quickly privatized after she assumed the presidency. Perhaps because there was such a lot to be sold off, the public service sectors of electricity, water and telephones were not slated for privatization in the first round. The World Bank, the IMF and the Inter-American Development Bank pressed for the privatization of these services also, and the government eventually sought to do so. But strong union protests blocked privatization during the Chamorro period (Close 2005: 212, 215).

It was not until 1998 that the National Assembly passed the laws[1] to reform the electricity sector and slate power-generating plants for privatization – or 'de-incorporation' as it was termed. Electricity distribution was also targeted for privatization, while responsibility for transmission was to remain with a state institution. Most of the existing power plants were privatized in the period 1999–2002. The notable exception were the hydropower plants of Hidrogesa, for which the bidding process was halted after intense popular protests led by the National Consumer Defence Network (RNDC, more on which below). From 1999, a number of private power plants were established. Electricity distribution was privatized in 2000, when the contract was given to the Spanish multinational Unión Fenosa.[2]

By the end of 2006, privately owned plants delivered just over 80 per cent of the total electricity supply (Lecaros et al. 2010: 64). Installed capacity reached 837 MW in 2007. However, owing to the age and deficient maintenance of many of the power plants, the figure of *installed* capacity was becoming increasingly irrelevant. *Effective* capacity was only 669 MW, while peak demand was 508 MW. This safety margin proved too small, and frequent power cuts were experienced, in particular from mid-2005 (Acevedo 2009: 11).

Sandinista power policies The government of Daniel Ortega reacted fast to address the energy crisis. With funding from Venezuela two new thermic power plants – Hugo Chávez I and II – with a total capacity of 217 MW were installed. Funding was also secured from Taiwan for another thermic plant of 27 MW. Thus, power shortages were largely overcome by 2008. The Ortega government has also reduced dependence on fossil fuel, and increased the share of electricity generated from renewable sources from 24–25 to 52 per cent. The plan is that the renewable share shall pass 80 per cent in 2018, and reach 90 per cent in 2020 (MEM, 24 January 2014). At the same time, electricity coverage has been greatly expanded. By the end of 2013, 76 per cent of Nicaraguan households had access to electricity, up from 54 per cent when the Sandinista government assumed power (ENATREL, 21 April 2014).

This effective response to the electricity crisis has been possible because of the massive support Nicaragua has received from Venezuela. Most of this support is channelled through the 'private' company Albanisa, in which the Venezuelan state oil company PDVSA holds 51 per cent of the shares and its Nicaraguan state counterpart PETRONIC holds 49 per cent. Among the subsidiaries of Albanisa is Alba Generación, which controls the Hugo Chávez and Taiwan plants, and Alba Eólica, which is planning to construct a 40 MW wind farm. This makes Albanisa the largest electricity producer in the country. In addition, Albanisa holds the bulk of the country's storage facilities for petroleum, the largest chain of gas stations, and the company that will construct the huge Venezuelan-financed oil refinery 'El Gran Sueño de Bolívar' (Bolívar's Great Dream), as well as a host of other enterprises, spanning TV and radios stations, a large hotel, cattle production, export of agricultural products, port services, transport activities, construction machinery, lumber production, and more. Thus, in the space of just a few years, Albanisa has become *the* dominant actor in the energy sector as well as the largest economic group in the country. While formally Albanisa is jointly owned by the Venezuelan and Nicaraguan states, the secrecy surrounding the commercial set-up has led to much speculation about who is the real owner – and about who would control this powerful economic group in the event of a change of ruling party in Venezuela or Nicaragua. It is unclear whether Albanisa really is a state entity, or a two-state joint venture, or whether the conglomerate is better understood as a private or party-controlled corporation. Many Nicaraguans assume that for all practical purposes it is the private property of the president.

Electric power to the people

On 22 August 2006, around one thousand demonstrators attacked the Masaya office of Unión Fenosa. Gates and doors of the office were broken down by the demonstrators, who were armed with sticks and stones and shouting: 'If we could throw out Somoza, why not Unión Fenosa!' Meanwhile, in Managua, there were demonstrations in front of parliament, demanding that the contract with Unión Fenosa be revoked. Three members of the bakers' association went on hunger strike. 'We cannot suffer this any more, better that they kill us straight away,' one of them was quoted as saying in an article in the newspaper *El Nuevo Diario* entitled '"Total war" against Fenosa'. In poorer barrios, streets were closed down, electricity meters destroyed, and workers of Unión Fenosa attacked (*El Nuevo Diario*, 22 August 2006). The special session in parliament, called to discuss President Bolaños' request to allocate $9 million to Unión Fenosa to alleviate the company's liquidity problems, was cancelled after demonstrators 'attempted to take over parliament' (*La Prensa*, 22 August 2006). Protests were also organized in other Nicaraguan towns: Ocotal, León, Granada Estelí, Juigalpa, Bluefields and Bilwi (Serra 2006: 231).

Why do people go on hunger strike and claim to be ready to be killed over the privatization of electricity distribution? While a large number of state companies had been privatized since the beginning of the government of Violeta Chamorro, none of those cases was met with such fierce and consistent opposition. The reasons for why the Unión Fenosa case generated such heated feelings are multifaceted, and have a lot to do with the way this process has intersected with key interests and discourses in Nicaragua's political sphere.

Privatizing electricity distribution One dimension relates to the opposing political fronts regarding privatization. Basically, this is an issue that pits 'the left' – the '*danielistas*' as well as those Sandinistas who are critical of Ortega's leadership – squarely against 'the right', the different liberal factions. The left's ideals of a welfare state crash directly with the liberal programme of reducing the state and giving the private sector the responsibility for tasks previously carried out by the state. In Nicaragua this gains added emotional significance owing to the ever-present context of the country's revolutionary past. In a certain sense, struggling over privatization means a rematch about the revolution and its meanings, legacies and costs. For many Nicaraguans currently above forty years of age, whose personal history and sense of identity were formed during the struggle to defend or oppose the revolution, such present-day struggles resonate with emotionally important ideas. This is particularly true for the (not insignificant) minority who were really dedicated during this period (Soto Joya 2009). But the sense of the privatization project as being part of the effort to dismantle the remaining legacies of the revolution adds, I believe, dimensions to the understanding of this project for very many others who experienced the 1980s.

From the perspective of those opposing privatization, it is significant that this is not only promoted by the Nicaraguan liberal presidents, but also by the international financial institutions. The whole restructuring and privatizing of Nicaragua's electricity sector was undertaken with heavy involvement and financing from the IMF and the Inter-American Development Bank. Further aid, as well as debt relief under the Heavily Indebted Poor Country (HIPC) scheme, was conditioned on the successful implementation of the process (McGuigan 2007: 6).[3] The role of the international lending institutions – promoting global capitalism 'at the cost of Nicaraguan electricity consumers' – has been a common theme in the criticism of Nicaraguan electricity sector reform (Acevedo 2005, 2009; Herrera Montoya 2005a, 2005b; McGuigan 2007). This is often linked to the fact that most of the investors have been foreign – Unión Fenosa being a case in point. Thus, in speaking about privatization in the electricity sector in general and Unión Fenosa in particular, in the presence of the Spanish prime minister Zapatero and King Juan Carlos, President Daniel Ortega said: 'It is a mafia. It is a Mafioso structure, gangster tactics within the global economy to which our countries are the victims because of the fault of our puppet governments ... The investors buy, through corrupt acts, the generating plants that are in good condition, from which they can sack the value ...' (*El Nuevo Diario*, 10 November 2007). In addition to the direct critique of 'global savage capitalism' contained in this quote, it is also worth pointing out an underlying theme resonating through many Nicaraguan political debates, concerning national sovereignty and the threat of outside forces seeking to take what riches the country possesses.

These themes of anti-privatization, anti-globalization, resource sovereignty and nationalism and their resonance in Nicaraguan political culture are important parts of the backdrop to understanding the strong opposition to Unión Fenosa. Yet by themselves they are far from a sufficient explanation. This requires delving further into the specifics of electricity provision in Nicaragua and how this service has evolved after Unión Fenosa took over. As with most privatization schemes, the underlying hypothesis was that state-owned electricity companies were inefficient and that the private sector would bring efficiency and transparency. This again would translate into increased coverage of electricity provision, better-quality services and reduced tariffs (McGuigan 2007: 5). Observers of different political shades concur that at least up to the end of the Bolaños presidency, the Nicaraguan reforms had failed on all counts:

- There were numerous and significant price hikes in the electricity tariffs.
- Blackouts and power rationing became increasingly common, reaching eight hours and more daily.
- Electricity coverage showed little increase and Nicaragua lagged further behind the neighbouring countries.
- There was little investment and maintenance of plants and grid, resulting in continuing high rates of energy loss.
- Customer service deteriorated drastically.
- Dependency on fossil fuels was increased.

Space does not allow an exploration of why the reforms failed to live up to their objectives (but see ibid., as well as Acevedo 2009, Herrera Montoya 2005a and Ripley 2010). Some causes may be briefly pointed out:

- Generating concessions were given on highly favourable terms, allegedly to tempt investors to enter what was perceived as a high-risk market. The terms did not stimulate reinvestment or much maintenance. As power plants received fixed capacity payments whether electricity was generated or not, even the incentive to produce was circumscribed.
- Distribution was given to one company, thus eliminating competition in this area.
- Unclear rules regulating the relations between electricity producers

and the distributor resulted in liquidity problems for Unión Fenosa. This again led power plants to operate at less than full capacity, to avoid the perceived risk of not getting paid by the distributor for the electricity supplied.
- Little attention was given to strengthening the market-regulating Instituto Nicaragüense de Energía (INE), which was never able to carry out its role with much capacity or authority.
- The policies did not include plans or targets for renewable energy development, and initially no incentives for investing in this area.
- The oil prices rose dramatically from 2002.

Privatized experiences How were the problems of privatization experienced by electricity users? As the electricity deficit increased, power cuts were becoming steadily longer and more frequent. Throughout most of 2006, there were daily blackouts of four to twelve hours. Furthermore, Unión Fenosa proved unable to provide a schedule of planned blackouts, as required by law. This impacted heavily on private businesses and on the population in general. Losses were great as electricity-dependent businesses had to adjust their hours of operation, and many were forced out of business. The Nicaraguan Chamber of Commerce reported that national businesses suffered a 20–25 per cent reduction in sales over several months in 2006 owing to the energy crisis, while Managua's huge Mercado Oriental announced that they would take legal action against Unión Fenosa owing to losses suffered by their meat and dairy businesses because of the power cuts (McGuigan 2007: 21–2). For the population at large, as described by Cupples (2011: 942), the blackouts meant food spoilt in the refrigerator, having to postpone work or studies, fear of walking home through unlit streets at night, sleepless and sweaty nights without fan or air conditioning, attending meetings over the sound of noisy generators, having to reorganize schedules owing to the rationing. Moreover, in many areas, water provision depends on electricity, and barrios of Managua and other cities were left without water for days on end. In hospitals, surgeries were halted, and in schools classes were cancelled.

In addition to the cuts, electricity consumers have experienced rising prices after privatization. While the tariffs charged are subject to approval by INE and calculated on the basis of the price of petroleum as well as other costs, they did involve new charges levied on consumers. Furthermore, as oil prices rose after 2002, so did electricity rates, in a series of adjustments. Between 2000 and 2005, the INE-controlled tariffs had risen 51 per cent ((McGuigan 2007: 26). This, however, had not been enough to cover the rise in production costs, and in 2005 the Energy Stability Law was approved, allowing for more rapid and flexible price adjustments whenever the oil price was above US$50 per barrel. In the six following years, there were reportedly thirty-nine price adjustments, while prices almost tripled (interview, RNDC, 2011).

Depending on your point of view, these price increases might be viewed as the natural and inevitable result of the rising oil prices on the world market; or, alternatively, as the effect of passing the costs of the favourable deals with the generating companies on to the consumers, along with the reformed system's inability to deal with inefficiencies and losses in the distribution network or to reduce the country's dependence on fossil fuels for electricity generation. Thus, you might end up with very different conclusions about whether the prices to consumers were the 'real' or 'just' prices to be paid. The complexity of the system and the lack of transparency about contracts and concessions do not favour informed and reasonable discussion.

In addition to the price hikes, there has been (and appears still to be) widespread over-charging on the part of Unión Fenosa. The Nicaraguan consumers' organizations report that they receive literally thousands of complaints yearly, from consumers who have suddenly, from one month to another, received huge jumps in metered electricity use. To illustrate the problem, McGuigan (ibid.: 18) recounts the story of a poor family that is charged for the use of 130 kilowatt hours (kWh) in October, an amount that jumps to 1,497 kWh in November and 1,673 kWh in December. Owing to the loss of the subsidy to small consumers, electricity charges actually increase by 2,736 per cent. But the case does not end there. In addition, as is standard practice, the company automatically assumes that the newly read usage is correct, and that this means that the consumer was stealing electricity in the past. Consequently, to the December bill is added 'non-registered energy' – what the household supposedly used during the last twelve months without being charged for. Thus, the family in question suddenly found that it owed the distribution company the equivalent of US$1,856, at a time when annual per capita income was US$890.

Fortunately, the amounts charged are extreme in this case. Still, it is quite representative in the fact that it tells a story that a great many Nicaraguan electricity users have experienced; being overbilled, and then automatically subjected to significant penalties. The RNDC reported that of all the complaints it received on any consumer issue, 76 per cent related to over-billing by Unión Fenosa. In 2007, according to the company's own statements, it received 54,000 complaints about bills, a number that equates to 9 per cent of its clients (Carrión Rabasco 2010: 16).

Complaining about the electricity bill is a difficult process.[4] First you must present a complaint in writing to the company within two weeks of having received the bill. The usual practice of Unión Fenosa is to respond to any complaint with a standardized letter stating that the original meter reading was correct, thus refusing the claim. The consumer then has another five days to present a new complaint. This time, if the company continues to refuse the complaint, the client must then present their claim to INE. In theory, INE should investigate and respond within twenty days, but in practice it usually takes longer. Often a customer will wait up to three months. For the persevering consumer, there is often a settlement reached at the end, whereby the original charge is reduced – although perhaps not by as much as requested. During the process, the bill in question is temporarily suspended and need not be paid. This, however, does not apply to the next month's bill, which must either be paid or an additional complaint must be lodged. If the customer fails to do so, she is immediately in default, meaning both that she will lose her energy subsidy, and that the company will be much less willing to reach a negotiated settlement. This is a trap many complaining customers have unwittingly fallen into – the company does not inform consumers about this.

This was the case for Doña Lilian, to whom I was introduced in the barrio of Batahola Sur. While her monthly electricity bill had been around 200 cordobas (C$),[5] it suddenly surged to 1,800 and 1,900 after Unión Fenosa had installed a new meter. Her neighbours point out how unreasonable the charge is – after all, she has a house with only two rooms, and uses electricity only for lighting and one electric appliance. She relates that she was unable to pay for three months. At one point she had managed to set aside C$1,000 and went to make a down payment. The company, however, did not accept any partial payment, and instead threatened her with jail. Several letters and a meeting with Unión Fenosa have not resolved the case, and Doña Lilian has had her electricity cut. Reinstallation seems to be out of the question, as she would have to settle her outstanding payments (including interest) as well as the installation charge. Her neighbours express their indignation that a respectable and honest old lady like her has been exposed to such treatment, and must suffer the indignity of sitting in the darkness in her house at night.

Doña Lilian is far from alone in being upset about the way she has been treated by Unión Fenosa. Visiting houses in the barrio, I was exposed to a number of stories from other customers. One lady had received bills that suddenly jumped from C$50–60 to 1,100. Her case had been resolved in her favour after going through the required steps until INE had become involved. A neighbouring man had had his charged consumption reduced from 245 kWh to 163 kWh, also after complaining to INE. A third lady had had her TV, fan and lamp all burn because of a power surge, for which the company was not willing to assume any responsibility. A fourth man tells a complicated story about how his meter had been damaged by a truck. Unión Fenosa installed a new one, which did not work, and showed zero after a month. A new and functioning meter was installed. However, for the month in which there had been no reading, the company estimated a usage more than four times his normal consumption – without offering a single reason for this. He complained, and had to go through all the steps until reaching INE before his claim that he should only pay the average of his electricity use before and after this month was accepted. In yet another case, a man had his electricity connection cut by mistake. Unión Fenosa admitted the mistake and reconnected him. In the next bill, he was charged for the installation and the new meter. Another man had to give up his meat-selling business as the unreliable electricity supply meant that meat in his freezer was spoiled several times – without Unión Fenosa accepting any responsibility.

In Batahola Sur, there are so many such cases that they have established a committee to negotiate collectively with Unión Fenosa. Altogether there are 145 electricity consumers who are contesting the amounts they have been charged. Three meetings were held between August and October 2011, between the committee and representatives of Unión Fenosa, as well as representatives from INE and the Ministry of Energy and Mines. According to the committee members, the agreements made in these meetings have not been respected by Unión Fenosa. In the first meeting, it was agreed that all cases should be properly revised, starting at a specified date, and that none of those complaining should have their electricity cut during the revision period. Moreover, representatives of Unión Fenosa should come to a public barrio meeting to explain and discuss electricity issues. However, nothing was done to review the cases, nor did representatives of Unión Fenosa come to the following meeting. Instead, with no warning, Unión Fenosa changed the meter installations to a type that is supposed to minimize electricity theft. This involves very much higher electricity posts, and mounting the meters 9 metres above the ground, thereby making them much more difficult to tamper with. A problem pointed out by the consumers is that they are now unable to see the meters, and thereby can neither check the readings made by the company nor use the meters to keep track of consumption and help saving energy. A second meeting was called, where the original agreement was approved once again. However, in the third meeting, Unión Fenosa said they were only going to review thirteen of the 145 cases, as the others were all in arrears in their payments (largely because of not paying the subsequent month's bill, after the disputed bill had been suspended). Subsequently, many of those in arrears have had their electricity connection cut.

Some of the members of the local committee for dealing with Unión Fenosa are adamant that they will not give up this struggle, and that, yes, they will pay their electricity bill, but 'only what is just!' Nevertheless, there is a strong sense of underlying frustration and perceived powerlessness. The huge multinational Unión Fenosa pays no attention to small consumers and their situation, takes no responsibility for mistakes or costs imposed on others, while the electricity user has to pay close attention to complicated requirements if she is to have any hope of getting a fair hearing for her case. From the complicated electricity bills with their host of different charges, to the many steps and pitfalls involved in presenting a complaint, Unión Fenosa seems surrounded by an impenetrable shield of bureaucratic rules and regulations designed to confuse the ordinary consumer. Batahola Sur is a poor barrio, where people have few economic resources, and most lack the knowledge required to deal efficiently with bureaucracies. Simply writing a letter of complaint is too large a challenge for many. Thus, there is an immense asymmetry in power and resources. Few really believe they will be able to change this situation.

Stealing electricity Seen from the other side of the table, the situation appears quite different, with other pressing concerns. Losses in electricity distribution are one of the great structural problems for the functioning of the sector. During the first years after privatization, as much as one third of the energy that Unión Fenosa bought from the generating plants did not reach paying consumers (McGuigan 2007: 15). Since then losses have slowly come down, to just under 29 per cent in 2006 and 21 per cent in 2011 (MEM 2011). Still, the figure remains the highest in Central America. As Unión Fenosa is only allowed to transfer losses of 15 per cent on to the bills of the consumers, the remainder cuts into the company's profits and is one reason why the company has claimed to be operating at a loss and is therefore in need of government support. While it has received such support in different ways over the years, the argument is nevertheless contested, since it is understood to be the responsibility of Unión Fenosa to reduce losses. If they don't do much about this, why should the Nicaraguan state come in and pick up the bill? The issue is important because it makes the electricity system inherently unsustainable and discourages new investments in the sector. Yet the set-up is one where the parties have tended to put all blame on the other actors – and it is impossible to settle 'objectively' how responsibility should be accorded.

The losses are of two kinds: technical losses, which to some extent are inherent in any transmission and distribution system, and electricity stolen by users illegally hooking into the grid. Technical losses may have amounted to up to 15 per cent in the period 2000–02, and have today been brought down to 8.5 per cent (interview, MEM, 2011), beyond which there is little potential for further reduction. Theft, then, accounts for the remainder of the losses, meaning that since 2000 between 20 and 12 per cent of the total electricity produced has been stolen. This has been an urgent problem for Unión Fenosa, and the billing and payment policies that have been described above are designed in response to this situation. Whenever there is an anomaly, such as a jump in measured electricity use, the assumption is that electricity theft has been discovered, and a charge for presumed previous unmetered electricity usage is added to the bill.

When picturing theft of electricity, the image that immediately springs to mind is of the illegal hook-ups from poor houses to the electricity network in new urban settlements. Such illegal connections are common in Nicaragua, and campaigns to combat the stealing of electricity have tended to focus specifically on this 'poor' category of thief. Yet estimates from different sources concur that this segment accounts for only a minor share of the electricity stolen – between a third and a fifth of the total (McGuigan 2007: 14; MEM 2011). The greater share is taken by larger consumers, mainly private businesses, from the very large to the very small. A typical example might be the owner of a small motel, making an illegal connection bypassing the meter, allowing guests to use air conditioning as much as they like without the owner being charged for it. But in spite of the knowledge that the poor residents of outlying barrios and recent settlements account for only a minor proportion of the stolen electricity, they have remained the group targeted when it comes to efforts to reduce non-technical losses, even under shifting political regimes.

It is unclear why this should be so. It may be that their illegal connections are easier to discover, and that the immediate payoff is therefore bigger by focusing on this group. However, it is also possible to suggest that whether from the point of view of a Sandinista state (with clear high-modernist aspirations – see Scott 1998) or of a neoliberal government (see, for instance, De Soto 2000), the marginalized dweller of unregularized settlements, outside the web of the state or the perceived blessings of the market, remains a particularly troubling and provocative figure. The businessman padding his profit by bypassing the meter, on the other hand, is an altogether more familiar and less threatening figure to both forms of governmentality. Thus, from either ideological perspective, it may appear most urgent to deal with the poor slum dwellers and their illegal hook-ups.

From the point of view of the inhabitant of a recent settlement where there is no legal electricity provision, the illegal connection is normally not considered theft in the full sense. It is rather seen as the only way of getting what is considered a right of everybody in urban areas – having electricity. In these settlements, just about everybody will eventually hook up illegally to the network. Still, they also insist that their wish is to have a legal connection and pay what is just for the electricity they are using. Such statements should of course not be taken at face value. Many undoubtedly also see a benefit in not having to pay. But this way of speaking is so common and so unquestioned that it undoubtedly reflects widespread moral ideas that this kind of illegal connection is not really wrong.

However, this claimed willingness to legalize is tempered by the policy of Unión Fenosa of charging back payments for illegal and unmetered use of electricity. Even if the company's estimate reflects correctly what the user has been getting without paying for, it nevertheless implies a very heavy and often impossible outlay for poor households living from day to day. And, knowing the reputation of Unión Fenosa, there will also be a strong likelihood that the retroactive charge will be much larger than is appropriate for the real usage. Thus, there are strong disincentives to formalizing one's connection.

To sum up, the protests of 22 August 2006 drew on widespread discontent. Power cuts, price hikes, unjust charges, arrogant treatment of clients, automatic assumptions of theft and an underlying sense of great powerlessness are all elements that have generated outrage against the power company among Nicaraguans. This discontent remains strong today. Apart from the end to the blackouts (not really due to Unión Fenosa), there have been few improvements in the company's services. Yet the protests and demonstrations did not continue after the high point in 2006. The reasons for this have to do with the change of government.

Changing opportunities for popular protest

Two days after the Sandinistas lost the 1990 elections in Nicaragua, Daniel Ortega gave a speech in which he admitted the electoral defeat and announced that the party would respect the result and hand over power. But Ortega added that FSLN would continue to fight and 'govern from below'. Time has shown that, through its extensive organizational apparatus, the Sandinista party has been able to wield considerable influence during the sixteen years of liberal rule. Although FSLN-affiliated organizations were weakened after the electoral loss, the party managed to retain considerable influence among unions, the organizations of farmers, cooperatives and farm workers, and among students. The many new NGOs that were formed after 1990, on the other hand, represented a somewhat different line: while most of them had their roots in the Sandinista revolution and were perceived to be on the left, the majority distanced themselves from what was seen as an authoritarian FSLN party structure. Thus, while Nicaraguan civil society has been dominated by the left, it has been a divided left. Generalizing broadly, one could say that while the FSLN retained the *poder de convocatoria* – control over grassroots activists and organizations that allowed mobilization for street demonstrations – 'the other left' gained a kind of intellectual hegemony as often media-savvy NGOs with professional expertise, high visibility in Managua and good access to donor funding.[6]

Under the last years of the presidency of Bolaños, a serious constitutional crisis – with the president and the National Assembly in deadlock – was further exacerbated by a string of popular protests, when demonstrators took to the street, set up blockades, stopped traffic, burnt tyres and fired rockets at the police. There were major demonstrations over proposed increases in bus fares (to cover the rising oil prices), as well as the demonstrations against Unión Fenosa described above. Clearly, the protests against rising bus fares and Unión Fenosa were expressions of deeply felt popular grievances. Just as clearly, it was in the FSLN's interests to foment these protests in the run-up to the presidential elections in November 2006, to make visible popular discontent and to underline the government's weak capacity for governing and maintaining order.

The Nicaraguan consumers' organizations were active in the campaign against Unión Fenosa, especially the RNDC under the leadership of Ruth Selma Herrera, a strong and outspoken critic of everything she saw as against the interests of ordinary Nicaraguans. RNDC gave practical help to electricity users who wished to complain about their bills, and ran media campaigns, initiated legal cases to promote consumers' rights, and arranged demonstrations. While the RNDC did not wish to be identified with any party, and maintained a critical distance from the FSLN, there were common interests between the two. RNDC was successful in mobilizing people for demonstrations against Unión Fenosa because they could draw on the organizational apparatus and grassroots activists of the FSLN. This 'strategic alliance' gave the RNDC a mobilizing capacity and an agenda-setting impact that it would not have had on its own.

During the election campaign in late 2006, Daniel Ortega took a clear stance against Unión Fenosa, threatening to cancel the contract with the multinational. Moreover, the Sandinista-controlled Supreme Council of the Office of the Auditor General declared the contract with Unión Fenosa invalid owing to the failure of the company to comply with its obligations to make infrastructural investments. Apparently expecting a hard line from the new government, Unión Fenosa requested payment of US$53 million from the World Bank's Multilateral Investment Guarantee Agency for the supposed expropriation of electricity distribution – a payment for which, if effected, Nicaragua would ultimately become responsible. This tough posturing from both sides was followed by drawn-out negotiations between the parties. Eventually, a Protocol of Understanding[7] between the Nicaraguan government and Unión Fenosa was signed in May 2008. This agreement stipulated the investment obligations of Unión Fenosa over the coming three years, as well as legal changes which the Ortega government was obliged to make (e.g. making electricity theft punishable), and included the transfer of 16 per cent of the shares of Dissur/Disnorte to the Nicaraguan state as settlement of debts Unión Fenosa had incurred.

Unión Fenosa and the Ortega government thereby became partners. In practice, the 'socialist and solidary' Sandinista government is continuing the privatized electricity distribution policy of its liberal predecessors.[8] This is not, it should be pointed out, any break with the government's overall economic policy. In spite of his radical rhetoric, Ortega has sought cooperation with the private sector, and the country has received praise for its economic policies from the international finance institutions.[9]

The space for civil society, however, has changed. RNDC can serve as a good example of this. During the Bolaños presidency it spearheaded the mobilization and protests against Unión Fenosa, but in the new political context it has had to make drastic changes in working methods and has given up public campaigns and mass demonstrations. A representative of the organization explained to me that since Ortega's election it has no longer been possible to draw on the Sandinista grassroots structures for popular mobilization. This is because the grassroots activists have become tied up with other tasks of the new government, for instance in the new neighbourhood organizations established by the Sandinista government. Also, one may assume, it is no longer in the interests of the Sandinistas to contribute to mass demonstrations when their own party is in power. The result, anyway, was that the RNDC was no longer able to mobilize. 'The irony', a staff member told me, 'is that while the current government is more supportive [of consumers] than the previous ones, this still leaves them with less possibility to mobilize and protest ... The consumers today are less protected than ever.'

Instead, the organization decided that its best chance of wielding influence in the current situation was through developing legal and technical proposals for ministries and the National Assembly party benches. The RNDC has, for instance, worked on finance issues and elaborated proposals for new laws on microcredit and regulating the use of credit cards. While the laws have not been adopted, there have been tangible results, such as new regulations on the interest rates that credit card companies are allowed to charge. Within the field of electricity generation and distribution, the focus was shifted to acquiring professional expertise and studying the current institutional, legislative and economic arrangements, as a basis for coming up with proposals for an improved regulatory framework. 'We are today the organization with the deepest knowledge of how Nicaragua's energy system and electricity market functions,' I was told in 2011. This shift in organizational strategy was also reflected in its staff. New, young and professional people were hired. At the same time, former key staff left, such as the leader Ruth Selma Herrera, who was given 'an offer she couldn't refuse', as director of the Nicaraguan Water and Sanitation Enterprise (which none of the liberal governments ever got around to privatizing).[10] Also, other employees were recruited into government positions. Whether this reflected a strategy of the new regime to co-opt critical voices – as some have alleged – or was simply a way of recruiting qualified people, the net effect has been that organizations like RNDC lost experienced staff members.

As is the case for many Nicaraguan civil society organizations, RNDC's activity is highly dependent on aid funding. Since 2007, however, almost all the European bilateral donors have closed down or dramatically reduced their aid programmes, including Austria, Denmark, Finland, Germany, the Netherlands, Norway, Sweden and the UK.[11] In terms of state-to-state aid, Nicaragua has been more than compensated by the support received from Venezuela. However, for Nicaraguan civil society organizations, the withdrawal of these international donors means a drastic drop in available funding, and competition for what is left is becoming increasingly hard. Many organizations have had to reduce their activity levels significantly, lay off staff, and sometimes stop operating completely. This was the fate of the RNDC. According to its former coordinator, the organization had been struggling financially since 2009, and had to close down in January 2012 (*El Nuevo Diario*, 30 March 2012).

While a special case, the rise and fall of the RNDC can nevertheless be seen as emblematic of some of the changes or reconfigurations that Nicaraguan civil society has undergone. The net effect is that the current regime has been relatively successful in curbing civil society protests. Grassroots activist structures previously supplying manpower to turbulent mass demonstrations have largely been transformed into the neighbourhood committees (CPCs) that perform local administrative functions, such as distributing different types of government support to needy target groups and carrying out health campaigns. The visibility and influence of the many Managua-based and outspoken critical NGOs, on the other hand, has waned as donor funds have become greatly reduced. The surviving NGOs have reduced their activity levels and/or shifted their work from lobbying and awareness-raising towards more service-delivery-type programmes.

Returning to electricity distribution, the new civil society configuration means that while dissatisfaction remains high, it does not have the organizational outlet and visibility it had during the Bolaños presidency. Of course, civil society engagement in the issue has not completely died away. The other national-level consumer organization, INDEC, continues to help individual clients and has attained some media exposure about the issue – but at a much lower level and in a more soft-spoken vein than was the case of RNDC. The Movimiento Comunal – a grassroots organization with a considerable national network – has held demonstrations against electricity tariffs – but succeeded only in mobilizing a couple of dozen participants. In sum, while problems with electricity delivery remain as before, public protest has largely been stifled.[12] The case illustrates how regime change can have significant impacts on civil society configuration and state–society relations – and that the dynamics of Nicaraguan electricity politics cannot be understood without an appreciation of this.

Conclusion

The Nicaraguan privatization of electricity distribution was an intensely contested affair, which failed to live up to its objectives. The material presented shows that what is sometimes considered a purely technological process – the electrification of a country such as Nicaragua – is fundamentally a social process which cannot be understood without considering its cultural and political dimensions.

The motivations behind popular protest are formed by the horizon of ideas, beliefs and experiences which give them meaning – in much the same way that technocratic plans are also shaped by the context of ideas, beliefs and assumptions in which they are developed. Neoliberal economic policies are based on the idea that market solutions ensure efficiency, and thereby result in reduced costs and better services for customers and users. Within its own frame of reference, this notion takes on the appearance of logical necessity. When electricity prices are set by impersonal market forces, they are 'right', in an objective sense, and cannot be contested – or such are the dominant claims. Yet the material presented shows that this technocratic claim is not as dominant as it likes to claim. It is actually hotly contested.

Electricity policies take on very differing characteristics according to the 'frames' through which they are understood and assessed. The neoliberal belief in the supreme efficiency of market solutions clashes with ideas of people's intrinsic rights to basic services such as electricity. Should electricity be a public service, or is the privatized market the best – if invisible – hand to provide this good? Seen from such different perspectives, the processes described will appear in a completely different light. In Nicaragua, with its revolutionary past, rights-based understandings have particular salience for many, as they relate to deeply held convictions and identity-forming experiences. This applies to the sizeable minority of people who played an activist role in the revolutionary project. But also for people active in opposition, or who simply lived through the turbulent decade of the 1980s, these issues can evoke strong positive or negative feelings.

Moreover, the revolution was a nationalist project, aimed against not only the country's dictatorship, but also against the perceived sacking of the wealth of Nicaragua by an international system dominated by the same USA that had kept the Somozas in power. The notion of national sovereignty is particularly strong in Nicaragua, and remains a potent argument against international financial institutions and transnational companies. This is seen in the criticism of Unión Fenosa, and was also a fundamental element in the successful campaigns that stopped the privatization of the hydroelectric power plant Hidrogesa and the planned mega-dam of Copalar,[13] to be built with Mexican capital and with the intention of exporting much of the produced energy. While many chapters of this book describe claims to resource sovereignty that challenge the state, these are examples where notions of resource sovereignty are co-extensional with – and therefore serve to strengthen – the idea of a national state.

There is also the widespread perception that electricity is a good to which people have an entitlement. Electricity as a right clashes with the idea that electricity is just another commodity in the market. Thus, the case of the respectable Doña Lilian who has had her electricity cut because she could not pay the charges that suddenly and inexplicably increased almost tenfold raises great indignation over the injustice and disrespect experienced by the poor lady, who has to suffer the indignity of 'sitting in the dark at night'. Similarly, for people in a recently settled barrio, still without electricity service, the use of illegal hook-ups to connect to the grid is considered normal. It is not understood as stealing, or even as unethical, but simply what you have to do if you are so unlucky as to live under such circumstances.

Furthermore, protests are fuelled not only by notions of the right to electricity, but perhaps even more strongly by the belief in the right to be treated fairly and with respect. Unión Fenosa's way of dealing with customers – its unwillingness to admit mistakes but readiness to brand customers as thieves, its practice of overcharging, its unapproachability for complaints, and its readiness to add extra charges and penalties – really generates outrage. And this is compounded by the sense that the playing field is not level: while the company may continue breaking the rules without any consequences – at worst they may have to repay what they have overcharged in a few cases – any formal mistake made by the electricity consumer inevitably leads to the loss of whatever valid claim she may have held.

Thus, energy politics is also a contest of ideas and values, where allegedly rational, objective and impersonal market logics confront a varied set of ideas and values, ranging from contrasting ideas of state responsibilities and notions of nationalism and sovereignty to deep-set ideas of entitlement to electricity and rights to fair and dignified treatment. But as shown, this ideational contest is not an open contest; outcomes crucially depend on the shifting constellation of social forces and political power, while state institutions, civil society and the interplay between them form the arena of the struggle.

Nicaragua demonstrates clearly that this is not a static arena: the boundaries between state, civil society and the market shift under the different regimes. The final years of the Bolaños government – with an inefficient market-based electricity sector and frequent blackouts that sparked popular protests and alienated parts of the private sector – could be dubbed a 'failed energetic state'. While this was certainly not *the* cause of the subsequent electoral loss to the Sandinistas, it was a contributing factor. In contrast, an aggressive state-driven policy for energy development on the part of the Ortega government has ended the energy shortages and made great strides towards unifying the nation in a national electric grid. While an unpopular privatization of electricity distribution has been upheld, outlets for protest over this have been stifled through the government's hold over grassroots organizing and general restrictions on independent civil society activism. One is tempted to believe that Ortega in his current 'pink socialist' incarnation still remembers Lenin's slogan that 'communism equals Soviet power plus electrification of the whole country'. At least, Ortega's Nicaragua is asserting its resource sovereignty and making great strides towards becoming an energetic state – while at the same time strengthening the party's hold on state power. In this reading, more thanks to the Sandinista government than to God.

Notes

1 Ley de la Industria Eléctrica (No. 272-1998), and Ley de Reformas a Ley Orgánica del Instituto Nicaragüense de Energía (No. 271-1998).

2 After a merger in 2008, Unión Fenosa changed its name to Gas Natural. In practice, however, Nicaraguans have continued to refer to Unión Fenosa, a usage I follow in this chapter. At privatization Unión Fenosa took over the two distribution companies Disnorte and Dissur. While these still exist and are formally the companies from which Nicaraguan consumers buy their electricity, I will refer to them as Unión Fenosa, as was common in Nicaragua in the period described. In 2013 Gas Natural Fenosa divested itself of its Nicaraguan engagement. Another Spanish company, TSK, acquired the holding.

3 When the privatization of the Hidrogesa hydroelectric plants was blocked by a National Assembly vote, the IMF responded by threatening that an agreement with Nicaragua would not be made unless the decision was reversed (*El Nuevo Diario*, 20 July 2002). Also, debt relief under the HIPC was used to seek to press Nicaragua to carry through the privatization of Hidrogesa, even though at the end a waiver to this condition was given (McGuigan 2007: 31).

4 The information given here on relations between Unión Fenosa and electricity users was collected in October/November 2011, in interviews with the RNDC and with electricity users in Batahola Sur. I was told that the current practices of Unión Fenosa were similar to those of 2005/06.

5 Around US$10.

6 See Borchgrevink (2006) for a more in-depth analysis of these processes.

7 Ratified by the National Assembly on 12 February 2009: legislacion. asamblea.gob.ni/normaweb.nsf/9e3148 15a08d4a6206257265005d21f9/de5ac72 7f7881850062575a700744bof?Open Document.

8 There are differences. The Sandinista government has fought hard and repeatedly over the years to make the international financial institutions accept subsidization of small electricity consumers.

9 Ortega seems determined to avoid the confrontations with the private sector, the Church and the multilateral institutions that led to civil war, economic crisis and the fall of his first government.

10 Ruth Selma Herrera resigned from the position in April 2010, owing to conflicts with the (Sandinista) union organizing the employees of the institution.

11 In most cases, the reason given for phasing out support has been a desire to shift aid towards the poorer countries of Africa. In some cases cuts have been made owing to dissatisfaction with the Sandinista government.

12 See Borchgrevink (forthcoming 2015) for a more detailed analysis of how the change of government in 2007 led to new state–society relations and restricted space for civil society activism.

13 Copalar was a grand hydroelectric scheme which the Bolaños government sought to get approved, but which was stopped after intensive protests by people from the area facing resettlement, in alliance with Managua-based and international organizations.

References

Acevedo, A. (2005) 'Crisis energética? Aquí está el detalle', *Envío*, 279.
— (2009) 'Sector eléctrico en Nicaragua', *Boletín Quincenal*, June.
Borchgrevink, A. (2006) 'A study of civil society in Nicaragua. A report commissioned by NORAD', NUPI Paper 699, Oslo: Norwegian Institute of International Affairs.
— (forthcoming 2015) 'Civil society under different political and aid regimes in Nicaragua', *Journal Für Entwicklungspolitik/Austrian Journal of Development Studies*, 31(1).
Carrión Rabasco, J. (2010) *La Ir-Responsibilidad Social de Unión Fenosa: Nicaragua, Colombia y Guatemala*, Nicaragua: Observatorio de la Deuda en la Globlización.
Close, D. (2005) *Los años de Doña Violeta: La historia de la transición política*, Managua: Lea Grupo Editorial.
Cupples, J. (2011) 'Shifting networks of power in Nicaragua: relational materialisms in the consumption of privatized electricity', *Annals of the Association of American Geographers*, 101(4): 939–48.
De Soto, H. (2000) *The Mystery of Capital: Why capitalism triumphs in the West and fails everywhere else*, New York: Basic Books.
ENATREL (2014) 'Nicaragua se aproxima al 80% de cobertura eléctrica', www.enatrel.gob.ni/index.php/noticias-y-eventos/1294-nicaragua-se-aproxima-al-80-de-cobertura-electrica, accessed 12 October 2014.
Herrera Montoya, R. S. (2005a) *Crisis del sector energético: Nicaragua apagándose?*, Managua: Red Nacional de Defensa de los Consumidores.
— (2005b) 'Nuestro sistema energético revela uno de los fracasos más grandes de nuestra clase política', *Envío*, 283.
Lecaros, F., J. M. Cayo and M. Dussan (2010) *Central America Regional Programmatic Study for the Energy Sector: General Issues and Options*, Washington, DC: World Bank.
McGuigan, C. (2007) *The Impact of World Bank and IMF Conditionality: An investigation into electricity*

privatisation in Nicaragua, London: Christian Aid.

MEM (Ministerio de Energía y Minas) (2011) 'Concol presenta informe de pérdidas en el sistema de distribución eléctrica', www.mem.gob.ni/index.php?s=1&idp=174&idt=2&id=76, accessed 11 September 2013.

— (2014) 'Gobierno Sandinista sigue haviendo patria con el cambio de la matriz energética', www.mem.gob.ni/index.php?s=1&idp=174&idt=2&id=745, accessed 12 October 2014.

Mostert, W. (2007) *Unlocking Potential, Reducing Risk: Renewable Energy Policies for Nicaragua*, Washington, DC: ESMAP, World Bank.

Ripley, C. G. (2010) 'The privatization of Nicaragua's energy sector: market imperfections and popular discontent', *Latin American Policy*, 1(1): 114–32.

Scott, J. C. (1998) *Seeing Like a State: How certain schemes to improve the human condition have failed*, New Haven, CT: Yale University Press.

Serra, L. (2006) 'Las luchas sociales en Nicaragua en el contexto electoral', *OSAL Observatorio Social de América Latina*, 7(20): 225–35.

Soto Joya, M. F. (2009) 'Nosotros? Sandinistas. Recuerdos de revolución en la frontera agrícola de Nicaragua', PhD thesis, University of Texas at Austin.

4 | WIND AT THE MARGINS OF THE STATE: AUTONOMY AND RENEWABLE ENERGY DEVELOPMENT IN SOUTHERN MEXICO

Cymene Howe, Dominic Boyer and Edith Barrera

Oaxacan wind power at a crossroads

On a windy day in February 2013, we encountered transnational green capitalism on the frontier of Mexican statecraft. This happened in the isthmus of Tehuantepec, in the town of Santa Rosa de Lima, where a state police checkpoint had been set up just before the road from Juchitán turns south-east towards Álvaro Obregón. 'Checkpoint' somewhat overstates the formality of the situation. There were two state police trucks idling by the side of the road, across the road from several weathered wooden canoes. The canoes and the gossamer of faded fishing nets near them announced to all passers-by that they had crossed into the fishing zone of the Laguna Superior. The police half-heartedly reviewed our papers after we asked Señor Tomás, our taxi driver for the day, to stop the car. A representative of the Oaxacan government had told us that morning that the situation in Álvaro was *muy tensa* (very tense). He said the police were reporting that trucks filled with armed men were driving around the village threatening violence. Señor Tomás, by contrast, said we had nothing to worry about. Driving through the verdant ranch land south of Juchitán, he explained how politics is always tied to money in the isthmus: 'This is how it always is, people make a lot of noise about this or the other thing. But it is just because they want to be paid. When they get paid, all this resistance will dry up. You'll see.'

What we did not expect to see in Santa Rosa, and the reason why we asked Señor Tomás to pull over, were two gringos talking to the local police commander. The tall one turned out to be someone, Andrew Chapman, we had been pursuing without success for some time. Chapman was part of the senior management team of Mareña Renovables, a consortium seeking to build the largest single-phase wind park in all of Latin America (consisting of 132 turbines with

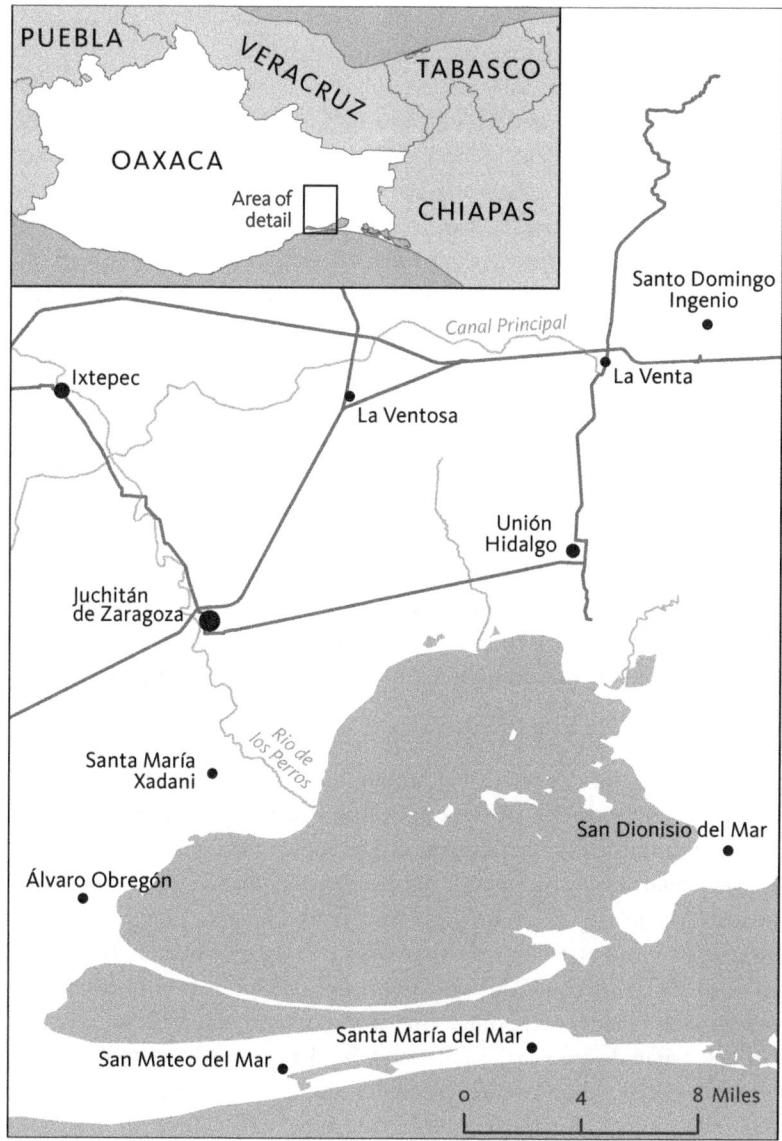

4.1 Isthmus of Tehuantepec, Oaxaca, Mexico

a production capacity of 396 MW). Mareña was, as everyone in the region knew, in deep trouble. Its park was designed to stretch across a sandbar at the southern end of the laguna from the *binnizá* (Zapotec) community of Álvaro across to the *ikojts* (Huave) communities of Santa María del Mar and San Dionisio del Mar.

Originally on schedule for completion by the end of 2012, the project suffered several delays before becoming the focus of intensifying resistance during the second half of 2012 from popular assemblies (*asambleas populares*) across the lagoonal region. The *asambleas* are, as discussed in more detail below, community-level organizations of self-governance united against governmental and industrial *megaproyectos* (mega-projects) in the region. By the end of 2012, *bloqueos* (blockades) had been erected in both Álvaro and San Dionisio to prevent the start of Mareña's construction. Indeed, the day before, the president of Mareña's board of directors, Jonathan Davis Arzac, had announced to the press that the project and what the company calculated to be an investment of 13 billion pesos in the region would leave Oaxaca unless 'rule of law' could be guaranteed. Davis claimed there were only 'twenty well-identified people' resisting the park, what he and the Oaxacan news media characterized as a violent and unscrupulous minority of political opportunists, holding communities to ransom and blackmailing project developers. Mareña was unwilling or unable to believe that large numbers of Istmeños were organizing in resistance to a project that the consortium viewed as hugely beneficial to all stakeholders.

Even if we had come to be suspicious about the public minimization and demonization of the opposition movement, we found it hard not to sympathize with Chapman. He spoke to us openly. Shouting himself hoarse over the wind, he seemed like a man desperate to be heard. 'My job is to go in there and try to open a dialogue and to go and listen. But I can't do that with threats of violence. If it's safe to send my people in, I'll send them in. ... The only way to change minds is to listen to people. But if you're not allowed to listen to people, what do you do? [Throwing his hands up in despair] We've got this project that I really believe is good for the planet, good for the region, good for the people down here.'

The people, Chapman felt sure, would come around.

> You can't help but be stunned by the beauty of this place. And then you see how the people are *living*. And I'm trying not to just impose my American values here but I don't think lousy medical care is a good thing, that lousy schools are a good thing ... So if

you can funnel resources into these communities to improve those services, imagine where they could be in five or ten years. They can still be fishing the lagoons but they'd have basic stuff, like electricity that is continuous, like transportation, like schools ... It may sound very idealistic but that's actually what we're trying to do. And to be confronted with this violence and with people who are essentially lying about what we're trying to accomplish ...

He trailed off; the resistance was wearying him and his investors. We asked how much patience he had left and he replied grimly, 'Not much.' Then, a moment later, he concluded,

I just find it frustrating, and sad, and the consequence is that the investor group that I represent ... they're sitting in their offices and they can put their money here, they can put their money there and they're just going to say to themselves, 'Why? I don't need these problems. I'm not actually in the business of saving the world, I'm in the business of earning money for my fiduciaries. And I need to do that in a low-risk way.'

More bad news came minutes later from the police – in the interests of his own safety there was no way they were going to allow Chapman to enter Álvaro that day. The risks were, as predicted, too high. The state police treat the Obregonian resistance with a great deal of caution. Memories are still fresh from the previous November when Mareña and the police last tested the Álvaro blockade during the Day of the Dead festival. With several protesters hauled off to detention in Juchitán and others subdued with pepper spray, the blockade was temporarily broken, allowing company workers to access the sandbar and begin topographic and vegetation removal work. But only hours later a much larger crowd of several hundred Obregonians rallied to chase the police and company off, overturning trucks and taking construction equipment hostage. We arrived in the aftermath to find the opposition more galvanized than ever. One of the leaders told us, machete at his hip, 'If they want to see blood, here we are, we're ready.'

As we finished our impromptu interview with Chapman, two gleaming white trucks glided up with other representatives of the company and the state government to collect him, and we parted ways at a crossroads that seemed designed by Hollywood as a symbol for the impasse in which wind power development in Oaxaca was increasingly finding itself caught.

An anthropology of failure

Construction on the Mareña park site has still not, as of the time of this writing, begun. Indeed, it seems increasingly unlikely that it will ever begin and that the project will have failed despite strong support from all levels of the Mexican government, generous financing from the Inter-American Development Bank, new national policy regimes favourable to both transnational private energy development and to renewable energy development, almost uniformly positive media coverage in the national and regional press, and a consortium of powerful international investors. The marriage of Mexico's aggressive renewable energy development campaign,[1] its high electricity tariffs and the world-class wind resources of the isthmus seemed to all parties to be highly auspicious. In a matter of less than a decade more than a dozen wind parks have come on-grid in the isthmus, according to industry experts the densest development of wind energy anywhere in the world. Investors continue to flock to the region while renewable energy advocates across the world have lauded Oaxacan wind power development for its positive contribution to climate change remediation. The Mareña project alone could avert the emission of up to 879,000 tons of carbon dioxide a year.

What we seek to explain in this chapter is how a project as diversely supported and ecologically timely as Mareña failed. As one might imagine, the vectors of this failure are highly complex, involving both new and old forces, relations and institutions, each of which signals contingent power relations and contested interactions between and among social and natural domains. To stay true to the complexity of the case, in lieu of a simplified argument for a single decisive causality, we explore a bundle of different issues and perspectives that help illuminate the challenges facing programmes of energy transition, especially in parts of the world like the isthmus of Tehuantepec that have long been marginalized or abandoned by their governing states. Renewable energy, very much like its carbon counterpart, provokes challenging questions about resource sovereignty, indexing the uncertain and uneven abilities of nation-states to manage territorial resources in the public interest (McNeish and Logan 2012; this volume). In the case of renewable energy, which is intended to prevent further climatological damage in the global pubic interest, and concerns about whose sovereignty, what territories and which resources ought to be state managed or locally controlled, surface dramatic ethical tensions between concerns for local and global well-being.

Of course it is no secret that large-scale projects of energy development are almost always politically complicated, especially when the intense energy needs of translocal governance and industry are perceived as threatening local interests. As critical institutionalist perspectives have shown (Cleaver 2012), it is very often the case that institutions tasked with managing natural resources are rarely explicitly designed to do so, and these disjunctive origins can easily result in ambiguous and 'patched together' development and responses to resistance. Recent anthropological studies have highlighted the complex and often contentious relationship between state- and industry-led energy development schemes and indigenous peoples, especially concerning rights to land and resource use (Colombi 2012; Smith and Frehner 2010; Turner and Fajans-Turner 2006; Westman 2006). Mexico has been no stranger to conflicts surrounding energy-related modernization schemes, especially in the areas of mining (Liffman 2012) and petroleum extraction (Breglia 2013). Until the 1980s, Mexico had the largest population displaced by irrigation and hydropower projects anywhere in the world (Robinson 1999).

But the isthmus also has a long political and cultural history of resistance to the hegemony of Oaxaca Valley elites and to the nationalist overtures and arms of Mexico City. Few histories of the isthmus, formal or informal, do not invoke or reinscribe an Istmeño tradition of struggle against external power (Campbell et al. 1993). Narratives often begin before the Spanish conquest with Istmeños depicted as the last true Zapotecs after the Aztecs made the northern Zapotecs their minions, assimilating them culturally and linguistically. Then one is reminded of revolt after revolt in the nineteenth century, of how national and Oaxacan hero Benito Juárez never broke the will of Juchitán even though he burned it. One hears how brave Juchitecos later fought valiantly against the French invasion of Mexico and helped secure the victory of the Mexican Revolution. But when that revolution turned corrupt, Istmeños rose again, with the Coalition of Workers, Peasants and Students of the Isthmus (Coalición Obera, Campesina, Estudiantil del Istmo – COCEI) helping to accelerate the dissolution of the PRI's (Partido Revolucionario Institucional) power across Mexico. In all such stories, the isthmus is frequently identified as the place where the sovereign powers of Oaxacan state and Mexican national governance end and where, as in Álvaro Obregón, the police fear to tread.

A local history of political resistance to national and state-level governmentality in the region surely explains, in part, Mareña's impasse. But the resistance also exists for reasons quite contemporary, illuminating deeper tensions and paradoxes in the dominant model of renewable energy development worldwide. As elsewhere in neoliberalism, market- and entrepreneur-oriented policies are trumpeted as the most effective and 'rational' solutions to anthropogenic climate change. Large-scale renewable energy projects like wind parks are highly capital intensive and thus become tightly bound to finance capitalism and to expectations for a positive return on shareholder investment. Thus, even though green energy projects typically speak a language of environmental sustainability, the growth imperative of dominant models of economic health and a rational-choice ethos carry over to them. Communities, perhaps especially indigenous communities, with resources like land, water and wind that can be converted to 'mega-project'-level renewable energy projects, thus often find their own interests compromised by the growth and profit motives of states and their transnational corporate partners.

In the fading light of the black sun

And there are more complications still. Oaxacan wind power development needs to be understood as a response not only to climate change but also to the vulnerability of the Mexican petrostate. Although peak oil fears have dissipated in many countries with the rise of shale oil and gas extraction, in Mexico such concerns remain both strong and empirically substantiated. Mexican heavy crude production fell by 46 per cent from 2004 to 2012. With its supergiant oilfield Cantarell running dry and the national oil monopoly Petróleos Mexicanos (Pemex) widely believed to lack the expertise and resources to effectively develop deep-water hydrocarbon resources, the Mexican petrostate finds itself at its own crossroads. In recent years Pemex has supplied as much as 40 per cent of the operating budget of the Mexican federal government, meaning that all aspects of Mexican statecraft depend critically upon revenue from oil sales. The crude production drop has been masked to some extent by a concomitant rise in international oil prices, which allowed Pemex to retain high revenue. But, one might say, the current financial model for the Mexican state is only one crash in the carbon energy market away from disaster.[2] President Felipe Calderón's push to aggressively develop renewable energy resources was formulated in this context. As an official in the Mexican environmental ministry explained to us, 'we need to diversify the sources of our electricity production. The hydrocarbons we don't use for our own energy consumption are hydrocarbons we can sell at a good price.' As anthropologists of oil have noted in other contexts, the combination of enduring societal dependency on hydrocarbon resources and the mounting difficulties of resource extraction and environmental impact are generating intense and sometimes unpredictable political and cultural effects across the world (Behrends et al. 2011; McNeish and Logan 2012).

In the fading light of what Reza Negarestani (2008) calls 'the black sun' of oil, new energo-political models are taking shape in petrostates to support the dominant growth model (Boyer 2014; Mitchell 2011). In Mexico, interest in developing the wind resources of the isthmus of Tehuantepec dates back to the early 1990s. However, it was only during Calderón's presidency that a serious campaign to develop renewable energy began. Crucial elements of this campaign were new legislation and a regulatory framework favourable to private–public partnerships in renewable energy development. The wind power sector skyrocketed, growing from two parks producing 84.9 MW in 2008 to fifteen parks producing 1.331 gigawatts (GW) by the end of 2012 (a 1,467 per cent increase that has made Mexico the second-biggest wind power producer in Latin America after Brazil). The dominant development scheme has been industrial self-supply (*autoabastecimiento*) partnerships in which a private wind developer contracts to produce energy for a large industrial client (examples include CEMEX, Walmart and Bimbo) over a period of several years or decades. These schemes are typically portrayed as win-win-win for the government, developers and industry. Companies can lock in lower-than-market energy prices for the long term, enjoy the financial benefits of *bonos de carbono* (emission reduction credits), and guarantee themselves a secure energy supply. Developers enjoy special access to green development financing through organizations like the Inter-American Development Bank and the UN's Clean Development Mechanism. States receive infrastructural development and economic multipliers without having to invest themselves. Communities are also frequently portrayed as winners in self-supply development in that they typically receive land rents and payments from usufructuary agreements.

But many Istmeños have come to have doubts about the benefits of *autoabastecimiento* wind development (Nahmed Sittón et al. 2011). Some have come to demand compensation for the use of their land beyond rents, some claim to have been tricked or pressured into signing contracts by agents of the government or developers, some deny the validity of mega-project-level development altogether. The Mareña project has helped to refine and to intensify resistance to the current paradigm of Mexican wind development for several reasons. First, the project is viewed by its critics as epitomizing a general lack of transparency in the development process. What is now known as the Mareña Renovables project has shifted names and forms several times since development began in 2003. Most recently, a Spanish energy firm, Grupo Preneal, which had signed contracts for land exploration and secured governmental permissions, sold the project rights (for what at that time were two separate projects) for $89 million to FEMSA, Latin America's largest beverage company, and Macquarie Group, Australia's largest investment bank. These two companies quickly fused the two projects together and sold part of their stakes to the Mitsubishi Corporation and to the Dutch pension fund PGGM, and signed a power-purchase agreement with FEMSA-Heineken for twenty years. Little if any of this information was communicated directly to the communities that would be impacted by the park.

Speculative activity was also quite common in the early days of the Oaxacan 'wind rush'. It is difficult to reconstruct precisely the behind-the-scenes politics of the period, but there is significant evidence that some type of cartel-like organization was organized or permitted by the Oaxacan state government in which wind developers were assigned exclusive negotiation rights over choice plots in the core wind zone. Many of these plots were 'flipped' at a profit and again without informing the communities that would be impacted. This cartelism had unfortunate legacies in that communities were never allowed to entertain competitive bids from different developers, which set remuneration rates at a low level by international standards. Also, speculators, knowing they would not have to see these projects through to operation, apparently frequently cut corners in terms of community relations, with many Istmeños and communities complaining later of having been given insufficient or inaccurate information regarding the benefits that projects would bring.³ In the Mareña case, a federal judge in Salina Cruz issued an injunction (*amparo*) to halt progress on the park project in December 2012 in order to further investigate opposition claims that communal land was being expropriated without prior information and the consent of the majority of the *comuna*.

A second criticism frequently raised against Mareña is that it has sought, like many wind developers, to advance its project through manipulation of local authorities rather than through consensus-building projects with whole communities. Companies fervently deny these claims. But critics contend that contracts for exploration and land use rights were facilitated by bribes paid to *presidentes municipales* (mayors) or *comisariados* (collective land commissioners) in the form of cash or trucks. Even when these authorities are not personally implicated in embezzling funds or resources, it is said that they share these resources only within their own political network, thus taking what ought to be a social good and privatizing it. Multiple high-ranking members of the Oaxacan state government singled out Mareña as the worst offender they knew in terms of these practices. One figure described Mareña to us as a 'clear case of how things ought not to be done'. Another wondered why Mareña would pay 'loads of money' to buy off local authorities who rotate every three years in the normal course of elections.

This points to a third general area of complaint – that wind development in the isthmus has been accentuating social inequality, political polarization and violence by deprioritizing general social benefits (e.g. the very schools and healthcare mentioned by Chapman) in favour of benefits to specific landholders and authorities. For example, although it has significant backing among most political parties in the isthmus, wind development is particularly associated with the PRI party's political network. The PRI are in turn closely associated with the construction unions that benefit directly from park contracts. Across the isthmus we have heard that PRI political authorities recruit groups of *golpeadores* (thugs) from the construction unions who are utilized to intimidate, threaten and in some cases actually attack those who resist or oppose wind park developments. This dynamic has been particularly evident in San Dionisio del Mar, which has been on the front lines of the Mareña conflict. Although there is some evidence that the wind park enjoyed bilateral support in earlier phases, as the project has advanced it has clearly exacerbated local political tensions between the long-reigning PRI and the leftist PRD (Partido de la Revolución Democrática). The PRD faction in San Dionisio has accused the PRI *presidente municipal* of signing a usufructuary agreement with Mareña to begin construction work without consulting them and of pocketing the fee for himself and his allies. Thus, as one San Dionisian student put it pithily, 'the whole town is divided. Basically, if you are a PRI family, you are for the park. And if you are in the PRD you are against it.'

Although many of the criticisms levied against Mareña are similar to those that all wind development in the region now faces, it is important to highlight certain singularities of the project that have helped make it a watershed case. When we asked one mid-level Mareña employee why she thought the project had generated so much controversy, she said, 'Well, for two reasons. It's a very ambitious project and it's the first one designed to occupy communal land.' The project is ambitious in that it would be the first one to impact multiple communities in the area (not only Álvaro and San Dionisio but also the many other lagoonal communities that have traditional rights of access to fishing near the sandbar). Likewise, it would indeed be the first park to occupy land actively administered in the *bienes comunales* land tenure system, which requires a majority of *comuneros* to approve any change to the use of the land. Although other wind park projects have occupied land that is formally communally owned, they have always been in cases in which communal organization had lapsed and/or a communal decision for land privatization had occurred previously, allowing current landholders the right to contract individually with developers.

Finally, we would also highlight a geographic singularity to the Mareña project: as apparently the only wind farm project in the world designed to occupy a sandbar, its environmental impacts are especially difficult to estimate. There is little evidence available to address burning questions such as whether the presence of the turbines would create vibrations and shed light that would scare fish away. The Environmental and Social Management Report published by the Inter-American Development Bank in November 2011, for example, noted the possibility of short-term 'economic displacement' from the disruption of fishing during the construction phase of the park (2011: 18–19) but curiously failed to discuss long-term impacts of the park's presence on local fish populations despite extensive analysis of possible impacts on bat, marine turtle and jackrabbit species.

Thus, to summarize, in certain respects Mareña amplifies doubts and criticism already being directed towards Oaxacan wind development (manipulation of authorities, cartelism, heightened social inequality). In other respects, the project represents a watershed event by being the first project attempted on communal land, the first project to impact multiple communities simultaneously, and the first in the region located near fishing communities instead of agrarian and ranching communities. These factors combined to create a context in which local resistance groups united in a regional and increasingly trans-regional network during the period of our fieldwork in 2012 and 2013. Since it has been by far the most immediate cause of the Mareña project's failure, we now turn to a deeper analysis of the 'anti-eolic' resistance movement in the isthmus of Tehuantepec.

Capturing the meter

Back at the crossroads of green capitalist aspirations and the barricades of Álvaro Obregón, one is struck by the sheer quantity of dust that arises as a truck rolls down the road; it is reason enough to wrap a handkerchief or T-shirt around any and all respiratory orifices. Men in Álvaro Obregón are often outfitted this way, bare chested with a thinning T-shirt shrouding nose and mouth to keep the dust out. Maybe the protective shirt is printed with the smiling face of a bygone candidate, maybe a rock and roll concert relic; in either case, the chronic cough that is audible all over Álvaro makes it seem as though this is a losing battle. Today, in front of the abandoned hacienda that the *resistencia* has appropriated as their meeting place, T-shirts have been fashioned into masks by a group of young men with a more symbolic purpose, signalling a touch of outlaw: part Zapatista, part universal gangster. As they jump down from the back of a white pick-up truck even the T-shirt masks don't conceal their smiles. They've just returned from an excursion to the site where Mareña Renovables has its test tower, a spindly metal steeple with a three-pronged wind vane to measure the quality, duration and force of the wind. The masked men have something in hand, a prize. The crowd, numbering seventy or so, soon gathers around, eager to see what it is. Passing the booty from hand to hand with care, the object finally gets close enough for us to see that it is a gauge of some kind, with settings and indicators in English and numbers on dials. 'It fell down,' they explained to us, 'from the tower.' 'It *fell* down?'

we asked, incredulous. Their grins grew perceptibly wider; they had decided the bluff was not worth pursuing. 'Well, it fell off when we pulled down the tower.'

The Mareña project may have a powerful set of allies and all the forces of transnational capital behind it. But it does not have the approval of the men in T-shirt masks. Protest against the Mareña project has political precedent from the COCEI movement and earlier anti-eolic protests in the isthmus to more pervasive concerns about mega-projects in general (Gómez Martinez 2005). Those in the resistance often connect their political affinities to these histories, such as their ideological links to the Zapatistas and an original (uncorrupted) COCEI. But the *resistencia* also shares affinities with a broader, contemporary set of political practices and protests against the status quo, from anti-globalization movements to Occupy. In the second half of this chapter, we document the antecedents of the anti-Mareña resistance, focusing on their political genealogies as well as their ideological commitment to collectivist, non-hierarchal models that aim to supersede historic ethnic and political rivalries. Often drawing from ideals of neo-indigenous horizontal organization and reinvigorated customary law, the *resistencia* has channelled opposition against the Mareña park for both its mega-project scale and its transnational financial ties. Posing an explicit critique of neoliberal forms of development and foreign financial intervention, the resistance has sought to foster collaborations between *ikojts* and *binnizá* populations and to encourage alliances across party lines, although each of these goals meets with uneven results. Finally, as we describe below, the *resistencia* has found political purchase not because it opposes renewable energy (it does not), but because it has brought to the surface concerns about the potential environmental and social consequences that might follow in the wake of the park's construction. The *resistencia* has codified a suite of concerns ranging from the displacing and destructive potential of mega-projects to worries about the loss of land and fish. *La laguna, la pesca* and *nuestra tierra* have become affectively aligned as both elements of quotidian practice and symbols of regional patrimony. The local resistance that has effectively halted the Mareña project has led us to understand that climate change mitigation measures have indeed, as Mike Hulme predicted (Hulme 2009: xxvii), fomented new opportunities for environmental consciousness and activism. However, the environmental critique levelled by the resistance contravenes corporate and state-sponsored sustainable energy projects that also claim to offer protections to our shared ecology. While transitions to renewable energy have the ethical potential to leverage a global climatological good, when they are seen to contravene local claims for rights, autonomy, environmental knowledge and ecological stewardship, they instead generate, as Mareña found, the conditions for failure. We have found that when renewable energy transitions and climate mitigation are coupled with neoliberal development schemes that mirror the logics of extraction, we are apt to see a reinvigoration of lateralist, collaborative and horizontal modes of activist response and resistance.

Rescuing the land from the wind

The office of the Asamblea de los Pueblos Indígenas del Istmo de Tehuantepec en Defensa de la Tierra y el Territorio is readily identified on the streets of Juchitán; it is the one with the anti-eolic art on its façade. As we sat down one Sunday afternoon with two of the founders of the *resistencia*, it was hard not to notice our intimate physical proximity in this tiny room decorated with images of past victories and heroes from Che Guevara to Subcomandante Marcos. Roberto P. began the conversation and proceeded to detail a vast historical narrative of the *resistencia* over the course of over an hour. Roberto is one of the primary voices of the resistance; he does not, however, like to be called a 'leader'. This is a designation that he associates with hierarchical, vanguardist and, ultimately, corrupt political forms. Roberto is a teacher by vocation and by nature, as became clear in his oration of historical events. The *resistencia* against Mareña, he explained, must be understood through a longer genealogy that spans many decades and locations. In addition to the insurrectionary politics of the isthmus – particularly the early COCEI movement – Roberto linked the *resistencia* to: the repression of the student movement in Mexico City in 1968 and a guerrilla *foco* in Chihuahua before that, the Chiapan rebellion of the Zapatistas following the North American Free Trade Agreement (NAFTA), the battle over the development of an airport in Atenco in the early 2000s, the teachers' strike and state violence in the capital of Oaxaca in 2006 guided by APPO (Popular Assembly of the Peoples of Oaxaca), and Maoism itself, with its agrarian peasant insurgencies and challenges to First World imperialism. Roberto's cartography of revolution and response to foreign domination, urban hegemony and rebellions against neoliberal development brought us to the origins of anti-eolic resistance in 2005, founded by a group of committed teachers. He and others had protested against the installation of the La Venta wind park in the 1990s, noting that Subcomandante Marcos himself showed up and spoke in solidarity with them. Beyond symbolic gestures of the Zapatista leadership, the *asamblea* could also claim several significant victories of its own. These included nullifying contracts across the region and 'rescuing' 1,200 hectares of land from being contracted and thus turned into wind parks.

Roberto gives credit where credit is due; he is faithfully citational in his rendering of the *resistencia*'s insurrectionary lineage. Originally, the Juchitecan arm of the resistance worked under the name Frente de Pueblos en Defensa de la Tierra y del Territorio. However, the designation '*frente*', Roberto explained, is overly encumbered by vanguardism, hierarchical leadership and a military etymology, all qualities that they hoped to surpass. By consensus it was decided that the title of '*asamblea*' better captured their ethos. An *asamblea* evokes, as Roberto put it, 'a more indigenous notion, that of community'. With an egalitarian order and their rejection of hierarchical leadership the *resistencia* has proceeded with their platform in place.

> That is how we began and how we have preserved the shape of this struggle ...
> Since then, and in a concrete way, we have defined the line we have held until now, and that is, legal defence, direct action, mobilization, constant information for the communities, and founding *asambleas*.

Roberto readily adds that 'We have traversed the entire historical process of the left in Mexico in order to be able to offer an alternative.' The communitarian spirit of lateralist leadership and collective consensus is emphasized in both name, *asamblea*, and spirit, rejecting leaders and, implicitly, a flock of 'followers'.

Dedication to the collective model is also manifest in the deeds of the *resistencia*, who have prioritized founding more *asambleas* as a central part of their mission. *Asambleas generales* have since multiplied across the isthmus in towns and villages supporting the anti-Mareña resistance. These working forms of *autonomista* protest and process call for collective decision-making in a reinvented use of *usos y costumbres* (literally, practices and customs) – a legal and social system operating in parallel to state governance. *Usos y costumbres* have generally been regarded as a counterbalance to indigenous people's marginalization in elite national projects throughout Mexico and in other Latin American countries. The *resistencia* has evoked similar neo-indigenist ideals in its organizational forms as well as discursively in its materials and pronouncements (Jackson and Warren 2005).[4] Lauding indigenous knowledge and evoking autochthonous environmental stewardship have been powerful and proximate logics for the resistance movement, even as they risk certain essentialist interpretations (Dove 2006: 195–8; Tsing 2003). Claims to environmental wisdom and indigenous sovereignty have also been tailored to a very specific, and novel, alliance between *binnizá* and *ikojts* communities.

The ideological traction of indigeneity draws upon millennia of *binnizá* and *ikojts* dwelling in the region, but the *resistencia* has reworked an ossified anecdote about interethnic conflicts between these two communities (Gómez Martinez 2005), regarding land, displacement and historic antagonisms. Challenging the construction of the wind park, for many people with whom we spoke, marks 'the first time in history' that *binnizá* and *ikojts* peoples have worked together collaboratively. Pan-indigenous movements, *neoindigenismo* and collaborative activism have a precedent in Mexico (Stephen 2002; Jung 2003), but the co-ethnic solidarity between these two populations is understood to be a singular and critical advance in co-rectifying corporate and state exploitation. Alliances between *ikojts* and *binnizá* communities, as several people in the resistance shared, still feel very new, emerging, and not fully complete. Their only common language is the colonizers', yet working in this idiom, the *resistencia* has crafted a collective agenda of indigenous rights in the face of renewable energy incursions. A *comunero* from San Dionisio, for example, described how autonomy for first peoples and their ability to continue to occupy the lands where they have resided hold a certain power of truth and right.

> Today in San Dionisio the struggle continues, just as it began with our heroes who came together to seek Mexico's autonomy; here also, we seek autonomy for first peoples, for indigenous people, because we are the true owners of the land and the territories of the seas.

The *resistencia* has fomented and fostered collaborations that traverse historical divisions between local indigenous populations, but they have also managed to navigate party political lines. Given the historical strength and chauvinism of party politics in Mexico it is no mean feat for PRI-istas in the resistance to share meals with the PRD and the COCEI. Political rivalries have not ended – especially since PRI-istas are believed to be the primary beneficiaries of the isthmus wind boom – but within the ranks of the resistance these divisions have been muted. The political machinations of the parties continue apace in many aspects of wind park development, but because the political parties, left and right, have largely taken an official stance in favour of the Mareña project, the parties themselves have lost favour in the eyes of those resisting wind park development. The anti-Mareña resistance has, in further bids for autonomy, gone as far as summarily questioning the validity of political parties as legitimate democratic entities. Antonio L., one of the founders of the *asamblea* and one of the key voices in the *resistencia*, speaking to a crowd gathered in Álvaro Obregón, affirmed, 'Today is a declaration of war against the political parties, against the government, against Mareña Renovables, against everyone who is allied with or affiliates with Mareña Renovables.' Antonio and others in the *resistencia* also publicly announced that no political party candidate would be allowed to campaign for political office in the upcoming municipal elections. In June 2013 they made good on this pronouncement by prohibiting the installation of voting machines. The parties and the park have become a combined menace for those in the resistance, and so both have been given a directive and a direction: *fuera* (out).

On fish and neoliberalism

In January 2012, San Dionisian *comuneros* were meeting in earnest to block construction of the Mareña park under a new appellation, '*los inconformes*' (the nonconformists). By April *los inconformes* had initiated a permanent occupation of the town's municipal headquarters. Allied with forces in Álvaro Obregón, Juchitán and other communities, as well as other indigenous rights organizations in the region, such as Union of Communities in the North Zone of the Isthmus (Unión de Comunidades de la Zona Norte del Istmo – UCIZONI), the San Dionisian protests found further collective force, impact and media attention. '*Fuera Mareña*' (Out with Mareña) was a slogan that was beginning to ring out, and ring out increasingly loudly, across the isthmus.

In the discourse of the *resistencia*, the potential for a '*despojo de nuestra tierra*' (being robbed of our land) has been a rallying cry and an ominous reminder of colonial histories. It also operates as a reference to the thirty-year (or more) contracts that landowners, *ejidatarios* and *comuneros* sign with companies for turbine and road placement. The land, much of which was bequeathed for collective use by the federal government over the course of the last century, has a powerful patrimonial significance as well as an economic role in many Istmeños' lives. While land has been a key concern regarding wind parks throughout the isthmus, water – and more specifically that which inhabits the water and those that subsist from the water – has been a critical subject for those resisting the Mareña project. Indeed, there has been a clear effort to 'conjure nonhumans' as potent forces in these political struggles (De la Cadena 2010).

The fishermen with whom we spoke in San Dionisio and across the isthmus were convinced that their lives and their livelihoods would be irrevocably impaired by the development of the wind park. The precise effect that the Mareña project might have on the fish or shrimp population was unclear, given the environmental impact report's failure to treat this aspect of the regional conditions and the project's unique placement on a sandbar. However, in part because of this absence of scientific analysis, fears abound. Ibrahim C., who would emerge as one of the key voices among the *inconformes* in San Dionisio, summarized the sentiments of many in the region.

> The wealth of our sea, of our people, of our source of work and nourishment is vital [...] If the wind project comes in we will be buying foreign products coming from other places which will make feeding ourselves more expensive [...] and so in a sense San Dionisio now sees itself, or has transformed itself, into a courageous town that defends its lands and teaches foreigners that our lands must be respected.

It is telling that Ibrahim begins his comments with the 'wealth of our sea', for over time, as the resistance to the project grew and spread, we began to hear, increasingly, the expression 'the sea is our bank' (*el mar es nuestro banco*). This was a canny spin on the evident presence, or imposition, of banking interests and multinational capital that have backed the Mareña project. But 'the sea is our bank' is also a factual statement to a degree. According to reports there are 5,000 indigenous families that rely upon fishing for their existence. Even if the number of fisherfolk in San Dionisio proper who survive exclusively by fishing is likely no more than a few dozen, many, if not the majority, of the population rely on fishing for a kind of security subsistence in conditions of economic and food insecurity; if all else fails the sea is there and you, and your family, can eat. Ibrahim is not naive, after all, to underscore how harm to *la pesca* will result in increased dependence on a market-based food supply and further insertion into a vast chain of imported products.

Many advocates, Ibrahim included, were clear that opposition to the wind park was not a refusal of its 'clean' capabilities and renewable energy aspirations. Rather, they wanted to mount a warning and protest against the prioritization of market-based growth models spreading across every dimension of daily life in Mexico. Berta C., one of the founders of the Juchitecan *asamblea*, underscored that it is not wind energy that is at issue, but the specific dangers of massive foreign capital investment in the region that the parks have portended. These fiscal threats come in the form of bribes, manipulation and payouts. Huge sums of financial investment, whether invested in land rents or secretly passed into the hands of local caciques, are viewed as denigrating local sovereignty, causing further 'tears in the social fabric'. If there is any parallel to be made between NIMBY (not in my backyard) complaints and the concerns being voiced by San Dionisian fishermen or those wo/manning the barricades in Álvaro Obregón, it is of a qualitatively different kind. NIMBY objections that are pervasive in places such as the United States and Europe – disrupting one's view, spoiling the look of landscapes – ring hollow in comparison to those of subsistence fishermen pleading for their ability to survive. It is especially apparent when they are able to articulate the multiple ways in which neoliberalizations have changed and, from the point of view of many, endangered their ways of life and livelihoods.

Conclusions: downed windmills

Andrew Chapman, in his baseball hat with the thrashing fish embroidered across the front, and the young man behind his makeshift mask with the absconded wind meter in his hand represent very different places, both metaphorically and physically. Chapman has long since returned to New York, and the young man with the meter, he is likely either swapping stories with his friends or out on the water hauling shrimp-filled nets from the lagoon by moonlight. There is no reason to make speculative comparisons between the very different lives of each of these men and the economic, social and cultural worlds they represent. But it is fair to say that they do have shared expectations and parallel hopes for the future. Each of them can claim a virtuous, ethical position: the American bringing development, fomenting markets for renewable energy use, slowing the creep of global warming and producing profit for green investors; the Istmeño fighting for his future on the lagoon, challenging foreign invaders and transitional capital, ensuring 'food sovereignty' for himself, his family and the region.

Since they have a climatological impetus, mega-projects of clean energy production would seem to have an ethical edge over other mega, extractive endeavours, such as mining or oil drilling. However, the very dimensions of a mega-project, 'clean' or 'dirty', are invariably worrisome and controversial because they consume such vast tracts of space, whether land or sea (Turner and Fajans-Turner 2006; Liffman 2012). Renewable energy production may be a dramatic improvement over its carbon cousin, but in the case of the isthmus, the injection of foreign capital appears to erase much of this environmental and social potential. In this sense, the Mareña project and its failure in the face of resistance reveals more than simply another development desire gone awry. Rather, it brings to the surface ethical tensions that position local economic and environmental health against global economic and environmental health (Howe 2014). The Mareña case is a political economic calculation that attempts to balance scale against compensation. However, it also challenges the deeper logics of energy transition by demanding responses to how putative benefits and remediations will be made now, and in the future, locally and transnationally. There is likely no one in Álvaro Obregón, San Dionisio or other communities in the resistance who would declare their outright opposition to preventing further climate change or, in the broadest terms, making the world and their environment more humane and more hospitable. However, to entreat them to sacrifice their land and fishing grounds on behalf of international global protocols and climate change mitigation mandates devised in Kyoto, Durban and Copenhagen, which, in turn, benefit investors and developers at least as far afield, is a proposition that does not sit well in the isthmus. In a place that has successfully thwarted outside influence and control for several centuries, development driven by private, foreign capital for the benefit of large corporate consumers seems like folly. Asking Istmeños to risk further precarity on behalf of global warming and an abstract 'greater global good' may appear preferable to re-enacting histories of corporate exploitation and extractivism, but this attempt to engineer climatological altruism has failed to gain much traction. As they seek to protect and defend the resource-rich places they inhabit, the *resistencia* reiterates a politics of territoriality. Their motives can be understood, in this sense, as another attempt at 'resource sovereignty', whereby autochthonous communities seek to manage their territorial resources – such as land, water and wind – with or without the sanction of the state (McNeish, Logan and Borchgrevink this volume: 3). The Mareña project has proved to be a critical referendum on the possibilities for renewable energy in Mexico, but it is not singular, nor will it be unique as renewable energy projects continue to expand in Latin America and around the world. It is a lesson, however, in how disjointed development and failed attempts at sustainability mirror other projects that have similarly taken market-based models as the only possible 'rational' solution to the threats of the Anthropocene.

Critiquing capitalist development and creating horizontal and collective political models, the anti-Mareña actions share an affinity with uprisings and protests from Tahrir Square to Zuccotti Park. Horizontal networks function in place of hierarchies, consensus democracy replaces top-down direction, and principles of decentralization are prioritized (Graeber 2002). Reacting to these projects of green neoliberalism, the *resistencia* has been able to recapitulate the roots of the anti-globalization movement(s) and direct actions that, for many, originated with the Zapatistas. Like the Zapatismo formed in the Lacandon jungle, anti-WTO protests in Seattle, protests against the G8 in Geneva, to, most recently, Occupy across North Africa, North America and Europe, the *resistencia* has rejected hierarchical orders and decision-making (Jung 2003; Muñoz Ramírez 2008; Razsa and Kurnick 2012; Stephen 2002). It has made definitive critiques of neoliberal policies and the ways in which private finance capital may endanger local communities' livelihoods and well-being. In this sense, we would argue, energy protest movements in the isthmus represent an 'outpost of the new opposition' (New Left Review 2001).

However, there are at least two important distinctions to make regarding the Istmeño response to *neoliberalismo verde* that make it stand apart from Occupiers, anti-globalization actions and neo-anarchist movements that have emerged in other parts of the world. The first is their explicit incorporation of a collaborative, neo-indigenous model coupled with an adherence to a communal *asamblea* ideology; these tactics have been seen in other Latin American struggles over sovereignty and resources, to be sure (e.g. Dove 2006; Turner and Fajans-Turner 2006), but anti-eolic protest in the isthmus signals an emerging logic of resistance that rejects both green capitalist aspirations and greenhouse gas reduction as justifications for territorial displacements (see Howe 2014). The second distinction to be made is that the *resistencia* in the isthmus is, unlike Occupy, not reacting to a global financial crisis, but instead drawing attention to protracted forms of marginalization exercised by state policies and green capitalist developers. In other words, they are reminding those that will listen that they have been 'the 99%' for quite some time. The *resistencia*'s critique is not aimed at the *failure* of global finance capital, but rather is a scathing commentary upon its '*successful*' propagation across the isthmus. In place of Wall Street banks that were too big too fail, the resistance has challenged a massive renewable energy installation that has also seemed – with all of its international development and corporate sponsorship – too big to fail. And yet, as of now, it has. Despite the heft of Mareña's clean energy aspirations, the occupations of roads and barricades have revealed their debilities and shortcomings. And in this sense, the refusal of the Mareña project is not simply a referendum on how renewable energy projects will proceed in Mexico, but a foreshadowing of potential resistances, North and South, where renewable energy projects may be, increasingly, objects of dissent.

Notes

1 Under the administration of Felipe Calderón (2006–12), Mexico made great strides towards combating climate change through *transición energética*. The country's General Law on Climate Change (signed by Calderón as he was leaving office) outlines a comprehensive strategy for climate-resilient and low-carbon economic growth. The scope of the legislation makes it one of the most ambitious climate remediation laws in the world and it has, along with previous policies to reduce greenhouse gas emissions and bolster conservation, made Mexico a world leader in climate change mitigation.

2 Indeed, as this volume goes to press, the Mexican government has

announced 8.3 billion dollars in budget cuts reacting to the downturn in oil prices in late 2014.

3 In our recent survey of La Ventosa, an isthmus town now surrounded by active wind parks, we were surprised by how many people reported having been led to believe that the wind parks would reduce their electricity costs.

4 Although they are understood to be a pre-Columbian inheritance, *usos y costumbres* have been modified over time and have experienced a resurgence throughout Mexico (Carlsen 1999: 2; Stephen 2002; Rubin 1998).

References

Behrends, A., S. Reyna and G. Schlee (eds) (2011) *Crude Domination: The Anthropology of Oil*, New York: Berghahn.

Boyer, D. (2014) 'Energopower: an introduction', *Anthropological Quarterly*, 87(2): 309–34.

Breglia, L. (2013) *Living with Oil: Promises, Peaks and Declines on Mexico's Gulf Coast*, Austin: University of Texas Press.

Campbell, H., L. Binford, M. Bartolomé and A. Barabes (eds) (1993) *Zapotec Struggles: Histories, Politics, and Representations from Juchitán, Oaxaca*, Washington, DC: Smithsonian Institution Press.

Carlsen, L. (1999) 'Autonomía indígena y usos y costumbres: la inovación de la tradición', *Chiapas 7*, Mexico City: Instituto de Investigaciones Económicas.

Cleaver, F. (2012) *Development through Bricolage: Rethinking Institutions for Natural Resource Management*, London and New York: Earthscan/Routledge.

Colombi, B. J. (2012) 'The economics of dam building: Nez Perce tribe and global capitalism', *American Indian Culture and Research Journal*, 36(1): 123–49.

De la Cadena, M. (2010) 'Indigenous cosmopolitics in the Andes: conceptual reflections beyond "politics"', *Cultural Anthropology*, 25(2): 334–70.

Dove, M. R. (2006) 'Indigenous people and environmental politics', *Annual Review of Anthropology*, 35: 191–208.

Gómez Martinez, E. (2005) 'Proyecto Perfiles Indígenas Diagnóstico Regional del Istmo de Tehuantepec', Oaxaca City: Centro de Investigaciones y Estudios Superiores en Antropología Social, Unidad Istmo.

Graeber, D. (2002) 'The new anarchists', *New Left Review*, 13: 61–73.

Howe, Cymene (2014) 'Anthropocenic ecoauthority: the winds of Oaxaca', *Anthropological Quarterly*, 87(2): 381–404.

— (ed.) (2015) 'Latin America in the Anthropocene: energy transitions and climate change mitigations', Special issue, *Journal of Latin American and Caribbean Anthropology*.

Hulme, M. (2009) *Why We Disagree about Climate Change: Understanding Controversy, Inopportunity and Inaction*, Cambridge: Cambridge University Press.

Inter-American Development Bank (2011) Annual Report, www.iadb.org/en/annual-meeting/2012/annual-report-2011,6410.html.

Jackson, J. E. and K. B. Warren (2005) 'Indigenous movements in Latin America, 1992–2004: controversies, ironies, new directions', *Annual Review of Anthropology*, 34: 549–73.

Jung, C. (2003) 'The politics of indigenous identity: neoliberalism, cultural rights, and the Mexican Zapatistas', *Social Research*, 70(2): 433–62.

Liffman, P. (2012) 'El movimiento de lo sagrado por Wirikuta: la cosmopolítica wixarika', *Ediciones MNA*, Museo Nacional de Antropología, www.mna.inah.gob.mx/index.php/ediciones-mna/articulo/180-el-

movimiento-de-lo-sagrado-por-wirikuta.html.
McNeish, J.-A. and O. Logan (eds) (2012) *Flammable Societies: Studies on the Socio-economics of Oil and Gas*, London: Pluto Press.
McNeish, J.-A., O. Logan and A. Borchgrevink (eds) (n.d.) 'Recovering power from energy: reconsidering linkages between energy and development', in *Resource Sovereignties: Converting Energy into Political Power in Latin America and Beyond*.
Mitchell, T. (2011) *Carbon Democracy: Political power in the age of oil*, New York: Verso.
Muñoz Ramírez, G. (2008) *El Fuego y La Palabra: Una Historia del Movimiento Zapatista*, San Francisco, CA: City Lights.
Nahmed Sittón, S. et al. (2011) 'El impacto social del uso del recurso eólico', Oaxaca City: Centro de Investigaciones y Estudios Superiores en Antropología Social Unidad Pacifico Sur.
Negarestani, R. (2008) *Cyclonopedia: Complicity with Anonymous Materials*, Melbourne: Re.Press.
New Left Review (2001) 'A movement of movements? The punchcard and the hourglass, interview with Subcomandante Marcos', *New Left Review*, 9: 69–79.
Razsa, M. and A. Kurnik (2012) 'The Occupy movement in Zizek's hometown: direct democracy and a politics of becoming', *American Ethnologist*, 39(2): 238–58.
Robinson, S. (1999) 'The experience with dams and resettlement in Mexico', Contributing paper, World Commission on Dams, siteresources.worldbank.org/ INTINVRES/214578-1112885441548/20480078/ExperiencewDamsResettlementMexicoSoc202.pdf.
Rubin, J. W. (1998) *Decentering the Regime: Ethnicity, radicalism, and democracy in Juchitán, Mexico*, Durham, NC: Duke University Press.
Sawyer, S. (2004) *Crude Chronicles: Indigenous Politics, Multinational Oil, and Neoliberalism in Ecuador*, Durham, NC: Duke University Press.
Smith, S. and B. Frehner (eds) (2010) *Indians and Energy: Exploitation and Opportunity in the American Southwest*, Santa Fe, NM: School for Advanced Research Press.
Stephen, L. (2002) *Zapata Lives!: Histories and Cultural Politics in Southern Mexico*, Berkeley: University of California Press.
Tsing, A. L. (2003) 'Agrarian allegory and global futures', in P. Greenough and A. L. Tsing (eds), *Nature in the Global South: Environmental Projects in South and Southeast Asia*, Durham, NC: Duke University Press, pp. 124–69.
Turner, T. S. and V. Fajans-Turner (2006) 'Political innovation and inter-ethnic alliance: Kayapo resistance to the developmentalist state', *Anthropology Today*, 22(5): 3–10.
Westman, C. (2006) 'Assessing the impacts of oilsands development on indigenous peoples in Alberta, Canada', *Indigenous Affairs*, 6(2/3): 30–9.

5 | OIL AND ENVIRONMENTAL INJUSTICE IN VENEZUELA: AN ETHNOGRAPHIC STUDY OF PUNTA CARDÓN

María Victoria Canino and Iselin Åsedotter Strønen

Introduction

This chapter explores how the poor fishing community of Punta Cardón in Venezuela historically became transformed by the development of the oil industry, and their contemporary struggle to be recognized as victims of social and environmental injustice caused by more than sixty years of local oil production.[1] The community of Punta Cardón is part of the parish of Punta Cardón,[2] located in the Gulf of Coro in Falcón state on the north-west coast of Venezuela. Since the mid-1940s, the community has become enclosed within the oil-refining complex called Paraguaná, today run by the state oil company PDVSA (Petroleos de Venezuela SA).

The chapter has dual aims. One is to provide an ethnographic genealogy of the historical encounter between this once-secluded fishing village and the international oil industry, as experienced from the point of view of local villagers. This exploration serves to sensitize us to, on the one hand, the subtle, gradual and far-reaching social changes that industrial development may generate, and on the other hand the dense social histories that contemporary local knowledge about oil's impact on nature and environmental conditions are built upon, having accumulated through generations. This historical narrative allows us to gain a deeper understanding of how oil production in Punta Cardón is far from being just an industrial, technological endeavour, but rather, from the local population's point of view, deeply embedded in their collective social histories, knowledge regimes and identities.

The second aim is to explore the current controversies between, on the one hand, the community of Punta Cardón, and on the other hand the Venezuelan oil company and state agencies, as the local residents try to achieve recognition for the socio-environmental damage they have suffered, both historically and in the present. This exploration shows the multiple ways in which the community's claims for substantial recognition and compensation are stifled, both by the power of authority and value vested in the oil company and the epistemological paradigm legitimizing its activities, and the deep-seated economic and sociocultural hierarchies characterizing the relationship between the community, the state institutions and oil authorities.

Through these explorations, this chapter will bring to light some of the contradictions inherent in the political ideology of Venezuela's so-called Bolivarian Revolution under the leadership of the late Hugo Chávez, and now President Nicolás Maduro. Indeed, as McNeish, Borchgrevink and Logan suggest in the introduction to this volume, it is paramount to capture how energy production in Latin America continues to generate social struggles within the context of self-proclaimed progressive governments. Venezuela, being one of the world's largest oil exporters and also the most politically radicalized country on the continent (except Cuba), is a prime example of how petroleum resources, as argued in the introduction to this book, are deeply entangled both in the historically conditioned structural ramifications that energy production entails, and, fundamentally, in multiple and often paradoxical struggles over meaning, value and relationships of power played out within and across different imagined and real social territories.

In Venezuela, as elsewhere (Sawyer 2007; Gudynas 2009; Gustafson 2010; Widener 2011), these tensions unfold at several levels, but perhaps most notably in those communities whose livelihoods, ways of living or mere existence enter into conflict with extractive projects governed by the state or state-supported private actors. Indeed, these tensions become overtly visible in the community of Punta Cardón, where the superior value attributed to oil extraction as a means for development within Venezuela's 'Bolivarian Revolution' silences the critique put forward by the local community over how they themselves and nature are 'sacrificed' in the name of national development and poverty alleviation. However, the analysis also speaks to McNeish, Borchgrevink and Logan's emphasis on critical institutionalism as a tool for understanding pathways of resource extraction. Through an illustration of how technocratic thought and practice in resource governance are both changed and reproduced throughout different ownership transitions and ideological conjunctures, we are offered an insight into how the epistemological power of extractive activities is inscribed into the everyday practices of state governance. Indeed, in this chapter, the deeply 'oil saturated' nature of the Venezuelan state and state imagery (Coronil 1997) is rendered visible as the notions of popular sovereignties and national sovereignties coalesce into a tense contestation over who the oil should benefit and at what cost.

Before the arrival of the oil company

Historians record the existence of the Punta Cardón community long before any kind of oil-related economic activity existed in the country (López 2002).[3] When construction work on the refinery (one of the two refineries that today make up the Paraguaná Refining Complex)[4] began in 1945, there were already three communities in the area. Elderly contemporary residents recall, among others, the 'La Puntica' or 'Punta la Barra' community, the 'Cerro' community and the defunct 'La Botija' community, where today the tanks and huge flare stack for this refinery have been erected. These earlier communities lived primarily from fishing as well as the farming of corn and goats. The whole community participated in fishing in one way or another, carried out on a seasonal basis. The fishing took place practically in front of their houses, about a hundred metres from the beach, through the *calado* technique.[5] Contemporary residents claim that the fish at this time were so abundant that they literally washed up onshore. Indeed, there were so many that they were not weighed as they are now, but were counted in catches of three to four thousand fish.

The Punta Cardón community, in short, subsisted from fishing and, according to Félix Sánchez, a local fisherman, 'men, women, the elderly and children went about the task of salting fish, weaving nets, hauling in fishing nets and grinding salt, in addition to making espadrilles, rag dolls, hammocks, *dulce de lechosa*, *debudeques* and rearing goats'. The fish were salted as a means of preserving them because the community had no electric energy or fresh water. Lighting came from lamps they made themselves, using organic fuel from the oil of jack mackerel (a fish indigenous to the area); and fresh water was provided through the construction of *aljibes*, *casimbas* and *jagüeyes*.[6]

The fish were sold in the closest town, Coro (capital of Falcón state), to which they travelled in boats. From the sale of fish they were able to purchase basic provisions for life in the community, including: *papelón* (solid sweetener made from sugar cane), corn, cloth for making clothes and shoes. Payment was not received immediately, so bartering was commonly used as a means of exchanging products. The fishermen and the owner of the boat (*el patrón*) also commonly used a system of bartering to settle exchanges of fishing equipment, food or other basic necessities. This explains why many residents, especially women and children, did not know the value of money until well into the twentieth century.

The standard diet was two meals a day based on fish and *arepa*.[7] Both sewage systems and drinking water provision were non-existent. Having no showers, they bathed in the sea, and because they had no toilets the forest was used for this purpose. Education barely reached third grade[8] and few people went to school. Some families kept goats, while others reared pigs. Meat was sold to neighbours upon request, and, as now, the pig's bones were used and lent among families to give more taste to the grains they cooked. All the men were engaged in fishing activities, and the art of fishing was transmitted from generation to generation as the boys accompanied their father and grandfather out on to the sea, where they were taught the art of handling the *bongo* (home-made vessel). The villagers did not have any means of transport by land, and moved by walking long distances, although some had the help of a donkey. This did not discourage them, however, from travelling long distances to fiestas at great distances from their community, for which they borrowed clothes and footwear.

In short, life in Punta Cardón was simple and precarious, and to a large extent shielded from political events and economic activities in the rest of the country. It was this semi-isolated community of fishermen, or a 'sleepy society' (Pérez and Rangel 1976: 6) blessed with an abundance of fish and a clean, transparent sea without noise or pollution, that people from Shell found when they arrived in Cardón in 1945.

The first contact with oil people

Contemporary inhabitants in Punta Cardón recall that the community looked with astonishment on the people who had arrived in their community overnight. They didn't know what to think, as the newcomers both spoke another language and dressed differently from what they had seen before. One woman, born in 1926 to local fisherfolk, recalls those first encounters this way:

> When the first boat came, all the habitants came out to see what was going on, and they were coming to look for people to cut

down the vegetation [where they were about to construct the refinery] ... they paid the workers and gave them two kilos of maize each. That made everyone want to work for the company, we were all very poor and to be paid just to cut vegetation was huge. We looked upon the men from the company with astonishment, all of them white, big and with hats, they crossed our patios to get to wherever they were going without asking for our permission, but as they were giving out maize, and later sugar and rice, no one said anything. And we didn't understand what they were saying, and that made us laugh, but we became acquainted with canned meat and soft drinks, so at least we got to eat something else, we were just used to eating fish and *arepa*.

As this quote also indicates, the villagers and fishermen didn't understand what the oil people were doing there when they first arrived. No dialogue or consultation with the community had been carried out, and no information had been issued either by the company or the state. The state had negotiated with Shell behind closed doors, and without taking into account the community that already existed there.

Thus, above all, the local villagers were astonished and also curious. Soon, however, tensions started to arise as the two life worlds converged without having a common language to communicate through. According to the interviewees, the company started to offer work and distribute bags of rice. A number of the new foreigners also tried to barter bags of sugar for sexual favours with young women. However, the women didn't understand, because of language barriers, that accepting the bags of rice implied an agreement to sexual contact. This led to several episodes where foreigners were beaten up when they tried to impose themselves on the women in order to get their part of the 'transaction'.

In the initial stages of their presence in the community, Shell contracted local men to clear the hill where the first refinery plant was to be built. The novelty and apparent ease of the work meant that many fishermen left fishing to go and work for Shell. It also appeared to be work that was well paid. Earning two bolívares for a day working on land was preferable to the uncertainty of earning the same or less from a longer period of fishing. The company started to build lodgings and bring in food for its employees. The products and materials were shipped by boat, predominantly imported from Europe. The foreigners were exclusively men, and none had brought their families with them. Shortly, a brothel was established; La Concha was visited by locals and foreigners for over fifty years, and closed only around ten years ago.[9] Stories of women being raped and abused were common. A significant number of babies were born whose foreign fathers returned to their own countries without acknowledging their offspring. The 'madams' governing the brothels were indifferent to the births of the babies, and the mothers had to leave their children with other women in the villages they came from. Likewise, women who went off to work as housemaids in the foreigners' homes had to leave their children behind. Some of these also got pregnant with foreigners who did not recognize the child.

The attraction of 'black gold'

The oil boom generated an appearance of progress and development, attracting a lot of people both from elsewhere in Venezuela and other countries in search of work with Shell and the activities generated around the company. The new residents began to settle in the surrounding areas of Punta Cardón, which grew in a sudden and disorderly fashion. Finally, with the grand opening on 7 May 1949, the Shell Oil Company kick-started the refinery operation in Punta Cardón;[10] it contained eight plants refining 100,000 barrels a day of crude oil brought from Maracaibo (there is no oil in Falcón). However, because of their low educational level, very few residents from Punta Cardón got work in the refinery. The best local men could hope for from the company was occasional manual piecework when extra hands were needed for expansion, emergency response or other physical jobs. Providing high salaries and other benefits, the oil company could attract the best-qualified and educated talents, and these skills were not found in the villages adjacent to the oil operations.

Environmental and social impacts

Contemporary residents recall that, as the years passed, the idea of progress and well-being that the company initially inspired turned to bitterness and the sense of having been passed over and deceived. In 1962 the first environmental impacts were felt in the community when contamination from the refinery destroyed the village of La Botija. The first incident was caused by inundation of seawater resulting from spillovers from the cooling tanks used in the plants. These tanks were constructed without internal reinforcements, which soon led to the water running through the walls of the *bahareque*[11] houses of La Botija, seriously damaging many of them. The company paid compensation, but many residents moved elsewhere when they constructed their new houses. The second major event occurred when 'alarmed residents found that oil was flowing out through their toilet bowls and latrines, everything smelled of oil', contemporary local resident Franklin Medina recalled, continuing: 'For the first time, people had the strange discovery of black, viscous liquid in the latrines, mixed with the human excrement deposited there.'[12] The affected families in La Botija were paid a small amount of compensation by the oil company, which soon tore down their houses and in their place built storage tanks and a flare stack of over two hundred metres on the site. According to local people, this ended the natural cycles of day and night, because it lit up the sky so much that 'it was impossible to see the films' at the community's cinema.[13] Given the immediate expansion of the refinery, many residents suspected that the oil spills into their septic tanks had perhaps not been an accident, but rather a deliberate leakage on the part of the company. Local residents still believe that this was the case, but they have never been able to prove it.

The people from La Botija who migrated moved to the villages of El Cardón Grande, La Puntica and El Cerro, leaving behind a gap in social relations and the spatial arrangements that had once defined the community as a collective. To Franklin Medina, these incidents constituted key moments leading to a rupture in local identity and in knowing how to live together in Punta Cardón. In his opinion, 'we made a sacrifice for a whole nation. We ceased to exist in order for the country to develop. We lived from the crumbs that fell from the table when the cake was divided up.' Manuel Arias, local poet, singer and social activist, expresses it this way:

> Botija wasn't only about [Shell] buying up some houses, it was the destruction of a pueblo, its streets, its culture, songs, baseball teams, bars, schools, friends and everything that makes life and gives life to a people and its traditions and customs ... mangroves, butterflies and snails abounded, and you could hear the little fishing boats leaving as they passed by the beach ... from the field you could hear shouting from baseball when Cheno-Cheno threw

the ball, Salvador Tremont announced the testament of the burning of Judas and all the inhabitants arrived to have a laugh; afterwards, Colacho came with his trumpet and Benjamín supported the singing girl with the guitar in his arms, while Fernando accompanied him with his violin.

Dialogue of the deaf

From 1947 to 1975, the refinery was operated by Shell. During the Shell era, the company commonly resorted to simple payoffs when they had caused some evident damage such as the inundation of the latrines, or when the complaints of the community had become too bothersome. For example, as oil production gained pace, local fishermen gradually realized that the oil activities had reduced the fish stocks. Shell had placed cooling and heat ventilation equipment close to the natural spawning site of some fish species, and the fish eventually disappeared. In the natural catch sites, quays had been constructed and pipelines that brought the crude oil from Maracaibo crossed the bay. This blocked the entrance to the bay of fish schools and filled the sea with noise and contaminants.

At the time, the fishermen had approached Shell to put forward their analysis of what was happening to the environment and their livelihoods. The company's response was to offer some minor individual economic compensation to the fishermen for *possible* damage that the company's procedures could have caused. However, they did not admit to the damage the oil production was causing to the community as a whole, nor commit to revising their operating procedures.

Even though the consequences of oil production were already starting to emerge when Shell was on site, the real problems didn't become evident until later, when the processing plant passed into Venezuelan hands. The Venezuelan oil industry was nationalized on 1 January 1976, converting Shell of Venezuela (a subsidiary of Royal Dutch Shell) into one of fourteen operators in the Venezuelan state oil company PDVSA. In Punta Cardón, it operated under the name of Maraven.[14] After the nationalization a new pattern of production emerged. The foreign oil companies had extracted most of the light crude, and new processing plants had to be built in order to exploit the heavier crude that was now becoming more important. These installations and procedures caused a lot more contamination than the lighter crude production had generated.

The conditions after nationalization changed a lot less than what local inhabitants had hoped for. Now being in the hands of Venezuelans, and having seen how the nationalization process was accompanied by a nationalist-developmentalist discourse, they had expected to be met with a greater will to conduct dialogue and awareness of the local environment. However, the results were the opposite; the levels of pollution increased as the refinery expanded, and the consequences for fishing were the same. The heavy crude production involved pollutants with a greater load of sulphur and heavy metals in general. Refinery production rapidly reached 300,000 barrels a day at the beginning of the 1990s. At the same time, as environmental deterioration and health problems became more evident, the community embarked on a long struggle to gain recognition for their problems.

Punta Cardón in the 1980s and 1990s

One of the gravest consequences of the shift to crude oil production was the increasing pollution caused by charcoal.[15] This was spread through the air as a fine, black dust covering the landscape, people, houses, water and practically everything within a large area surrounding the installations. At the end of the 1980s and the beginning of the 1990s, serious health problems in the community became more frequent, and unusually high numbers of 'anencephalic'[16] children were born. These circumstances finally managed to bring together the community in a concerted front, organized as Junta de Vecinos de Punta Cardón (the Punta Cardón Neighbourhood Association), and later as the Comité por la Defensa de la Salud y la Vida (CODESVI) or Committee for the Defence of Health and Life. A series of protests were staged outside PDVSA offices and the refinery; they went to the press to complain and tried to make allies with political offices and other institutions. However, their desire to have a constructive dialogue with PDVSA was fruitless. Eventually, the Ministry of the Environment intervened, demanding that PDVSA conduct a study of operating procedures and environmental consequences. PDVSA withheld the report emerging from this study from the Punta Cardón community, but handed it over to the ministry, which then elaborated a series of recommendations and demands to PDVSA about how to clean up their operations. The community of Punta Cardón, represented by CODESVI, put pressure on the Senate, demanding that the results of PDVSA's report be made known. The report has been withheld from the public to this day, but eventually the Senate produced its own report based on it, and also included a series of qualitative investigations in the community. The Senate's publication, entitled 'First report, incidences of pollution generated by the oil industry in the Paraguaná peninsula', identified a long list of problems and breaches of legal provisions, including:

- Illegal dumping of industrial effluents into the sea with raised concentrations of fats, oils, phenols, mercaptans and sulphates.
- Atmospheric pollution generated by burning sulphur dioxide, an acid gas that contains sulphur, vanadium, oxides and peroxides that are carcinogenic and pyrogenic.
- Existence of a toxic open waste pit, 'El Muladar', adjacent to the Punta Cardón community. In addition to rubber and catalyst batteries, the pit contained more than fifty metal drums in bad condition holding, among other toxic substances, vanadium, asbestos, glass fibres and tetraethyl lead.

The report established that the deterioration in the health of this community between 1988 and 1992 was significant and showed a progressive increase of approximately 35 per cent in morbidity associated with diseases linked to air quality, such as colds, pharyngitis, tonsillitis and otitis. Furthermore the Senate Committee asserted that: 'today we have no doubt that the contemporary way of living of the inhabitants in these areas is in stark contrast to their normal pathological state, and this situation would bring, as a consequence, different changes to their usual behaviour reflected in social, psychological, health problems, etc.'.

The committee also investigated the complaint concerning the birth and subsequent death of several anencephalic children between 1990 and 1993, concluding that this problem was not natural but of an anthropogenic order.[17]

The report further evidenced that there were large oil spillages in the sea in front of the refinery, polluting the fishing grounds formerly used by artisanal fishermen. It also revealed that the 'sulphur washer' was very close to the sea, and that illegal dumping was taking place, causing a progressive impact on the coastal environment and its surrounding areas. The report also detailed that the oil refining company was using seawater for its cooling plants (entering the refinery at 28 degrees and once heated to 350 degrees circulated and dumped back into the sea) and that the water temperature at sites close to the discharge was 10 degrees above the initial temperature. This could explain why some of the fishermen said that 'the sea is now a soup' and that they were unable to cast their nets and leave them overnight to collect them in the morning as they used to do because 'we find the fish parboiled'. However, even with this extensive evidence of environmental pollution and substantive damage to the life and health of local residents, nothing more came out if it. The report was put in a drawer, and business continued as usual.[18]

Punta Cardón today

After more than sixty years of uninterrupted operations in the Paraguaná Refining Complex, local inhabitants today still remain in a situation of poverty, inequality and inequity (Bebbington and Bebbington 2010; Fraser 1997). According to the Venezuelan National Statistics Institute (INE 2013) there are today almost 96,000 inhabitants in the parish of Punta Cardón. According to censuses undertaken by the local Communal Councils,[19] approximately 25 per cent still do not know how to read and write.[20] Their investigations also indicate that the unemployment figure stands at 54 per cent and a large proportion of employment (approximately 46 per cent) is informal or uncertain. Many local men still fish, but the incomes it generates barely allow people to cover their day-to-day necessities and to survive precarious situations in the short term. In the absence of storage centres or distribution networks that pay them a fairer price, local fishermen continue with the same practice of delivering fish to buyers in coolers. As a result of not being able to accumulate surplus capital, they generate debts for petrol and other equipment for work. Although continuing to claim that 'fishing has real value', the local men often rent their boats for 500 bolívares a day to PDVSA for different operations at sea because this is seen as 'secure money'.

Public services in the local community of Punta Cardón are overall of a poor quality. A significant percentage of the population have no garbage collection service. There is no regular drinking water supply owing to deficiencies in coverage; there is a shortage of electrification in many areas, and a substantial number of houses still use the old latrines and septic tanks because they do not have the necessary connections for sewerage. There is also a shortage of houses, and up to three families can be found living in the same house. There is just one medical centre in the community, without an ambulance. The public transport service remains irregular and the entire road infrastructure inside and outside the community is in a bad state. Some of the inhabitants have recently benefited from the governmental decree granting all citizens a social security pension upon reaching the age of sixty-five (men) and fifty-five (women), though many older men still fish in the hope of adding to their meagre income.

The environmental situation today

Despite the fact that the PDVSA has applied some bio-remediation policies during the past decade, many of the problems recognized long ago remain. Indeed, open-cast charcoal deposits continue to pollute, aided by the strong wind that characterizes this peninsula, carrying dust to the sea and elsewhere. Serious health problems also continue to be caused by this, and many of the inhabitants suffer from colds, hypertension, virus infections, dermatitis, diarrhoea, bronchial asthma, anaemia and diabetes.[21] Recently, there have been a number of cases of cancer, heart problems, strokes and other cerebrovascular diseases, and there are a significant number of disabled people living in the community. In interviews, the local residents expressed bitterness over how their health was deteriorating owing to the combined toxic mix of accumulated poverty and industrial pollution. As one man expressed it: 'We have the right to health, and to have a place to live which can coexist with the requirements of industrial development, not the vast pollution of our environment in which the possibilities of life are diminishing with each passing day ...'

Indeed, many of the residents feel that they have been sacrificed in the name of development and prosperity, but that barely any of the benefits have reached them. In spite of it being separated from the oil refinery by just a wall, the social programmes that the government has financed with oil revenues are by and large poorly implemented in the community. Adding to their bitterness, the middle-class communities near by, primarily inhabited by oil workers, have prospered and grown, sporting all the necessary public services in addition to a shopping mall, supermarkets, schools, banks and recreational areas. In spite of these complaints and apparent disparities, the inhabitants still agree with the principle that the oil industry is necessary for national development. Thus, they are not opposing the oil complex per se, but they want to be compensated for the damage they have suffered, have their claims about environmental and human damages recognized, and get a share of the benefits derived from oil extraction too.

Diverging epistemologies

The Venezuelan governments of Hugo Chávez and now Nicolás Maduro have been characterized by a political discourse emphasizing 'popular power' (*poder popular*) and political participation at community level as key components in the transformation of Venezuelan society. Indeed, the Bolivarian Revolution, as the political processes in Venezuela are often termed, has been carried forward by a significant upsurge in grassroots activism in both rural and urban areas. Moreover, the Bolivarian Revolution has been underpinned by cultural politics emerging from below, challenging the legacy of racism and classism in Venezuelan society, and a discourse that constitutes *el pueblo soberano* (the sovereign people) as legitimate political protagonists within the vision of a participatory political model (Strønen 2014). Respect for popular knowledge, popular modes of socio-political organization and popular conviviality has also been a key tenet in both political and popular discourse, and indigenous people, peasants and artisanal fisherfolk are in official discourse and cultural politics presented as representatives of 'the real Venezuela'. However, as we will see in this last part of this chapter, the community of Punta Cardón has yet to be recognized as a victim of environmental damage, or to substantially reap benefits from the oil production taking place in their immediate vicinity. Indeed, what is revealed in this case is a pattern similar to what is described by Gudynas (2009), whereby local inequalities are generated in sites of extractive economic production, at the same time as local claims for recognition or compensation are trammelled in numerous ways.

Memories discarded

Deteriorating fish stocks have long been a topic in Venezuela. Fishermen along the whole coastline agree that they have to go farther out to sea to get their catches, and that they have also had to change their methods and equipment. This problem has also been a grave concern for the government, and since 2009 the practice of trawling has been suspended as a measure to allow fish stocks to recover. Thus, the question at stake is not whether fish stocks have been reduced, or not in general, but, as in the case of Punta Cardón, the extent to which oil production has contributed to the diminishing fish stocks and general environmental deterioration.

The contemporary community organizations in Punta Cardón, including the two Fishermen's Councils in the area, the Communal Councils and the still-existing Committee for the Defence of Health and Life (CODESVI), are all arguing the case for oil production being partly responsible for the depleted fish stocks and environmental damage locally. But they run up against a broad body of scientific reasoning countering their claims, advocated by PDVSA, the National Institute for Agricultural Research (INIA),[22] the Socialist Institute of Fishing and Aquaculture (INSOPESCA in Spanish)[23, 24] as well as some universities doing research in the bay, such as the Central University of Venezuela (UCV in Spanish) and Simón Bolívar University (USB in Spanish). PDVSA's official stance is that it operates in accordance with internationally recognized parameters. They regularly refer to the findings of their own scientific teams as well as apparently 'objective' studies commissioned from national scientists to corroborate these claims.

Over the course of the research for this chapter, numerous interviews were conducted with local community members and fishermen, many of them of advanced years who were able to recall local flora, fauna and environmental cycles before the oil industry was settled. They explained and made drawings of where the pipelines, quays and effluents are, as well as the route of the disturbed fish, and made sketches of how and where fishing was conducted before the pollution, as well as the equipment they used to use. They repeated their testimonies and illustrations time and again, recalling their practices and histories, and showed photos of what Punta Cardón used to look like before the environmental changes really set in. Today, they also use these memories of reproductive cycles and fish stock movements to try to assess how to best access the dwindling fish population. They have a solid layman's understanding of the correlations between the refineries' installations and procedures, and how these are interfering with the ecological balance at sea. Yet their mode of reasoning is based on their historical and primarily orally transmitted knowledge of the bay, and they cannot back up their claims with any form of statistics, or 'hard science'.

INIA and INSOPESCA, the two state institutions working directly with fisheries in the area, are in their institutional design meant to support local fishermen and promote sustainable fishing policy. However, there is a considerable divergence between these institutions' assessment of the environmental situation in the bay, and that of the fishermen. Following a series of measurements in the bay on the reproduction of some species, INIA claims that there is a 'sufficient' level of reproduction and that different species of young fish are still thriving. The institution also holds that there is a sufficient increase in species in the Gulf of Venezuela, situated just in front of the Gulf of Coro, where all the pipelines and infrastructure are situated. However, unlike local fishermen, INIA does not consider the conditions in the bay thirty to fifty years ago, which is the fishermen's point of reference. Moreover, the technical protocols on which the INIA technicians base their research establish very narrow parameters for measuring what kind of species are reproduced and the baseline that the current level of reproduction is measured against. Moreover, they also conducted their measurements outside the traditional breeding grounds, sometimes against the fishermen's advice. In anonymous interviews, some of the INIA technicians recognized these methodological errors. Additionally, all the local INIA technicians were new to the area, and did not therefore have proper knowledge of how the stocks had changed over time.

Blaming the fishermen

However, though INIA staff to a certain extent recognized that there could be something to the fishermen's qualitative assessment, this knowledge nevertheless fell outside the institution's official knowledge base. This has several and complex reasons that cannot fully be explored here. Suffice to say that in spite of an official political ideology legitimizing 'popular' in Venezuela, the encounters between popular knowledge and scientific knowledge regimes are still filtered through hierarchies of power and knowledge that ultimately mute the fishermen's arguments. These processes also have deeper and subtler antecedents, such as the ways in which scientific institutions historically have served to legitimize oil activities in Venezuela (and elsewhere) (Lander 2012). However, currently the articulation between the state institutions and the fishermen of Punta Cardón has also been jagged along a different, yet interrelated, dimension. For the past several years, the fishermen have been engaged in a tentative collaborative process with the state institution INSOPESCA (advised by INIA), developed with the purpose of creating fishing regulatory policies. Through this cooperation, the institution aims to collect the fishing statistics on a daily basis from each fishing vessel, including the name of the boat, engine horsepower, site of operation, techniques used, crew members, catch, species, weight, and so on. To that end, the institution has a template it applies to each fishermen's council.[25] When the fishermen return to port, they must fill in this template and send it to the office of INSOPESCA, located in the city of Punto Fijo. However, to get there, the fishermen of Punta Cardón must take two forms of public transport, and then the same route back to return. This journey commonly takes three hours, and the fishermen are therefore often willing to pay a manager to collect the templates from all the fishermen and take them away. On some occasions, INSOPESCA sends some of its officials to collect the template in the communities; however, since this places a heavy toll on the institution's already limited staff capacity, it prefers to delegate this responsibility to the fishermen.

These challenges have several consequences. First, the complicated and time-consuming way of working makes the update of statistics difficult in general. And secondly, the fishermen do not fully embrace the task of gathering and delivering statistics, simply because the procedures take important time and energy away from more pressing tasks. Significantly, precise data collection is difficult to reconcile with the hustle and bustle taking place at the beach as the fishermen arrive onshore and sell their catch on the spot to local merchants. Afterwards, the fishermen gather and recollect the size of their catch 'from memory', and this then has to be delivered to INSOPESCA in one way or another. However, they have no computers with internet access or means of transport that could make this task easier for them. Moreover, the abstract character of the effort seems far removed from their daily practical concerns, and INSOPESCA staff members also confide in anonymous interviews that the fishermen have received limited instructions and support that could help consolidate the cooperation between the institution and the fishermen.

Confronted with the difficulty of persuading the fishermen to gather data more systematically and meticulously, a discourse has arisen within INSOPESCA that links the weakness of their statistics to the 'cultural traits' of the fishermen. They allege that the fishermen are sloppy and unreliable, a stance that also subtly discredits the fishermen's claims about diminishing fish stocks and environmental degradation. This is interlinked to INSOPESCA's claim that the fishermen's declarations about smaller catches are a deliberate strategy to avoid the repayment of loans to the state. Moreover, both INIA and INSOPESCA argue that another reason for diminishing catches is that there are more fishermen working at sea, owing to the government's credits for boats and equipment.

The politics of payouts

Seen together, multiple processes are at play here. One is that the fishermen's claims about long-term environmental degradation are derived from collectively accumulated oral knowledge that is emasculated in the face of PDVSA's and the state institutions' claims to scientifically based arguments. The other process is one of reproduction of deep-seated prejudices against the fishermen based on deeply rooted economic and sociocultural hierarchies. This latter process also needs to be explored along an additional dimension, namely the ways in which the relationship between the oil company and the community historically has been mediated by the politics of economic compensation, and how this is played out within the current political conjuncture.

Over recent years, contact between the community of Punta Cardón and PDVSA has been primarily channelled through the two fishermen's councils operating in the area. Through these fishermen's councils, the fishermen have on several occasions approached the company to claim economic compensation when oil spills or other damage have directly affected their fishery. PDVSA, wary of the fishermen's capacity to create hindrances to their operation through, for example, blocking the entrance to the bay, has on these occasions paid out compensation to the fishermen through the fishermen's councils. These dynamics have nurtured a social imagery whereby the fishermen are commonly described by the oil people and the state institutions as '*interesados*' – an expression that in Venezuela describes someone who is always on the lookout for an opportunity to enrich themselves through taking advantage of a situation or other people. This imagery is embedded in a more deep-seated political culture of assistentialism and clientelism that cannot be fully explored here (see Strønen 2014). Suffice to say that the imagery of the 'scrounger' who wants to live off the state is a powerful collective one that in the current political conjuncture is reproduced in an ambiguous manner alongside the ethos of the Venezuelan state's obligation to attend to social needs. This label has not only been attached to the fishermen, but also in an ambiguous manner to the community at large, who are heavily dependent on PDVSA for the collective benefits that do trickle down. Through the ethos of using oil revenues for social development, and supported by legislation promoting and giving guidelines to Entrepreneurial Social Responsibility, it is PDVSA that is now directly responsible for social welfare programmes and community initiatives in the communities adjacent to its oil installations. Indeed, requests for support or resources are commonly addressed directly to PDVSA, and not to the local municipality. However, the other community organizations in Punta Cardón, such as the Communal Councils and CODESVI, incorporating a wide range of community members from different organizations and professions, have in recent years tried to initiate a dialogue with PDVSA that goes beyond these material endowments. Rather, reflecting a more socially and ideologically radical ethos that characterizes many grassroots organizations in contemporary Venezuela, they want to be recognized by the authorities as a partner for de facto dialogue and simultaneously address in depth the socio-environmental consequences that six decades of local oil production have generated. However, so far, their attempts to reshape the relationship between the oil company and the community have been met with indifference at PDVSA. Faced with a lack of 'hard scientific data' that corroborates their arguments about historical and persistent socio-environmental damage,[26] they have so far been unable to muster public support for their cause, and since they do not pose a direct 'threat' to the company's immediate operations (like the fishermen), it has been easier for PDVSA to stifle their attempts to initiate contact.

Finally, it is important to recognize the marked sociocultural hierarchy that characterizes the relationship between PDVSA staff and local community members. PDVSA workers have always been regarded as 'elite workers' in Venezuelan society; slightly set apart from the remainder of society. This exacerbates already existing levels of deep-seated educational, economic and sociocultural hierarchies, reinforcing PDVSA staff workers' perception of their own superior knowledge of the realities surrounding their oil production. Moreover, through the practice of supporting the local community with socio-economic benefits, and bolstered by a political discourse emphasizing how oil wealth is now 'trickling down', PDVSA is habitually able to legitimize its presence and benevolence with a discourse that reiterates that 'we are helping them', 'we are setting up this project', 'we are giving them this and that'. Even though it is indeed positive that socio-economic benefits do to a certain extent trickle down, this discourse and practice simultaneously foreclose substantial recognition of the damage suffered by the community, and reproduce the dynamic of 'paying out' and 'paying off' that has characterized the relationship between local residents and the oil industry ever since the first Dutch oilmen set foot in Punta Cardón more than half a century ago.

Punta Cardón and the Bolivarian Revolution

In Venezuela, oil production has always been doused with nationalist and patriotic sentiments, seen as the source to the nation's exceptionality, strength and path to prosperity (Coronil 1997). With Chávez and the Bolivarian Revolution, PDVSA and oil wealth itself were vested with an additional ideological layer. After being 'renationalized' at the turn of the century through legislation putting a halt to privatization processes initiated by Chávez's predecessors, PDVSA became a symbol for the refounding of the nation through social development, deep state reforms and poverty alleviation. Indeed, the parallelism of the slogans 'PDVSA is now for everyone' and 'Venezuela is now for everyone' serves to accentuate how the association between the 'new, socially inclusive' Venezuela and PDVSA is intimately intertwined.

PDVSA's revenue currently represents at least 50 per cent of the national GDP, and the company also directly finances the majority of the so-called missions constituting the central axis of the current government's social policies,[27] as well as more indirect social inclusion policies such as old-age pensions, social security benefits and public investments. The widespread identification of the national oil company as part of the 'Socialism for the 21st century' makes it very difficult for the inhabitants of Punta Cardón to criticize the company publicly. Moreover, the majority of the residents in Punta Cardón are poor people who voted for Chávez and subsequently for Maduro, and who have periodically benefited from the missions implemented by the government. Indeed, even if the collective improvements in the form of social programmes and social security benefits are not sufficient to lift the community out of poverty, it is the first time in history that the community has been taken into account by the state at all.

Thus, given all these factors seen together, it is very difficult for the community of Punta Cardón to draw attention to the negative consequences of oil production. Those who dare to do so have to politically and symbolically manoeuvre through layers of strong ideological sentiment about oil production as the road to national prosperity, at the same time as they, given the politically polarized climate in the country, run the risk of being accused of being opponents of the 'revolution'. Simultaneously, they are faced with an oil company that rather than view itself as a representative of its citizens has acted as if it were a mere representative of capital (Lander 2012) in its quest to maximize production. Moreover, PDVSA has its own scientific-technical teams and the unlimited financial capacity to finance and commission research from scientists in the main national universities, which is then presented as 'objective science' and used to legitimize its operation (ibid.). Complementing the picture, the state institutions involved in the controversy locally are viewed through the lens of realpolitik and dominant knowledge regimes, too intimately incorporated in the extractivist state to be able to challenge the modus operandi of the oil industry. Finally, it is also important to take note of the fact that the Paraguaná Refining Complex will have an important role in the expanded exploitation of heavy crude in the Orinoco belt, established as an important part of the country's development process anchored in the government's five-year plan, 'Plan de la Patria 2013–2019'. Additionally, exploration for and production of gas are already under way in the bay, producing a situation whereby PDVSA has much to lose from initiating a process looking into how their oil installations and operations are affecting local environments and livelihoods.

Conclusion

As McNeish and Borchgrevink write in the introduction to this volume: 'friction commonly arises between those living in the areas where the energy resources exist and those wishing to extract or harness them'. Likewise, the historically evolving socio-environmental patterns in Punta Cardón are similar to what Gudynas describes as 'extractivist enclaves' in the form of a process of industrial development which externalizes social and environmental effects (Gudynas

2009), often creating 'pockets' of inequality in sites of extractive activities. These processes not only have local ramifications, but form part of a broader scenario of resource struggles reflecting contested resource sovereignties (introduction, this volume) mobilizing diverging notions of power and knowledge, wealth and values, grievances and entitlements.

As is carefully documented and argued in this book, these traits are not necessarily reversed within the current political conjuncture of progressive governments in power in Latin America. Rather, in some cases contestations and conflicts are aggravated as energy production is presented as a necessary means to national development, and damage may be presented as 'necessary sacrifices' in the name of collective progress. As the account from Punta Cardón illustrates, these processes are filtered through dense layers of historically conditioned socio-political relations and state practices, showing how 'resource governance' always needs to be analysed as relations of power and struggles over meaning that are dialectically formed between global histories of extractive activities and particular configurations of the local. Indeed, only when the 'technocratic' exercises of extractive power are put under scrutiny can we fully understand how deeply contentious and ideological the idea and practice of energy production actually are.

However, a final remark is needed. The processes that are currently taking place in Latin America, whereby the ethos of extractive capitalism has been forcefully challenged from below – but also in some respects at the level of the state – have illuminated both its fragility and solidity. Fragility, because there is a simmering and growing recognition of the urgent necessity to scrutinize the truths upon which this model is based, and to critically rethink the relationship between humans, society and nature. Solidity because, as the case from Punta Cardón illustrates, the economic and epistemological power of orthodox energy production, as well as entrenched cultural and socio-economic hierarchies, are still-dominant patterns across the continent – and indeed across the world. Yet we should not think that the struggle waged by the inhabitants of Punta Cardón is theirs alone. Rather, their efforts to draw attention to socio-environmental damage caused by oil production in their immediate vicinity constitute both the centre and frontier of a global process, whereby the consequences of extractive activities are increasingly also felt outside the oilfields.

Notes

1 The chapter is based on several periods of fieldwork in the area and more than thirty interviews conducted by the first author with local residents in 2010/11, as well as reading of historical and contemporary documents and interviews with public authorities relevant to the case. Many of the interviewees were elderly people living in the community at the time of the arrival of the oil, who provided detailed information on the history of the community.

2 In the 2013 census carried out by Instituto Nacional de Estadistica (INE), the parish of Punta Cardón, located in the Municipio Carirubana, is registered with 95,829 inhabitants. The community (pueblo) of Punta Cardón itself, comprising the 'capital city' of the parish of Punta Cardón and a multitude of informal communities (barrios), has an estimated 40,000 inhabitants.

3 The first Venezuelan oil company, La Compañía Minera Petrolia del Táchira, was founded in 1878. The first oil well, Zumaque I, gushed open on 31 July 1914, beginning the continuous production of oil in Venezuela up until this day.

4 The Paraguaná Refining Complex (CRP) is the second-largest refining complex in the world, after the Jamnagar Refinery in India. It emerged from the merger of the Amuay, Bajo Grande and Cardón refineries in 1997. At present it has the capacity to refine 940,000 barrels a day, which represents 65 per cent of the Venezuelan refining capacity. The refinery is owned by the Venezuelan state oil company, PDVSA.

5 The technique consists of installing a labyrinth of nets in the fish route, which is normally situated near the coast. The process begins when the fish, during emigration, are passing and enter the bay. This technique is used where there is really an abundance of fish. Some watchmen positioned on stakes nailed into the seabed observe when the shoals enter and issue the warning, and at that moment all the inhabitants participate in pulling in the nets.

6 *Aljibe* is a kind of tank built with a mixture of wood and grass to catch rainwater. *Casimbas* are wells that are opened wherever shallow water accumulates, and a *jagüey* is a kind of large pot sunk into the ground that is also used to collect rainwater.

7 Typical Venezuelan corn-based staple food.

8 The first school was established in 1932 in La Botija, the second was established in 1937 in the La Puntica area, and the third was built in 1960. The most important thing was to learn to read and write.

9 La Concha was part of what was called 'a tolerance zone' made up of a range of bars in which, at the time of its peak, women of all ages and nationalities worked, brought first by the company and subsequently by traders who specialized in the trafficking of women. These women were often trafficked into the country illegally with the promise of decent work, and found themselves in a situation of semi-forced sexual slavery as they were left with few or no other options than to prostitute themselves. This enforced prostitution was perpetuated as the '*patrón*' of the women kept the lion's share of the money that they generated.

10 In 1950, very close to the Shell Refinery in Punta Cardón, the Amuay Refinery from the Creole Petroleum Corporation was opened in the Amuay area, in Paraguaná. When oil was nationalized in 1976 in Venezuela, both ended up being run respectively by the new Maraven and Lagoven companies – subsidiaries of PDVSA – until 1997, when all installations were unified under the

current Paraguaná Refining Complex (CRP in Spanish).

11 A construction of interwoven sticks and mud.

12 A latrine is a well especially designed for defecation.

13 Interview with local resident Enna Medina, 2011.

14 The company operated for twenty-two years under the name of Maraven, until PDVSA decided in 1997 to change its functional structure, thereby eliminating operating subsidiaries and developing a new integrated operational structure based on business units. It is also also worth mentioning that the foreign oil company continued its presence and profited heavily from Venezuelan oil production after the nationalization in 1976. This was organized through lucrative technical assistance agreements, in Spanish called *Contratos de Asistencia Técnica* (CAT).

15 Charcoal is a carbonaceous solid derived from coking units in an oil refinery and from other cracking processes. It is deposited in an open pit forming huge mountains, and the dust is then swept away by the wind.

16 This is a condition whereby children are born without a brain.

17 Over a three-year period, twenty-two cases were recorded and another fourteen were not reported officially; their frequency varied between 0.5 and 2 for every 1,000 births, which is high for such a small population.

18 The report was furthermore corroborated by previous reports from the Ministry of the Environment and of Renewable Natural Resources (MARNR) in which operational irregularities and breach of current environmental regulations by the oil company had been demonstrated.

19 Communal Councils (*Consejos Comunales*) are locally elected, territorially based community groups regulated by law, which since 2006 have played an increasing role in public and political affairs (see Strønen 2012).

20 However, there are forty-one educational centres in the community, which according to the Communal Councils' own estimates are attended by almost one million people. These comprise the full range of governmental educational missions, from primary school for adults to work training programmes.

21 Information about the inhabitants' current health problems is based on interviews with doctors and nurses in the local health station, the leader of the health committee in Punta Cardón, and the censuses carried out as part of the governmental social programmes 'El Amor Mayor' and 'Hijos de Venezuela'. These programmes are providing economic support to economically disadvantaged elderly and poor families with smaller children, respectively.

22 Instituto Nacional de Investigaciones Agropecuarias (INIA) is a state institution that forms part of the National Agrarian System. It is devoted to agro-food innovation in accordance with the political ethos of 'Bolivarian' ideology, including recognizing and promoting ancestral, traditional, formal and informal culture.

23 Instituto Socialista de la Pesca y Acuicultura (INSOPESCA) is responsible for generating policies regulating fisheries and aquaculture throughout the national territory.

24 It is worth noting that we made formal and informal contact with the aforementioned institutions repeatedly. However, it was not possible to obtain written reports from them, and we obtained some data only in interviews that could not be recorded. To no avail, we were promised formal reports and interviews 'at the next appointment'. Only INSOPESCA gave us electronically

the data it had on fishing. The university researchers contracted by PDVSA work on contracts stating that PDVSA is the owner of such work and that sharing results remains at their discretion.

25 In Punta Cardón there are two fishermen's councils: Consejo de Pescadores Abdías representing approximately 210 fishermen, and Consejo de Pescadores de la Barra, representing approximately ninety-five fishermen.

26 In 2013, CODESVI approached Instituto Venezolano de Investigaciones Cientificas (IVIC), a state-financed but independent research institution based in Caracas, soliciting help to conduct a thorough investigation of their socio-environmental challenges. Currently, IVIC has initiated a process of putting together a multidisciplinary team responding to this call. Please note that the first author of this chapter forms part of this effort.

27 Among the missions PDVSA directly finances are the Ribas educational mission, the Barrio Adentro health system, the Gran Misión Vivienda Venezuela housing mission and, more recently, Misión Hijos de Venezuela, which caters to the children of very poor adolescent mothers, and Misión Amor Mayor, which is aimed at giving older adults a social security pension equivalent to the country's minimum salary.

References

Bebbington, D. and A. Bebbington (2010) 'Extraction territory and inequalities: gas in the Bolivian Chaco', *Umbrales*, 20: *Hidrocarburos, política y sociedad*, Bolivia: CIDES-UMSA, pp. 127–60.
Comisión Permanente del Ambiente y Ordenación Territorial del Senado de la República (1997) 'Incidencia de la contaminación que genera la industria petrolera en la Península de Paraguaná'.
Coronil. F. (1997) *The Magical State. Nature, Money and Modernity in Venezuela*, Chicago, IL: University of Chicago Press.
— (2002) *El Estado mágico. Naturaleza, dinero y modernidad en Venezuela*, Consejo de Desarrollo Científico y Humanístico de la Universidad Central de Venezuela – Nueva Sociedad.
Fraser, N. (1997) *Justitia Interrupta: Critical reflections from the 'post-socialist' position*, Santa Fé de Bogotá: Siglo de Hombres Publishers.
Gudynas, E. (2009) *Diez tesis urgentes sobre el nuevo extrativismo. Contextos y demandas bajo el progresismo sudamericano actual*.
Gustafson, B. (2010) 'La soberanía en los tiempos del gas: territorialidades y tácticas en el sureste boliviano', *Umbrales*, 20: *Hidrocarburos, política y sociedad*, Bolivia: CIDES-UMSA, pp. 161–88.
INE (Instituto Nacional de Estadísticas) (2013) *Reporte Ambiental*, Gobierno Bolivariano de Venezuela.
Lander, E. (2012) '¿Un nuevo periodo histórico? Crisis civilizatoria, límites del planeta, desigualdad, asaltos a la democracia, estado de guerra permanente y pueblos en resistencia', Paper presented at the Thematic Social Forum, Porto Alegre, January.
López, S. (2002) *Semblanzas de mi pueblo (Punta Cardón)*, Alcaldía del Municipio Carirubana, Punto Fijo: Ediciones Mar Afuera.
Pérez, J. and D. Rangel (1976) *El Desastre – Venezuela*, Vadell Hermanos.
Sawyer, S. (2007) *Crude Chronicles: Indigenous Politics, Multinational Oil and Neoliberalism in Ecuador*, Durham, NC: Duke University Press.
Strønen, I. Å. (2012) 'Money and power to the people: development from below and oil money from above', in J.-A. McNeish and O. Logan (eds),

Flammable Societies. Studies on the Socio-Economics of Oil and Gas, London: Pluto Press, pp. 133–55.

— (2014) 'The Revolutionary petro-state. Change, continuity and popular power in Venezuela', Unpublished PhD dissertation, University of Bergen.

Widener, P. (2011) *Oil Injustice: Resisting and Conceding a Pipeline in Ecuador*, Washington, DC: Rowman & Littlefield.

6 | 'EVERYTHING MOVES WITH FUEL': ENERGY POLITICS AND THE SMUGGLING OF ENERGY RESOURCES

Cecilie Vindal Ødegaard

Introduction

McNeish and Logan (2012) argue that the term 'resource curse' may contribute to under-communication of the struggles and conflicts of interest involved in natural resource extraction, for instance by 'naturalizing' its problematic consequences. This is an important point, and reflects the significance of asking how such conflicts of interest may be related to struggles about the establishment and definition of value (Graeber 2001), and to different understandings of 'development'. In my approach to energy politics in this chapter, I therefore seek to go beyond the idea of 'conflict of interest' in a conventional sense, by understanding energy politics – in a wide sense of the term – as a struggle about value. I will approach this issue by focusing on the smuggling of energy resources from Bolivia to Peru, and by discussing these countervailing forces in the light of structures of inequality. While the smuggling of combustible energy resources actualizes questions of il/legality and the blurring of legal and illegal practices, which is a rising concern in anthropological studies of neoliberalism and its effects in post-colonial contexts (e.g. Nordstrom 2004; Comaroff and Comaroff 2006; Aguiar 2010), I will discuss these practices as actualizing questions of state sovereignty in struggles of establishing and defining value.

My argument is twofold. First, I argue that the smuggling of combustible fuel reflects long-standing inequalities and a deep sense of local autonomy in border areas characterized by an ambiguous presence of state. A focus on smuggling thus serves to illustrate how controversy and contestation over energy may be responsible for uncovering and renewing long-standing social cleavages. Indeed, the smuggling reveals not only controversy over energy resources, but also how state legitimacy is negotiated and questioned. This negotiation of state legitimacy is informed by the ways in which border trade is socially and spatially embedded (see Polanyi 2001 [1944]), and underpinned by local forms of sociality, circulation and ritual exchange. Secondly, and related to the first point, I argue that the smuggling illustrates the limitations of an equation between energy extraction and development. In so doing, I reveal how people at the margins of these economies relate to questions of prosperity and circulation in a way that differs from dominant notions of development as facilitated through resource extraction.

Mitchell (2011) argues that no nation escapes the political consequences of our collective dependence on oil. By showing how energy extraction and the development of political forms have been mutually constitutive in the course of history, he argues that the dependency on oil shapes the body politic and creates both the possibilities and limitations of modern democracy. In this manner, Mitchell argues that the control and contestation of the transformative power of energy have become quintessential for the modern power over life. As noted by Boyer (2011), Mitchell's argument makes it worthwhile asking whether sovereignty is vested not only or primarily in territory or territorial borders, or in body politics (Foucault 1976; Agamben 1998), but in particular commodity constructs. I therefore find Mitchell's argument useful in studying smuggling as an inroad into understanding the struggles to transform energy into political power. Such a focus is especially interesting since the smuggling takes place beyond the institutional and infrastructural framework for circulation and distribution, where claims to sovereignty are conventionally made. As noted in the introduction of this book, energy politics actualize a complex series of interdependencies and social contracts, and this demands a more pluralized discussion of 'sovereignties' (Blom Hansen and Stepputat 2001). Indeed, a focus on the smuggling of combustible fuel demands a pluralized and multilayered understanding of sovereignty, and may contribute in questioning rigid accounts of the relationship between energy and development. In this regard, the chapter is concerned with how long-standing inequalities continue to represent an issue in governments' attempts to legitimize their energy politics.

As I will illustrate, governments' preoccupation with the smuggling of combustible fuel has indeed instigated new and intensified attempts to control the borders in order to reduce the extent of these practices. These new mechanisms of control can, if we follow Chalfin (2006), be seen as a sheltering of state sovereignty in a commodity construct. She argues that new mechanisms of border control can thus be understood as free-floating signifiers of state authority, as a sort of sheltering, or outsourcing, of sovereignty in a commodity construct (ibid.: 262). So, by focusing on energy resources as commodity and commodity construct within and beyond institutional and infrastructural frameworks, this chapter is concerned with how different claims to sovereignty become vested in a commodity construct. Such a sheltering of claims to sovereignty in a commodity construct is actualized also in state responses to popular protests against energy politics and extractive enterprises, and is illustrated by the lengths to which governments are willing to go in protecting state/corporate interests.

The chapter is based on anthropological fieldwork among traders and *contrabandistas* primarily on the Peruvian side of the border, mainly in the city of Arequipa. Arequipa is a regional distribution point for trade between highland and coast, and for the trading of goods brought from Bolivia and Chile. Indeed, people come to Arequipa even from Lima to buy contraband goods. Fieldwork was conducted in January and February 2011, and the chapter also builds upon previous fieldworks in 2007, 2004, 2003, 2001 and 1997. The fieldwork involved open-ended interviews with traders and *contrabandistas* as well as with public functionaries, in addition to participant observation at one of the marketplaces in Arequipa, more specifically at the Mercado Nuevo.[1] This market association involves about nine hundred members, many of whom have a background in the Andean highlands and are bilingual Aymara or Quechua and Spanish speakers. A research assistant was also hired to conduct interviews and observations in Puno, La Paz and El Alto, as well as in the border areas between Peru and Bolivia, with a focus on the routes and circuits of the smuggled fuel. My assistant's name is Magaly Cardich, and she has an MA in anthropology from the Catholic University in Lima. Her fieldwork was undertaken in January and February 2012. Before I describe and discuss the smuggling of fuel, I will give a brief outline of energy politics and increasing popular protests against extractive enterprises. This is in order to create a background for understanding the contested nature of energy extraction, as well as the practices of smuggling.

Politics and prices

Owing to the subsidization of the energy sector in countries such as Bolivia and Ecuador, energy prices in Latin America vary quite significantly from one country to another. In Bolivia, the nationalization of the energy sector by Evo Morales in 2005 involved a subsidization of energy prices for the national market, causing a significant price difference[2] between Bolivia and neighbouring countries. As a result, the illegal exportation of energy resources from Bolivia became widespread. A *gasolinera* (vendor of combustible fuel) interviewed in El Alto described how during several periods of the year 2010, the vendors sometimes ran out of fuel, since so much was being illegally exported.

The illegal exportation of fuel not only represents economic loss[3] for countries like Bolivia, where energy prices are subsidized, but also a questioning of the legitimacy and territoriality of the state. The smuggling may in this regard be considered a double challenge to state sovereignty in the countries involved, involving denials both of the state's legitimacy in controlling the flow of commodities, and of the territorial boundaries themselves. In this regard, it is not my intention to romanticize the smuggling, taking into consideration, for instance, how these practices may affect the Bolivian economy, or undermine tax revenues that could have contributed to improve social welfare. What I find important is understanding the rationale for and background to these practices. In examining this, I hope to throw light on the challenges, contradictions and dilemmas involved in realizing and establishing legitimacy for the politics of resource extraction.

As I will return to, the government in Bolivia has introduced several measures to reduce the smuggling, especially since 2009/10. These measures were introduced partly owing to the increasing difficulty of meeting delivery requirements for the countries with which Bolivia has entered export agreements, such as Argentina and Brazil. The question of illegal exportation also created the backdrop for the Bolivian government's decision to remove the subsidies in December 2010, causing the prices to almost double. In protest, people in Bolivia took to the streets and refused to pay bus fares, and after three days prices were reduced again. In the aftermath of this situation, the prices of fuel in Bolivia nonetheless remained somewhat higher than before, since they were regulated with the rest of Latin America. Although the intensity of traffic was somewhat reduced after this incident and the strengthening of controls, the traffic nonetheless continued.

Some of the countries to which oil and gas from Bolivia are being illegally exported, such as Peru, are also extracting their own energy resources. Indeed, Peru is currently accommodating a search for and exploitation of new gas and oil reserves. In contrast to Morales' attempts to nationalize the sector, recent governments in Peru have held a liberal line in relation to foreign investments and interests in resource extraction. These policies have been reinforced by Peru's signing of free trade agreements with a range of partners during the last few years (the USA, Canada, China, Japan, the EU, Korea, etc.). The idea that natural resource extraction will lead to economic growth and development is prevailing in official discourse, and represents a model which in Peru is supported by state–corporate partnership. The extraction of oil in the country was started in 1971, by PetroPeru, which was privatized in 1993. With Alberto Fujimori's presidency in the 1990s, economic policies were thus increasingly liberalized. This included allowing foreign actors to search for and extract energy resources along the coast as well as in the Amazonas; as much as 72 per cent of the latter has been opened for oil concessions. Peru is still a net importer of oil because production has been lower than expected. When it comes to natural gas, the country has increased its exports, especially since 2010, for instance to countries such as Canada, Mexico, Chile and Argentina. The major area for gas extraction opened in 2004, located in the department of Cuzco and referred to as Lote 88, where extraction is led by the state/corporate company Camisea. While the state still controls most downstream production, pipelines and most refineries, over half of the crude production is controlled by Argentina's Pluspetrol. Other foreign actors include the USA's Hunt Oil and Spain's Repsol.

Protests and responses

In recent years, there has been a rise in popular protest against extractive industries in Peru, owing to concerns about environmental effects in local communities, a common disregard of laws about prior consultation,[4] and limited employment opportunities for local populations.[5] In 2009, there was an uprising in Bagua against the extractive industry, first initiated in rural areas. The protests were a response to the opening for oil concessions in the area, fuelled by local opposition to a series of decrees to implement a Trade Promotion Agreement with the USA, involving the elimination of tariffs and trade protection for farmers. Later the protesters were supported by people from the town of Bagua, and the issues of energy exportation and energy prices were included in the complaints. During the intervention in the uprising on 5 June 2009, thirty-five people were killed (both protesters and police).

The protests in Bagua indicate how popular criticism in Peru has not exclusively been related to the consequences of extractive industries in local communities, but also to the prices of energy in the national market. Indeed, energy prices represent a continual concern among many, and most people seem to follow the prices of gas very closely. As a result of protests such as those in Bagua, agreements were made during 2010/11 to reduce the exportation of gas.[6] Owing to price adjustments and the continued rise in prices, protests nonetheless went on, and during my fieldwork in 2011, people took to the streets in Arequipa on 27 January, followed by similar protests in Cuzco, Puno and Tacna.[7] In Arequipa, I witnessed hundreds of people arriving in the Plaza de Armas from the shanty towns surrounding the city. They were protesting against the rising prices of gas.[8] In support of the demonstrators, the bus companies also stopped work. One of my interlocutors, Juan, who works as a taxi driver, said: 'This is a serious problem and therefore people protest; when the gas price rises, the prices of everything else rise along with it. Transport, food, water; everything becomes more expensive, because everything moves with fuel.' Indeed, since energy prices affect the general price level, most people, and especially the poor, are significantly affected by price adjustments. This is reflected in their willingness to mobilize and protest.

In Peru, Arequipa is known for the inhabitants' willingness to revolt against the central state, a fact that many of my interlocutors are clearly proud of. Indeed, current protests are often seen in relation to a massive strike and demonstration in Arequipa in 2002, against the privatization of water and electricity, the so-called 'arequipazo'. This demonstration initiated a series of popular protests not only in Arequipa but also more generally in Peru.[9] Many interlocutors in the shanty towns of Arequipa participated in these demonstrations, among them Arturo, who carefully stressed that both he and his wife participated in the *arequipazo*: 'In Arequipa we have a tradition for protesting, and politicians are afraid of us. When Arequipa rises in protest, the rest of Peru rises too.' There are two points that I would like to stress with regard to this series of protests. First, it is important to underline that the protests in recent years primarily involve people from the popular classes, or *gente popular*, and not *gente que tiene* (people who have). For instance, while my interlocutor Luis agreed with the criticism behind the demonstration in Arequipa in 2011, he did not participate himself. He always used to take part in such demonstrations, though – that is, when he and his family were still constructing their house in a shanty town in the outskirts of Arequipa and before they got water and electricity. Now that they have both water and electricity, he commented that: 'It is not so important to me any more. Those who participate now are those who live farther up, those who lack installations of water and discharge.' In this manner, it is common that *los que tienen* (those who have) are less likely to take part in demonstrations compared to those 'who don't have'. A second pertinent point is that although the protests primarily involve people from the popular classes, this does not imply that people's criticism reflects concerns only for their most immediate life situation. In their criticism, people also express concerns for the more general implications of a liberalization of resource extraction, and thus question the dominant equation between energy extraction and development in public discourse.

While energy has increasingly come to be regarded as a basic commodity or service, a widespread opinion is that energy prices should not be defined by the market alone. People often used the term *vende-patria* (vendor of the nation) to refer to politicians who are seen to be selling Peru's natural resources to foreign interests. As Luis and I were discussing the candidates before the elections in 2011, he said that all the candidates were *pura rata* (pure thieves). He argued that before the elections gas prices were rising as part of an exchange between the government and the enterprises: 'They make the prices rise in order to receive the support of the enterprise owners.' Luis continued to speculate that the prices were rising since Alan García was soon to leave his position as president anyway, so therefore he did not care any more, but was amassing money to take with him on leaving office. This and similar speculations illustrate a deep sense of mistrust, as people perceive that the authorities have priorities other than what is best for the *pueblo* (people). This sense of mistrust increasingly fuels people's willingness to protest, especially when they see that decisions are made for the benefit of foreign interests. In this rejection of the authorities' energy politics, the notion of oil and gas as national patrimony is at the heart of the protests. It illustrates Gledhill's (2008) point about the importance of oil in popular imageries in Mexico, as closely tied to particular forms of nationalism and claims to national sovereignty against perceived US imperialism.

One of Ollanta Humala's promises during the presidential campaign in 2011 was that there should be no more deaths resulting from conflicts between extraction companies and local populations. But in 2013, there had nonetheless been fourteen deaths in such incidents following his assumption of office. Before the presidential elections, Humala was regarded by many as representing an alternative to the neoliberal politics of previous governments, pledging to extend the benefits of Peru's economic growth to all Peruvians. The fact that little has changed since he took office, however, seems to further feed into the continuing protests related to extractive industries, for instance in Tacna, Paita, Cajamarca and Espinar.

On 21 May 2012, inhabitants in Espinar initiated a protest against the mining company Xstrata, after it was clear that they would not get their claims through. Central to the complaints was the copper mine's pollution of the community's drinking water, and the fact that the company did not provide employment opportunities as previously promised. When the police became involved to protect the company's property on 28 May, clashes between police and protesters led to two deaths and nearly fifty people injured (both protesters and police). For the second time in six months, Humala declared a state of emergency in response to protests against extractive industries.

The clashes in Espinar illustrate how far the government is willing to go in order to protect the interests and investments of extractive companies. In the debate that followed, the pervasiveness of the official discourse – that extractive enterprises will lead to economic growth, no matter whether the people most immediately affected are consulted or enabled to take part or not – is evident. In the front page of the Lima-based newspaper *Correo* (22 May 2012), the violent repression of the uprising in Espinar was described in the following way: 'Now they have their dead bodies. Police had to hit two and hurt 20 people in order to defend the mine in Tintaya.' In a similar vein, Congress member Lourdes Alcorta said: 'Unfortunately, there will have to be some deaths.' Commenting on this use of language, the blogger Eduardo González (30 May 2012[10]) notes that the killing of citizens is in this manner made into a technical question, in the sense that, in specific circumstances, there is no alternative. In other words, the logical consequence of such conflicts is that the police force will be used and protesters killed. This language further serves to transfer responsibility for the violence to the demonstrators, and to demonize social protest. It involves the logic that: 'Nobody has the right to oppose the politics of the state or of an enterprise, and the one who protests is a public enemy. The police is a killing machine, so it is your own fault if you place yourself in front of it' (ibid.). What thus appeared as an acceptance of these deaths in public discourse involved an acceptance too that these people are not Peruvians, not citizens, but natives, agitators, subversives – and in this manner reflect the position of some people as second-class citizens, in a social hierarchy defined by race and class inequalities. Resource extraction was represented as somehow inevitable – and thus made the violent interference in social protest appear inevitable too. In what follows, I focus on the smuggling of fuel, and subsequently discuss how these practices are socially and spatially embedded. In so doing, I relate the smuggling of fuel to questions of state sovereignty and local forms of sociality.

Contraband

For centuries, inter-regional barter and trade have played an important role among people in the Andes (Murra 1980). Indeed, contemporary cross-border trade in southern Peru follows the same routes as trade during the colonial period – that is, along the routes of mule drivers who brought goods between highland towns such as La Paz, Puno and Arequipa, and towns along the Pacific coast, to Chile. While thus building on a long-standing history of trade, these activities increased significantly during the 1980s and 1990s, especially among people from the highlands, but also among people from the urban middle class who were losing ground in this period.[11] This increase in self-employment was a result of the economic downturn, inflation and rising unemployment in the 1980s, and was reinforced by the structural adjustment programmes imposed by the IMF, privatization and the reduction of the public sector. The need for alternatives to waged labour was further reinforced by the continued migration of people from rural areas to the cities. These changes combined to make self-employment, e.g. through trade, the only alternative for many, often taking place at the margins of the formal economy. Especially since the government of Fujimori in the 1990s, the involvement in trade has also been promoted by NGO and state agencies, e.g. through microcredit programmes and courses, often directed at women in particular. In these programmes, the value of entrepreneurship is promoted alongside the expectation of reliance on family and kin networks. Free trade zones have also been established close to the borders of both Chile (in Tacna) and Bolivia (in Puno), the latter being established in 2007/08. These free trade zones are located along the historical routes of cross-border trade, and can be considered a way to incorporate, or make legible (Scott 1998), already existing practices of trade. Several of these arrangements are examples of how policy-making under neoliberalism often involves a rhetoric about inclusion through and by market principles, in line with the thinking of the Peruvian economist De Soto (1989). So, while the activities of these small entrepreneurs can be seen to represent a 'non-hegemonic globalization' (Mathews et al. 2012), and as central for local responses to market fundamentalism, it is important to note that these small-scale entrepreneurs are also the perfect neoliberal subject; hard-working and self-made, and continually adapting their own quest for social mobility to growing demands of flow and consumption.

There is and has been a majority of women in these forms of cross-border trade in the Andes, and particularly so in the south. This can be understood in light of the historical traditions for women to be involved in barter and trade in this part of the Andes, and a gender complementarity whereby women manage money and economic matters, while men manage politics (Harris 2000). In response to my question of why the majority of *contrabandistas* are women, the *contrabandista* Olinda replied: 'This is a women's area. If a man gets involved, people will say he is doing women's work.' Most of the trade in the border areas involves goods brought from Bolivia to Peru, owing to the lower price level in Bolivia. The merchandise being smuggled is not necessarily defined as illegal, but generally involves basic products of consumption. It is therefore first of all the undocumented importation of goods that makes the cross-border trade illegal, including that of fuel. Concerned about the extension of these activities, the legal assessor of the Chamber of Commerce and Industry in Arequipa claimed that almost 90 per cent of inhabitants in the border areas are involved in *contrabando*, providing the whole country with consumer goods: 'It is difficult to do anything, because people get angry, and will ask for an alternative – something which does not exist.' He added: 'It is a very complicated and difficult situation, and the authorities are afraid of interfering.'

The smuggling of fuel and other goods takes place in border areas characterized by a strong sense of local autonomy, infused by a history of marginalization and the ambiguous presence of state actors. Especially in villages close to the border, many people are engaged in contraband businesses and think about the delineated border not primarily as a delimitation of the nation-state, but as representing a possibility to earn money. These perceptions are also related to the many cross-border connections between Peru and Bolivia both historically and currently among inhabitants in the area. This situation illustrates how space and spatial delineations cannot be taken as ontologically given, but rather are created and contested as a result of relational processes (Massey 2005).

Being located just at the delineated border between Peru and Bolivia, the town of Desaguadero is an important location for the transport and distribution of contraband. Crossing the border in Desaguadero, you are not requested to produce documents of identity, and many *contrabandistas* lack such documents. There is a bridge in Desaguadero where one officially crosses the border and where *contrabandistas* pay *cuota* or *coyma* (contribution, or bribe) to the customs officials to pass with their goods, generally 2–3 soles to enter Bolivia, and 10–20 soles on the way back. You can cross the bridge either on foot or by bicycle, and those who walk are primarily women, wearing *polleras* (many-layered skirts) and carrying merchandise in *mantas* (rugs) on their backs. At the end of the bridge are the exchange brokers with their small wooden tables and radios. About 200–300 metres away from the bridge it is possible to cross the river by boat, and people who bring greater quantities of merchandise often use this route, hiding contraband goods under the other cargo. On the Bolivian side of the border, a port is located just in front of the police station, where the police officers stand at the door observing the traffic of people and goods. In the area around the bridge, there are people who offer to bring people's goods across the border, for a payment of 2 soles. Referred to in the media as *pulga* (lice) or *hormigas* (ants), they may cross the border many times in order to bring a customer's merchandise across. Many houses in the town are also rented as storehouses for contraband goods.

Desaguadero is often referred to as *tierra de nadie* (no man's land), and people come here from all across southern Peru to buy contraband goods, including fuel. This term points to the absence, or arbitrariness, of state control in certain areas, where trade is *already* institutionalized. In this respect, official interference in trade at the margins of the legal economy in Peru has generally been infrequent or random, and the authorities on both sides of the border have more or less closed their eyes to this traffic – that is, until the smuggling of energy resources began to be considered sufficiently problematic. A customs agent in Desaguadero said about the area: 'It has been taken by the inhabitants … It is difficult to interfere there since people become aggressive and defend in any way they can the continued functioning of the place.' One day they had sent one of their officials down there to implement customs controls, but as people responded with aggression, the official threw himself in the river – apparently in order not to be captured and beaten – and the other officials went out in a boat to rescue him. Police officers interviewed in both Bolivia and Peru similarly said that it is difficult and even dangerous to intervene in these areas of traffic and commercialization – and that they prefer to implement their controls primarily along the roads.

The routes and the goods On the highway from El Alto to Desaguadero, there is a petrol station almost every ten kilometres, and the number increases as you approach the border to Peru. On the bus to Desaguadero from the Peruvian side, you can hear people talking on their mobiles, planning their trading for the day. Among those involved in the traffic of fuel are people both from the Peruvian and the Bolivian side of the border, and owing to the lower prices in Bolivia, the fuel being trafficked is generally bought at petrol stations. While some *contrabandistas* sell the fuel on the streets, others deliver it to specific addresses in Peru for further sale, e.g. at petrol stations or in shops. In Peru, the fuel trafficked from Bolivia is said to be used primarily for three purposes: ordinary consumption and transport (gas, petrol), the production of cocaine (diesel, oil) and informal mining (diesel, oil). Along Cuenca Suches, the river that runs through Desaguadero, there are, for instance, several unauthorized mining businesses that have increased in numbers in the last few years, relying on illegally imported fuel from Bolivia.

In Peru, it is primarily women who sell the fuel to the public on the streets, generally offering cylinders of 2, 3 or 5 litres. Arriving in Desaguadero from the Peruvian side of the border, you soon see the vendors of gas and petrol, set up in front of the houses along the road. Some vendors live in Desaguadero and sell close to their houses, bringing out only a little fuel at a time and keeping the rest at home. Indeed, many houses in the town serve as storehouses for merchandise. The majority of these vendors are women, as already mentioned, who sit waiting for customers next to their cylinders of fuel, stacked on top of each other. Sometimes the vendors are knitting as they wait, with a parasol to protect them from the sun, and sometimes they are accompanied by their daughters. When pouring the petrol, they use nozzles similar to those in petrol stations, and several taxi drivers in Desaguadero reported that they fill up with fuel at the houses along the road.

Some *contrabandistas* transport the fuel across the border by foot, or bicycle, keeping the cylinders close to their bodies when crossing. This is especially common among women. There are also *contrabandistas* who transport the fuel by bus, placing it on the roof, among other luggage. The *transportistas* (drivers) involved in the smuggling are generally men who combine their work in transport (bus or car) with the traffic of contraband. Some drivers are involved also in the '*acopio de combustible*' – that is, accumulating fuel little by little and storing it, before transporting it across the border. There are even '*casas de acopio*' – houses where people stockpile the fuel in order to take it across the border by night. While many border villages are characterized by their poor housing, there are a surprising number of cars, often big and expensive *camionetas* (vans). Some of these vans have a double tank, or *falso piso*, which facilitates the accumulation of fuel. In Bolivia, it is, however, illegal to have a double tank, so many drivers hide the fuel in their cars. In and around Desaguadero, it is said that the fuel is transported across the border primarily by night or in the early morning, since confiscation of this merchandise causes major loss compared to that of other products.

In other areas along the border, grand-scale *acopio* takes place also using cars or trailers. Indeed, there are many different routes for the trafficking of fuel across the border, such as through Moho and

Huancané, where trailers go on towards Juliaca, sometimes accessing the roads leading all the way to Cuzco. Long lines of trailers may pass the borders by night (called *culebra* or *serpiente*, or snake), and the drivers are said to leave sacks of merchandise for the communities along the way, in return for help in sustaining the traffic. As many as six trailers may pass the border in a row, often without number plates, and sometimes leaving a bag of money at the police station in order to pass without interference. Fuel is also brought into Peru from Bolivia via the Amazonas, given the tax exemptions in this region on imported goods.

The city of Puno is also an important location for the commercialization of illegally imported fuel, where it is quite openly being sold on the streets, as part of everyday life. The traders greet the police officers when they pass by, and the police officers generally let traders continue their business. In order to let the big trailers pass the control points around the city, they are said to accept between 5,000 and 15,000 soles. In the Avenida del Sol, Puno, the *gasolineras* come out at sunrise to sell their goods, placing gallons of petrol or gas under a table in order to sell to drivers passing by, often combining this with selling other merchandise such as fruit. In and around Puno, it is said that the police cooperate with the *contrabandistas*, especially the major-league ones. While illegally imported fuel is sold in Arequipa too, the commercialization there takes place in a more hidden way. Some interlocutors reported that, owing to increased controls, the main distribution points in Puno are being moved farther away from the town.

Eufemia is one of the fuel traders in Avenida del Sol, Puno, selling her goods just in front of the hospital, behind a newspaper pitch. After settling down with some blankets, she often starts knitting while waiting for customers. Eufemia sells three types of fuel, she explained; one is green, with more octane, and is therefore the best (60 soles per gallon). The second is white, with less octane (59 soles). The third is Peruvian, which is the most expensive (68–70 soles), but she sells this for a lower price than in the petrol stations, so people come to buy this Peruvian product too. She buys the Peruvian fuel from police officers or hospital employees, owing to an arrangement in Peru where people in certain public positions receive fuel as a bonus. Eufemia has only a few gallons at her pitch, since the rest is placed in a rented location close to the avenue, where people run small businesses hiring out space for vendors. She has different kinds of clients, generally motorcyclists but also taxi drivers and private drivers. Her business is not paying off as it did before, however, owing to the rise in prices in Bolivia. Nevertheless, she still earns enough to keep her daughters in bus tickets and school equipment.[12] In this regard, Eufemia's business indicates how the smuggling of fuel is sustained by the involvement of small-scale actors who take care of distribution and sale to the public. For Eufemia and others, this is a way to secure a basic income.

There is a widespread idea in the Peruvian media that the smuggling of fuel is carried out by well-organized cartels, sometimes hiring individuals to transport the goods by foot (referred to as lice or ants). The strengthening of border controls during the last few years can be seen in the light of this idea that the traffic in contraband, and especially of fuel, is organized by cartels making use of advanced technological equipment to monitor official interference.[13] The case described below, however, indicates the existence also of medium-sized enterprises where people work independent of any cartel. Indeed, the case illustrates how the smuggling must be seen in light not only of its social and spatial embeddedness in kin and social networks, but also in relation to people's situation of poverty and need.

Since 2009, Olinda has earned a living bringing fuel from Bolivia. She used to buy the fuel from other vendors in the border village of Casani, and then brought it to Puno to sell at petrol stations, where the fuel was mixed with ordinary products. Olinda used to bring about fifty gallons at a time, travelling on the big buses and placing her goods on the roof, hidden underneath other packages. She started in the fuel business having amassed sufficient capital by bringing in other contraband *goods*, which was facilitated by her travelling for free in the car of her husband, who works as a taxi driver in the border areas. After she used this capital to invest in fuel, she started travelling on the big buses instead. In order to facilitate her traffic of fuel, Olinda used to rent a room at a relation's place in Pomata, a small town located close to the border. This gave her a good opportunity to accommodate the merchandise without being disturbed. Kin and contacts in the border areas of the highlands thus represent an important asset for many traders, not only for the accommodation of goods, but also for the supply of goods and information about prices and availability.

Olinda is originally from Arapa, a community located farther away from the border, close to Asangaro, which Olinda and her sister Julia both left while they were young in order to look for work in nearby towns. During my fieldwork in 2001, I came along to their home community in order to meet their mother, who was looking after Olinda's three children while she was working in the border areas. Their house at that point was an old and simple one made of *adobe*, and Olinda's children were obviously suffering from the limited food and care their old grandmother could give them. Particularly, I remember an incident soon after we arrived, as Julia had brought some bread from Arequipa to offer to the mountain spirits and share with her nieces and nephews. We then all had some potatoes to eat, but owing to the bad harvest that year, many of them were in a condition that I perceived as spoiled. While eating, I removed these parts and ate the rest, and felt embarrassed to see the children throwing themselves at my leftovers. During my fieldwork in 2011, in contrast, Olinda's children were attending good schools in the city. She had also managed to have two three-floor houses constructed, made of brick, to replace their old houses in Arapa. Olinda was able to make these changes thanks to her earnings from the fuel business.

Olinda's story illustrates how a situation of poverty and need often motivates people's involvement in the smuggling of fuel. For many, this involvement is and continues to be a precarious means of making ends meet, but others – especially the major-league *contrabandistas* – earn significantly more than the minimum salary, and have managed to expand and make new investments. Smuggling may therefore represent the potential for social mobility, and for disturbing the class- and race-based hierarchy in Peru.

Olinda got caught just before I arrived in Arequipa in 2011, however, as she tried to cross the border with fuel worth about 10,000 soles. All her goods were confiscated, and in addition she had to pay 3,000 soles to the customs officials (in *coyma*, not in fines) so that the bus would be allowed to carry on. After she got caught she cried for a month, she told me, and apparently her husband will no longer let her travel with him in his car, so she will have to build up capital again on her own, without the benefit of free transport. She described the first period after getting caught and losing all her capital as a terrible time, not having money even to buy food, but happy to receive some potatoes from a friend. She will not become involved in fuel smuggling again, she said, and added that fuel is more expensive in Bolivia now anyway. Having decided not to bring in fuel any more, Olinda has started to bring in other kinds of merchandise instead, such as olives and clothes, especially jackets.

The way in which Olinda chose to switch to textiles and olives, after having failed in her last attempt at bringing in fuel, indicates that for some of the traders involved, fuel represents 'just another commodity'. So while it is common for the *contrabandistas* to suffer when their goods are confiscated, they generally start up anew a short time after, although sometimes having downsized the quantity, or changed to another commodity. For instance, as we were discussing things while waiting for customers at the market pitch of Olinda's sister, Olinda and her sister agreed that 'either you win or you lose, you just have to continue'. In this regard, Olinda's story is typical of many *contrabandistas*' experiences and accounts about their encounters with the authorities. The case of Olinda and others further indicates the involvement not only of major- or minor-league actors in the smuggling of fuel, but also of medium-sized entrepreneurs who work independently. This does not mean that the cross-border trade is not socially and spatially embedded in important ways, as I will illustrate in the following. Indeed, the embeddedness of border trade is central to the negotiation of state legitimacy and territoriality that these practices involve.

The social embeddedness of cross-border trade

It is not uncommon for *contrabandistas* to criticize official interference for being immoral and illegitimate, questioning what right the officials have to interfere in people's attempt to make a living. Regarding the smuggling of fuel more specifically, it is interesting to note that people often legitimized this smuggling by particular reference to energy politics in Peru. For instance, people referred to the higher prices in Peru, and pointed out that their own energy resources are also being exported. Typical of people's responses when I asked about their opinions regarding the smuggling of energy resources was the following from Luis, as he said, shrugging: 'Since it is cheaper there. We also have gas, from Camisea in Cuzco, but we export this to other countries – at cheaper prices even – while we [that is, Peruvians] have to buy expensively.'

In general, when unauthorized economic activities are acted upon by the authorities, there is a common phrase that people utter with great contempt: 'They prevent people from working.' This phrase illustrates the value that people ascribe to this kind of work, independent of its degree of formalization. People thus often referred to poverty and unemployment when questioning the legitimacy of official interventions. Regarding the experience of a colleague whose contraband goods were confiscated, a woman called Rosaria said: 'They ought not to confiscate if this is how she works. It is not that they [*contrabandistas*] steal or anything, since they buy with their own money. Some even with borrowed money. This is the only kind of work there is ... As if they were the authorities? [i.e. those who confiscate goods] And who stays with the goods? Well, they share it between themselves of course! [that is, the officials].' State legitimacy is in this manner questioned and challenged, and a form of parallel or overlapping sovereignty seems to characterize this cross-border trade. As illustrated below, this form of parallel sovereignty is based on the social and spatial embeddedness of economic activity and relations of cooperation and exchange between kin and colleagues. Roitman (2006) makes similar observations in her research on illegal trade in the Chad basin. Although this trafficking may be illegal – since it departs from the codes and regulations of official law – she notes that people do not necessarily understand it as illicit. She suggests the term 'ethics of illegality' to refer to the values ascribed to forms of trade that are seen as both economically strategic and socially productive despite their often illegal character (ibid.: 264). Against this backdrop, Roitman makes an important argument that I find relevant also in the Peruvian context, namely that the forms of resistance associated with the trading practices must be seen as generated out of states of domination.

It is common that whole households or groups of kin get involved in the business of contraband, e.g. children, partners and parents/grandparents, since cross-border trade demands long working hours and a lot of travelling. Central to the realization of these forms of trade is therefore their social and spatial embeddedness, being based in networks of cooperation through household and kinship relationships. This reliance on household relations and kin also involves older relatives often making investments for younger relatives, for instance in merchandise or land for a market pitch. In this manner, people often expand their businesses, turning them into linked enterprises and thus spreading the risk (see Roberts 1995). In some cases, there is also a division of labour between household members who travel to bring in contraband, and household members who sell from a market pitch. Many vendors have learned the skills of trading from their ties with household, kin or partner, and it may be difficult to enter these businesses without such connections. Indeed, many *contrabandistas* are said to be *celosas* (jealous) about their business, referring to their unwillingness to share information beyond household relations and kin. With reference to the cooperation with her sisters-in-law who all bring in contraband, one of my interlocutors said that she learned everything she knows from them, and that they continue to cooperate: 'We are a chain of kin.' At the borders, kin as well as local farmers often support and help the *contrabandistas* carry out their operations, providing information and supply or storage of goods, as illustrated in the case of Olinda. These relationships also provide important connections between urban markets and rural areas along the trading routes (see also Smith 1989).

Indeed, success in these businesses depends on relationships of trust and cooperation, and there is an intense cultivation of social relations among vendors and *contrabandistas*, through sharing and giving, and the establishment of godparenthood and relationships to customers and suppliers. As I have demonstrated elsewhere (Ødegaard 2008), many *contrabandistas* also offer ritual payments to the powerful earth beings as a way to maintain good relations with the sources of well-being, health and prosperity. These payments take place through the offering of alcohol, food, herbs or coca to the *pachamama* (earth spirit) or the *apus* (mountain spirits), or to particularly prosperous virgins or saints. Such payments are meant to reproduce the sources of prosperity and improve success in business, and major-league vendors have a particular responsibility to serve as sponsors for festivals or parties. Particularly among major-league *contrabandistas*, it is also common to make pilgrimages to luck- or wealth-bringing places to secure success for high investments and risky businesses. Best known is Copacabana in Bolivia, a pilgrimage goal for merchant people from all over the Andean region. The virgin of Copacabana, Candelaria, is supposed to be extremely prosperous, and many *contrabandistas* and traders gather here each year to celebrate her and make offerings to promote prosperity in business. To serve as a sponsor for religious or other festivals and parties may contribute to securing prosperity in business. This illustrates the significance of a relational understanding of prosperity in this context, and of prosperity as dependent on reciprocity, circulation and exchange (see also Harris 2000). Harris has argued that, according to this logic, consumption (through sharing and offerings) and reproduction are not contrasted but closely interlinked, since offerings to the natural surroundings involve both consumption and the reproduction of the sources of prosperity.

Jorge and Norma both work as *contrabandistas*, and they generally travel to make payments to the virgin of Copacabana during her anniversary in August, expecting this to benefit their business. In 2003, they celebrated the patron saint of their village of origin instead. They had been appointed as *padrinos* for this festival, and were responsible for covering expenses on costumes, food and drink. Compared to the prosperous virgin of Copacabana, the saint of their home village is poor, and they did not really expect their input to strengthen their economic success. They nonetheless accepted the appointment in order not to get punished by the saint and risk the failure of their business. This was only one of the many other commitments to kin, neighbours and colleagues that they try to meet regardless of the expense, even if they have to borrow money. As a quite successful couple, their involvement as ritual sponsors illustrates the responsibility of the wealthy to secure the prosperity of others by sponsoring parties or festivals and maintaining a good relationship with the sources of their prosperity, e.g. *pachamama* or the saints. If they shirked their responsibilities, they also risked exclusion and the loss of the support, work help and gift-giving that these relationships provide. In this manner, people understand prosperity and progress as being maintained through reciprocity, exchange and circulation. Such understandings not only differ from the public discourse of development as facilitated by natural extraction, but also serve to legitimize contraband businesses. Indeed, *contrabandistas* are generally respected for their ability to make and manage money, and for their independence whether their work is illegal or not. They are seen as important suppliers of merchandise, performing a valuable social service by travelling to bring in goods that other vendors can sell. There is thus a general appreciation of the often hard-working *contrabandistas*, in a way that reflects the appreciation of circulation and exchange as a cultural value in this context (Ødegaard 2008). From this perspective, those who engage in trade are seen to perform a valuable social service by making money 'give birth' (Harris 2000: 61).

While relying on kin and social networks, cross-border trade is achieved also through the involvement of public functionaries. There are also public officials who legitimize their non-interference in the same way as the *contrabandistas* themselves, arguing that there are no jobs to get anyway. Many *contrabandistas* have brought goods across the borders for the duration of their working careers; on some occasions having their goods confiscated, but not often enough to prevent their businesses from flourishing. This is facilitated by the payment of bribes. In some cases, these monetary interchanges seem to be automatized and normalized. At one point, for instance, a woman behind a trolley was moving towards one of the boats along the river in Desaguadero, passing a Bolivian police guard. Upon passing, the woman gave him her hand containing a 20-sole note, apparently in the quickest way possible, upon which she continued towards the boats, without expecting any further interruption. The woman and the police guard did not exchange words, only looks – and thus the monetary interchange seemed automatized. Through this charging of bribes to let people pass, there seem to exist networks of complicity on both the Bolivian and Peruvian sides of the border between public officials and *contrabandistas*. In this way, the relational character of trade constitutes a means by which state officials at different levels are able to make money from the extralegal activities of *contrabandistas*. The form of overlapping sovereignty that characterizes cross-border trade is thus realized by the involvement also of public functionaries, who often receive bribes in return for letting the *contrabandistas* pass, or otherwise just letting people continue about their business. Economic flows in these *tierras de nadie* can be seen to involve an exception to the general rule in Agamben's (1998) terms, reflecting how the rule itself is constituted by an exception, creating a zone of indistinction between 'bare life' and political order, where 'bare life' is included only through an exclusion.

During my last fieldwork trip, several *contrabandistas* complained that the border and inter-city controls were increasing, especially along the roads before entering Arequipa. In contrast, they perceived the controls before entering Puno as less strict, and observed that customs officials there were still willing to accept bribes. According to the *contrabandista* Norma, for instance: 'There [in Puno] they always respect you, but not here in Arequipa. They do not respect you and they do not want to accept bribes.' Norma's formulation indicates how people see the reception of bribes among official functionaries as a mark of respect, a respect for their efforts and for their right to earn a living, now diminishing owing to the stronger controls. In this regard, the *contrabandistas* underlined the fact that the small and medium-sized actors are more affected by the increased controls than organized cartels, since the really major-league actors always manage to pay their way anyway. As the controls have intensified, the officials are apparently demanding higher bribes, which only the grand *contrabandistas* can afford.

Sovereign loss, sovereign things While fuel may be 'just another commodity' for the *contrabandistas* involved, this is certainly not the case from the perspective of the authorities. Being capital-intensive rather than labour-intensive – and demanding advanced technology and infrastructure – the extraction and distribution of energy is highly institutionalized, and represents commodities of high demand in the global market. In this regard, the illegal exportation of fuel not only represents economic loss for countries like Bolivia where energy prices are subsidized, but also a questioning of borders and state legitimacy. Indeed, the smuggling of fuel can be seen to challenge state/corporate interest and investment. Through the smuggling, energy resources are made into a petty commodity, deinstitutionalized and informalized. This transformation of the commodity represents a challenge to state investment in its extraction and distribution.

Especially on the part of the Bolivian government, measures have been introduced to reduce the smuggling of fuel and prevent the national economy from bleeding. Not only has the border control system been modernized and strengthened, but sentences have also been increased (from two to ten years). Especially since 2011, a range of new laws and arrangements has been introduced, and it is no longer permitted to sell fuel piecemeal (that is, without a car). Furthermore, cameras and police have been placed at petrol stations, the price of fuel for cars registered abroad has been doubled, and petrol stations have quotas for how much fuel they can sell. In addition, President Morales has introduced campaigns in order to increase awareness of the negative effects for the national economy, and rewards are given to people who denounce *contrabandistas*. There has been a suggestion too that number plates of cars should be recorded when taking on fuel. Military patrols have also been used to undertake controls, mainly by checking cars and trailers, and often being positioned along the roads in order to stop vehicles before they enter the border villages. Being targeted at energy resources in particular, these measures indicate how state territorialization, through the intensification of border work, is directed at some commodity constructs more than others. As noted by Chalfin (2006), such new mechanisms of border control can be seen as a sort of sheltering, or outsourcing, of sovereignty in a commodity construct. From the perspective of Bolivia in particular, the specificities of energy – as capital-intensive and demanding advanced infrastructure – appear to intensify the sheltering of sovereignty in a commodity construct.

Although less so than in Bolivia, the border controls have increased in Peru too during the last few years, in an attempt to reduce the expansion of contraband activities. Control stations both for customs and taxes have also been introduced along the main intercity highways, e.g. before entering bigger cities such as Puno, Arequipa and Ica. These internal controls represent an addition to those at the delineated border – which between Peru and Bolivia is difficult to control not only because of the somewhat challenging topography, but also, as I have illustrated, because of the way in which border trade is socially and spatially embedded in networks of kin and cooperating border communities. In addition, routines for the registration (and destruction) of confiscated goods have been introduced, in order to reduce the problem of functionaries who keep the confiscated goods for themselves. Peru's strengthening of border controls has been initiated largely in response to the Trade Promotion Agreement with the USA from 2009. As part of this agreement, the USA agreed to assist Peru in limiting the exportation of narcotics, modernizing the equipment and procedures of the National Police and Customs Agency, and strengthen the rule of law. While Bolivia is strengthening controls owing to the illegal exportation of fuel, Peru is strengthening controls in response to the demands of international partners. The Peruvian situation illustrates Chalfin's argument (ibid.) that the development of new customs policies and standards is not necessarily controlled by national governments, since extranational agendas and entities may endow governments with new tools and objects of operation and control.

On both sides of the border between Peru and Bolivia, the

intensification of controls has become increasingly important as a way of asserting national boundaries and state sovereignty. The new forms of border control involve a rationalization and implementation of forms of territoriality that follows pre-constituted political divisions (nation-states, departments, provinces, municipalities), divisions which the *contrabandistas* both contest and make use of to make a living. In this regard, the smuggling of fuel has intensified the ways in which energy is made the object of different claims to sovereignty.

Conclusions

My argument in this chapter has been twofold. First, I have argued that the smuggling of fuel reflects long-standing inequalities and a deep sense of local autonomy in border areas characterized by an ambiguous state presence. In so doing, I have illustrated how the smuggling involves a negotiation and challenging of state legitimacy and territoriality. *Contrabandistas* in Peru legitimize the smuggling of fuel by reference to unemployment and lack of other opportunities. In addition, they legitimize the smuggling by reference to energy politics and the state-authorized exportation of Peru's own energy resources. The smuggling thus reveals some of the controversies and contradictions involved in transforming energy into political power (Mitchell 2011), as different claims to sovereignty are made by state actors and the *contrabandistas* working in the border areas. Such claims to sovereignty in the border areas are illustrated also by state attempts to reinforce control over commodity flows. The smuggling thus illustrates the dilemmas of establishing legitimacy for energy politics in a highly pertinent way. In this regard, I have tried to go beyond the idea of a conflict of interests in a conventional sense, and have rather been concerned with some of the struggles involved in establishing and defining value.

Secondly, and related to the first point, I have argued that the challenging of state legitimacy in this context is informed by the way in which border trade is socially and spatially embedded, and underpinned by local forms of sociality, circulation and exchange. Through my analysis of these dimensions, I have illustrated how people at the margins of these economies relate to questions of prosperity in a way that differs from dominant notions of development as correlating with resource extraction. Indeed, the smuggling is informed by a relational understanding of circulation and ritual exchange as important for the creation and maintenance of the sources of prosperity. In these relational notions and practices of circulation lies a claim to territorial and cultural autonomy too.

In this manner, I have argued that the smuggling gives expression to an understanding of development and prosperity that goes against the equation in public discourse between energy and development. The challenges to the state that the smuggling involves illustrate how development through resource extraction cannot be reduced to a technological fix. Rather, resource extraction actualizes structures of inequality and feelings of local autonomy in new ways. In this manner, the possession of hydrocarbon reserves is not, as indicated in the introduction to this book, in itself a source of power – nor is it destined to create social pathologies as indicated in the 'resource curse' thesis. Rather, there are other socio-economic forces at work to compound the political value of energy resources, such as competing claims to sovereignty and the mistrust towards state agencies, both dimensions being related to structural inequalities.

Notes

1 The name and characteristics of this market have been altered in order to protect the identity of my contacts, including people's names and villages of origin.

2 Gas in Bolivia was about a third of the price in Peru.

3 Estimated at US$450 million per year (*La Razon*, La Paz, 27 January 2011).

4 Based on the Native Communities Law (signed in Peru in 1974) whereby the state recognized indigenous rights over land, and which later, at least on paper, has been reinforced by laws about participation and prior consultation.

5 Being capital-intensive, and not labour-intensive.

6 In September 2010, an agreement was made between the government and the private syndicate for the production of gas (Camisea) to limit the exportation of natural gas – in response to protests and production blockades. This specifically involved the gas extracted from Lote 88 in Cuzco.

7 Also in July 2010, people in Cuzco, Puno, Tacna and Arequipa protested against the exportation of gas from Camisea.

8 From 14 to 15 soles per gallon of gas, while 95 per cent gas rose to 20 soles.

9 While water and electricity have been privatized in most other cities, this is not the case in Arequipa. This was a result of the '*arequipazo*'. As one of the interlocutors said, 'Toledo did as they told him and decided not to privatize in Arequipa.'

10 Blog of Eduardo González Cueva: latorredemarfil.lamula.pe/2012/05/30/sus-muertos-y-los-nuestros/EduardoGonzales.

11 Thanks to Peruvian economist Oscar Ugarteche for pointing out to me these historical continuities.

12 She pays 55–58 soles for a gallon and sells it for about 70 soles.

13 See, for instance, *La Razón*, La Paz, 6 April 2011.

References

Agamben, G. (1998) *Homo Sacer: Sovereign power and bare life*, Stanford, CA: Stanford University Press.

Aguiar, J. C. (2010) 'Stretching the border. Smuggling practices and the control of illegality in South America', New Voices Series, 6, Global Consortium on Security Transformation, pp. 2–21.

Blom Hansen, T. and F. Stepputat (eds) (2001) *States of Imagination: Ethnographic Explorations of the Postcolonial State*, Durham, NC, and London: Duke University Press.

Boyer, D. (2011) 'Energo-politics and the anthropology of energy', *Anthropology Today*, 52(5): 5–7.

Chalfin, B. (2006) 'Global customs regimes and the traffic in sovereignty: enlarging the anthropology of the state', *Current Anthropology*, 47(2): 243–76.

Comaroff, J. and J. Comaroff (eds) (2006) *Law and Disorder in the Postcolony*, London and Chicago, IL: University of Chicago Press.

De Soto, H. (1989) *The Other Path: The invisible revolution in the third world*, London: Taurus.

Foucault, M. (1976) *History of Sexuality*, vol. 1, New York: Random House.

Gledhill, J. (2008) 'The people's oil: nationalism, globalization and the possibility of another country in Brazil, Mexico and Venezuela', *Focaal*, pp. 57–74.

Graeber, D. (2001) *Towards an Anthropological Theory of Value*, New York and Hampshire: Palgrave.

Harris, O. (2000) *To Make the Earth Bear Fruit. Ethnographic essays on fertility, work and gender in highland Bolivia*, London: Institute of Latin American Studies, University of London.

Lazar, S. (2008) *El Alto, Rebel City. Self and citizenship in Andean Bolivia*, Durham, NC, and London: Duke University Press.

Massey, D. (2005) *For Space*, Los Angeles, CA: Sage.

Mathews, G., G. Lins Ribeiro and C. Alba Vega (2012) *Globalization from Below. The world's other economy*, London: Routledge.

McNeish, J.-A. and O. Logan (2012) *Flammable Societies: Studies on the Socio-Economics of Oil and Gas*, London: Pluto Press.

Mitchell, T. (2011) *Carbon Democracy: Political Power in the Age of Oil*, London and New York: Verso.

Murra, J. (1980) 'The economic organization of the Inca state', *Research in Economic Anthropology*, Supplement 1.

Nordstrom, C. (2004) *Shadows of War: Violence, power, and international profiteering in the twenty-first century*, Berkeley: University of California Press.

Ødegaard, C. (2008) 'Informal trade, contraband and prosperous socialities in Arequipa, Peru', *Ethnos Journal of Anthropology*, 73(2): 241–67.

Polanyi, K. (2001 [1944]) *The Great Transformation: The Political and Economic Origins of Our Time*, Boston, MA: Beacon Press.

Roberts, B. (1995) *The Making of Citizens. Cities of peasants revisited*, London: Edward Arnold.

Roitman, J. (2006) 'The ethics of illegality in the Chad Basin', in J. Comaroff and J. Comaroff (eds), *Law and Disorder in the Postcolony*, Chicago, IL: University of Chicago Press.

Scott, J. (1998) *Seeing Like a State: How Certain Schemes to Improve the Human Condition Have Failed*, New Haven, CT: Yale University Press.

Smith, G. (1989) *Livelihood and Resistance: Peasants and the politics of land in Peru*, Berkeley, Los Angeles and Oxford: University of California Press.

7 | THE CONTINUOUS NEGOTIATION OF THE AUTHORITY OF OIL- AND GAS-DEPENDENT STATES: THE CASE OF BOLIVIA

Fernanda Wanderley

A range of different authors within the social sciences have been interested in explaining the factors that influence the shape of development models in countries that are highly dependent on non-renewable natural resources such as oil and gas.[1] They are asking a number of questions: How can we explain the common occurrence of political turmoil, lower economic growth and increased levels of poverty in countries that are rich in non-renewable natural resources? How can we explain the failure of the majority of oil- and gas-exporting states to efficiently control and use the income from their territories' rich natural resources for the long-term benefit of public interest?

These concerns apply to reflections on the historical conditions of the formation of nation-states and the patterns of formulation and implementation of public policy. In these studies, two main interpretive directions can be noted. On the one hand, there are theoretical frameworks that assume a rationalist and technocratic position regarding the formation of state and public policies, and on the other, there are theoretical frameworks that consider state construction and decision-making as eminently political processes that are fraught with contradictions, struggles and internal and external responses to the state apparatus. Both approaches start from assumptions about the historical evolution of nation-states, but also aim to provide suggestions and proposals for reform.

The rationalist and technocratic approach assumes that the decisions are relatively unconditioned by social and economic structures and that the outcomes of political decisions depend more on strategic choices and the qualities of specific leaders. Therefore, the importance of factors such as the continuing formation of social classes and their relationship with the formation of states, political cultures and the legacy of collective meanings and cognitive and ideological frameworks in the decision-making process are disregarded. These factors become exogenous parameters in the analysis because they are not considered important explanatory factors.

The political approach to public policy starts from the opposite premise: i.e. it opposes a rationalist and technocratic approach. This means that government actions occur in frameworks of interaction between government actors and political and economic actors. Indeed, these are also informed by historically constructed formal and informal rules.[2] In other words, it is through state–society interaction that objectives, instruments and political actions are defined. From this perspective, public policy represents the outcome of a model of interaction between the state and the society that is dependent on interpretive processes, conflicts of interest and logics of collective action. Therefore, the state is defined as a conglomerate of organizations, each one with its own partially articulated objectives and dynamics. The intense political dispute between different bureaucratic bodies and civil society suggests an organizational structure that is continuously designed and redesigned and in which the functions, hierarchies and dependencies are constantly modified. These transformations do not necessarily respond to rational and technical requirements and may be influenced by power relationships that are both external and internal to the state apparatus.[3]

These two theoretical orientations also touch on another critical debate in social sciences, i.e. the concept of the actor and his/her relationship with social structures. On the one hand the rationalist-instrumental approach gives pre-eminence to individual actors who are conceived as relatively autonomous with regard to the social relationships in which they are immersed. Their actions are viewed as a priori rational and instrumental, based principally on a cost–benefit analysis in which stakeholders assess possible courses of action in relation to the objectives sought.[4] On the other hand, sociological and anthropological approaches view the actions of individual and collective actors as socially constructed because they necessarily take into account the actions of other individuals to guide their own action. From this perspective, the motivations of actions are plural a priori and are configured by the meanings that individuals and groups of individuals give to them. Therefore, social actions are, to a lesser degree, rational-instrumental and may be motivated by habits, traditions and different values.[5]

This theoretical debate can also be seen as underpinning a very important methodological question: What is the status of the meanings, ideas and interests of social and political actors in explaining development paths and of the efficacy and efficiency of public policies? According to the sociological and anthropological perspective, the behaviour of actors may not be deduced solely from assumptions about universal preferences and incentives. In order to identify what they are, empirical investigation aimed at elucidating the collectively constructed meanings and the struggles of interests that structure political fields and their dynamics is essential.[6]

With these theoretical discussions as background, the current chapter draws on an interdisciplinary approach combining anthropological, sociological and public policy perspectives to analyse the political and institutional processes behind the development paths of non-renewable natural-resource-exporting countries. To exemplify this concretely, the chapter focuses attention on a particular course of events, i.e. the failed attempt to reform hydrocarbon revenue distribution policy in Bolivia. The choice of a specific event and its outcomes is based on the assumption that micro-analysis can provide a privileged window on to the disputed meanings and ideas that structure policy areas and their dynamics. This kind of analysis offers specific inputs on how government actions and their outcomes are interwoven in historically constructed social and political relationships. Furthermore, it enables an exploration of how much leeway governments really have, given the limitations imposed by the social, economic and political structures in which they are embedded. In this sense, we argue in favour of understanding government actions, their scopes and limits as ongoing processes of negotiation, dispute and interpretation in areas populated by stakeholders with multiple and constantly evolving interests, perceptions, values and behaviours.

From this perspective, the study of the Bolivian case illustrates the difficulties of constructing legitimation for collective agreements on how 'to govern oil and gas'. It also enables us to understand the political dynamics that construct institutional relationship frameworks between society and the state in countries dependent on natural resources extraction. In this way the unsuccessful government measure of reducing fuel subsidies in 2010 provides us with a window through which to analyse these political dynamics, where there is visibly always space for interpreting formal and informal rules, for criticism of what is appropriate and, therefore, for the struggle, negotiation and dispute over policies driven by governments. From this perspective, institutions are understood as collective constructions, the stability of which requires continuous affirmation.[7]

The chapter begins by reviewing the theoretical framework of a constructivist approach to public policy decision-making processes that characterize countries that export non-renewable natural resources, specifically oil and gas. The chapter proceeds to introduce the Bolivian hydrocarbon resource redistribution policy in an unprecedented context of tax revenue increases. Based on these elements, we reconstruct the process of the attempt (failed) by the government of Evo Morales to reduce hydrocarbon subsidies in 2010. In this section, we will first put forward the official justifications for choosing this measure and analyse its empirical support. Secondly, we reconstruct the views, perceptions and interpretations that structure public debate and the social mobilization that defined the limits on the state's authority to control and use the revenue from non-renewable natural resources. The chapter concludes with some discussion of the relevance of the interdisciplinary perspective to understanding the energy sector and the dilemmas of oil- and gas-producing countries within the context of a boom in international prices.

The theoretical ground – power, institutions and agency

In this chapter we start from an interdisciplinary approach combining anthropological, sociological and public policy perspectives. The central assumption of our interdisciplinary discussion is that the formulation of public policies, their criticisms and responses always involve a complex correlation of interests, perceptions and meanings, and that as such policies are historically situated symbolic constructions. The analysis of ideas, perceptions, meanings and interests that motivate the formulation, response or acceptance of public policies permits an exploration of the difficulties in constructing social and political legitimacy in decision-making and, therefore, the continuing challenge to the authority of the state on the use of non-renewable natural resources. It is therefore very important to understand the social processes that define the feasibility and effectiveness of policy choices. This understanding furthermore requires empirical study aimed at the interaction between organized social groups, social expectations, the meanings and interpretations in dispute, political behaviour and forms of society–state relations.

Public policies may be defined as decision-making processes whereby governments address problems of public order, that is to say, 'pursue goals considered to be of value to society or to resolve problems, the solution of which is considered to be of public interest or benefit',[8] such as, for example, situations of inequality, poverty and loss of food security. In this sense, public policy constitutes a specific subset of government public action characterized by being intentional and based on causal hypotheses. These hypothetical constructions inform which actions are the most appropriate to change the 'problem of public order'. Thus, public policies involve a narrative construction in the sense that they are built on analytical logic based on information (which is expected to be appropriate), which results in a history (causal model) concerning the factors that supposedly configure a situation considered to be problematic. The degree of appropriateness of this analysis (well-constructed information and a plausible causal model) is an important element for the effectiveness of actions and policy reforms.[9]

However, the appropriateness of the analytical or rational logic (which is always based on values concerning the 'desired' society and, hence, from a political positioning) is not enough for policies to be feasible, effective and efficient. It is essential that the group of social stakeholders share the narrative of what is problematic and that the cause-and-effect relationships which supposedly created the problem are considered 'correct' and that the solution (measure or reform) is also perceived as 'relevant'. Creating sufficient political support to neutralize economic and political interests that are opposed to the common good is equally important. This means there is a need for institutional frameworks that enable minimum agreements on the just distribution of resources, especially if they are non-renewable natural resources.

From a policy perspective, public policy decisions must be legal and legitimate under democratic principles, such as

> respect for the freedoms of their citizens, the production of outcomes of public interest and benefit to which markets contribute indirectly with their productive activities (employment, income, consumption ...), vigilance to prevent private interests seizing

public interest, damaging the rights of third parties and destabilising co-existence, in addition to deciding robust social policies, aimed at creating and developing citizens' abilities so that they decide and realise their own life options and remove the various kinds of obstacles that hinder it.[10]

However, from the point of view of public policy, the 'must be' perspective does not sufficiently recognize the fact that there are no optimal policy principles for all societies. Rather, the principles on what public policies 'must be' are collective constructions, whose meaning, importance and translation into practice vary from society to society, and even within societies. Therefore, one cannot assume a set of principles as static and exogenous parameters. As we will see in the analysis of the Bolivian case, collective meanings on the necessary role of the state and on the fair use of revenue from non-renewable natural resources are built on standardized practices between society and state, which have had a long historical evolution within the country.

Nor may it be assumed a priori that public policies are coherent, systematic, feasible, efficient and prepared to improve and evolve. Contradicting the view that 'what distinguishes public policy is the fact of integrating a set of structured, stable and systematic actions that represent how government permanently and steadily performs public roles and deals with public problems',[11] what is observed in many countries, and Bolivia in particular, is the predominance of contradictory, inefficient, improvised, non-consensual and highly contested actions. One of the features of many states dependent on non-renewable natural resources is precisely that decision-making, its implementation and the exercise of influence by social actors do not occur within an institutional framework characterized by plural and transparent deliberation that favours legitimate and stable agreements on hydrocarbon policy.[12]

Therefore, the analysis of public policy must be framed in the long history of decision-making in which formal and informal rules on the relations between state and society are reproduced. Thus, no public policy is an isolated and unprecedented decision. On the contrary, policies and measures are embedded in decision paths that the state and society have taken throughout their history. This is true even for government administrations that are proclaimed revolutionary and foundational, such as is the case of the Movement for Socialism (MAS in Spanish) government that began its first administration in 2006.

Public policies in countries dependent on non-renewable natural resources

In the case of post-colonial countries dependent on non-renewable natural resources, specifically oil and gas, similarities in the forms of performance of states and governments may be identified. Financial dependence on non-renewable natural resource revenue, mainly from oil and gas, has significant effects on state structures, decision-making systems and government actions which explain the difficulties in overcoming old models and restructuring development paths. The political fate of these countries is strongly associated with the performance of their main economic sector. They are states dependent on alternating cycles of prosperity and scarcity defined by the international prices of raw materials.[13]

In countries that export non-renewable natural resources, specifically oil and gas, the state is a central actor in the process of accumulation. This fact is mainly based on control of and dependence on the revenues from an economic sector that drives growth.[14] In the case of Bolivia, it is calculated that around 50 per cent of the state's financial resources comes from non-renewable natural resource revenue. States such as Bolivia are, therefore, rentier and distributive states. However, their leading role does not involve autonomous decisions given that their decision and implementation process occurs within highly anti-establishment frameworks of interaction between society and state.

Hence a paradox emerges in these countries mainly in cycles of boom: the greater influx of resources from non-renewable natural resource revenue results in greater power for the state owing to the increase in its financial base. On the other hand its authority to control and use those resources might be called into question owing to the reinforcement of economic, social and political dynamics by the boom and the relative improvements in living conditions. The greater power in cycles of boom results in very similar government actions in countries dependent on non-renewable natural resources, i.e. an increase in state jurisdiction (scope and degree of state intervention in the economy), and changes in the role of the state (through mega-economic projects, new redistributive mechanisms from the revenue via social programmes, greater investment and public spending, a salary increase not necessarily associated with the increase in productivity, greater employment opportunities mainly in state bureaucracy, and an increase in national consumption mainly via an increase in imports of products). These processes generate new cycles of accumulation for some economic sectors, a generalized feeling of improvements in living conditions and an increase in political legitimacy for incumbent governments. This seems to be the process experienced by many countries in Latin America, and Bolivia in particular, in the first decade of the twenty-first century.

The social and cultural effects of dependence on natural resources are equally significant. The restructuring of social classes and strata, the emergence of new economic elites, new notions about what the state is and its intentions, and political and economic expectations are deeply associated with the revenue from natural resource extraction. Precisely because the power of the state and its political authority rest on its capacity to extract revenue and distribute it internally, the state is seen as an apparatus, the principal function of which is the redistribution of revenue from natural resources within a framework of a rentier political culture. The problem lies in the fact that redistribution in times of prosperity without overcoming financial dependence on international prices may have dramatic consequences in downturn cycles, when social benefits or employment created by economic prosperity can no longer be financially maintained. Once society has acquired new benefits and rights, it is very difficult to go backwards without social conflicts that are profound and dramatic.

From a more structural perspective, this circumstantial situation may hide Dutch disease, i.e. the presence of new resources or the increase in prices of a sector – gas or oil – cause negative effects in other economic sectors, such as, for example, rapid and distorted growth in services, transport and other non-tradable goods, while national productive sectors such as food and manufacturing experience a slowdown. In the short term, this situation is not felt by the population as a 'problem' owing, in many instances, to the increase in employment, greater monetary flows via social policy and good short-term macroeconomic and social indicators. In this context, governments have little political incentive to confront this structural problem, which has medium- and long-term consequences.

Comparative analyses on the features of the political economy of countries dependent on non-renewable natural resources raise the question of whether the Dutch disease dynamic is inevitable, making government actions to counteract it impossible or, on the contrary, whether these outcomes depend to a greater or lesser extent on government decisions and policies. From our analytical framework, government policies always have some leeway demarcated by political and economic fields populated by diverse social actors and, therefore, by interests, beliefs, practices and strongly entrenched expectations. In this regard, government policies can be used to reinforce or change these practices and expectations. Therefore, governments are not exempt from responsibility for policy choices and their outcomes, although they do not have total control over them.

Before proceeding with the analysis of the specific event of the failed attempt to change hydrocarbon policy in Bolivia, it is important to show briefly the antecedents and characteristics of this policy.

Hydrocarbon policy and revenue redistribution mechanisms in Bolivia[15]

The proclamation of Hydrocarbon Law 3058 of May 2005 represents a turning point in Bolivia's hydrocarbon policy, i.e. a reform that permits demarcation of two significant historical moments. At the conceptual level, the new precepts introduced with this law closed a cycle marked by an aggressive liberal policy of investment attraction – inaugurated with Law 1689 in 1996 – and, at the same time, opened up a new phase marked by a series of reforms that gradually characterized a model of state management of the oil and gas sector.

Within this framework, the arrival of the government of President Evo Morales Ayma in January, far from consolidating the bases of what would be a new model of sectoral structuring, inaugurated a period of reshaping of hydrocarbon policy. In this context, the 'Heroes of the Chaco' Nationalization Decree No. 28701 was issued in 2006, deepening the scope of the reform inaugurated a year earlier and – under it, in October 2006 – new terms for the relations between the state and private investors were signed under the guise of operating contracts. This new policy model emphasizes state monopoly management of hydrocarbon resources, but also relies on foreign private investment.

Continuing the chronology of structural changes, in February 2009 a new Constitution was proclaimed that endorsed the fundamental principles of the Nationalization Decree. The Constitution validates state monopoly management of the hydrocarbon production chain and strengthens the principle of industrializing natural gas as a strategic pillar for national development. Private investment is provided a subsidiary role subject to a service provision system. The Constitution also introduces a series of guidelines for the scope of the industry and the economic system in general.

Under national regulations, the hydrocarbon sector must not only ensure energy security on the basis of a new model of consumption reliant on natural gas, but must also generate the surplus that will finance social policies and productive development. With the perspective of 'sowing gas', the National Development Plan of 2006 states that the objective of the sector is 'to maximise the economic surplus from strategic sectors and optimise its use for economic diversification and an increase in well-being in a context of equilibrium with the environment' (p. 91). In addition, the revenue generated by mining activities must be reinvested to continue exploration and exploitation activities that are essential to reproduce them, and furthermore be channelled for the industrialization of natural gas on Bolivian territory, seen as the principal strategy for changing the primary exporter model. Given that these are capital-intensive sectors, the plan acknowledges that the viability of these strategic lines requires the consolidation of substantial long-term investments, technology and specialized labour and, for this reason, 'the country requires, on the one hand, strategic alliances with the national and international private sector, as well as with other countries, and on the other, performance criteria to generate virtuous circles between the state and transnational companies' (p. 93).

Within the Energy Development Plan of July 2009, sectoral policy guidelines are reinforced by the National Development Plan, which, as already stated, gives the hydrocarbon sector a strategic character and the role of surplus generator. Along these lines, the Energy Development Scenarios presented by the plan seek to address the following objectives (p. 8): (i) guarantee national energy security and consolidate the country as a regional energy centre, giving priority to supplying the internal market; (ii) explore, exploit and increase national hydrocarbon potential, by means of strengthening the state-owned oil company, Yacimientos Petrolíferos Fiscales Bolivianos (YPFB), to operate in the upstream; develop, increase and quantify reserves; and develop new and existing hydrocarbon fields in order to increase production; and (iii) industrialize hydrocarbon resources in order to generate added value: industrialize natural gas – petrochemicals, liquefied petroleum gas (LPG) plants – and increase liquid hydrocarbon production capacity (refineries).

Since the application of Law 3058 of 2005, the country has witnessed a spectacular and historically unprecedented growth in the amounts raised by the state from hydrocarbon revenue. While Law 1689 of 1996 was in force, revenue via royalties and equity at around US$290 million reached their highest point in 2004; by 2009, these items plus the new direct hydrocarbon tax (IDH in Spanish) had created financial revenue of around US$1,347 million.[16]

It is nevertheless important to stress that although the growth in revenue responds largely to the tax reform of 2005, it is also explained by the upward trend in international oil prices and an increase in production levels made in order to meet the growing demand for natural gas from both the internal market and sale to the Brazilian market. This scenario, unprecedented in the history of the sector and encouraging (even though there are doubts regarding its sustainability in the medium or long term), generated an intense parallel dispute over these resources. Indeed, the structuring of the redistributive mechanisms and their relative weights was marked by pressure from social and political groups.

In general terms, four key mechanisms can be identified in the structure of hydrocarbon revenue distribution. The first of these is financing from the state budget at central level. The second, which represents the bulk of the allocation, is the revenue-sharing system of sectoral taxes inaugurated during the neoliberal period and reinforced by Law 2058 of 2005. This law strengthens the pressure on the state to raise hydrocarbon revenue directly by means of the direct hydrocarbon tax, while expanding the distributive base of this tax for the benefit of subnational levels of government and other entities such as universities, armed forces and police, indigenous and native peoples and rural communities, among others. The third mechanism, also inaugurated in the 1990s, and expanded starting in 2006, is a programme for direct transfer through the following mechanisms: 1) adjustment of universal old-age pension – 'Bonosol', now known as 'Renta Dignidad' (Dignity Payment); 2) the 'Juancito Bonus' programme (since 2006) for children in primary education; 3) the 'Juana Arzuduy Bonus' programme (since 2009) for pregnant women with no social security, payable at every medical check-up until the child is two years old; 4) an allowance for former Combatants of the Chaco War; and 5) a provisional bonus for public officials of the 2009 administration. Of this package of direct transfers, the first two would be financed by revenue from hydrocarbons.

The fourth mechanism refers to the universal fuel subsidies. A special feature of these subsidies is that they are financed, to a large extent, through freezing prices either as direct energy – if we refer to the residual natural gas[17] input for thermoelectrical generation or the gas distributed to industrial consumers and the domestic sector, via networks – or, in the case of condensate and natural gasoline (petrol), as an input to obtain refined products.

The *gasolinazo* and government justifications for reducing fuel subsidies

The significant increase in tax collection has been characterized as one of the most important achievements of the government's administration, with intensive media campaigns that highlighted the increase in financial reserves mainly in revenue from natural resources, with no clear information on the overall situation of the hydrocarbon sector. Government pronouncements generated high expectations regarding the possibilities the bonanza in the sector could produce. This included the improvement of socio-economic indicators and the sustainability of the proposed development model.

In strong contrast to this 'fiscal illusion' scenario, the government announced the decision to reduce subsidies on some fuels. On 27 December 2010, the price of a litre of petrol was increased by 72 per cent and of diesel by 82 per cent. The announcement, without prior notice, of the rise in fuel prices had been a recurring practice throughout previous administrations. They resulted from an institutional vacuum in terms of managing disputes among agendas, interests and perspectives, and of conducting a plural and informed public debate for consolidation of collective agreements on the use of revenue from non-renewable natural resources. Their persistence can also be explained, to a large extent, by an ongoing situation of change in sectoral management rules and the confusion between optimum levels of investment and production on the one hand, and of collection and use by the state of economic revenue on the other.[18]

The liberal restructuring of 1996 resulted in a significant flow of investments that increased productive capacity and reserves, consolidating energy as the most dynamic sector of the national economy. However, application of this model did not produce similar levels of success in terms of a better interpretation of the social mandate, tax collection and expected social outcomes (quality employment and a reduction in poverty and social inequality). Rather, this was a process that unleashed social mobilization and profound political upheaval in the country from 2000. With the reforms in this sector, and specifically those in 2005 and 2006 which had wide social support and acceptance, historically unprecedented tax revenue collections were achieved in the country. However, this period also saw the start of a phase of stagnation in investments, thereby putting at risk the chance of reproducing the surplus and with it the policies of redistribution and energy self-sufficiency. One possible interpretation of the reduction in fuel subsidies in 2010 is that it was an attempt to reverse the downward trend in investments.

During the first hours of the subsidy removal and price hike in December 2010, spontaneous protests began breaking out on the streets of all the cities in the country. In Cochabamba, the unemployed, who gathered daily on the street to wait for work, protested, and they were driven back by the police; heavy transport drivers also marched, announcing an increase in freight tariffs. In La Paz and El Alto, spontaneous protests from residents, public transport passengers and market vendors expressed their rejection of the measure. In different parts of the country, various sectors called town meetings, marches and protests. The process escalated to become an aggressive mobilization that led the president to back down after three days. This measure, known as the *gasolinazo* of 2010, had political and economic repercussions that were not possible to reverse in the following years, marking a new post-*gasolinazo* context with significant political consequences.

Government justifications for the need to reduce subsidies after six years of frozen prices were based on three pillars: (i) the need to eliminate smuggling to neighbouring countries such as Peru, Brazil, Chile, Argentina and Paraguay; (ii) the need to stimulate declining oil production; and (iii) the regressive nature of this redistributive mechanism. According to the authorities, the measure 'levels fuel prices in Bolivia in relation to prices in neighbouring countries, and aims to eliminate all forms of petrol and diesel smuggling, and encourages oil production'.[19] Alongside the economic reasons, the social nature of the measure was defended with the argument that the resources removed from the petrol and diesel subsidy would be reinvested in socially significant programmes.

The unexpected nature of the measure due to the lack of previous debate with broad sectors of society on the financial and sectoral issues justifying the measure, and a context of rather high optimism concerning the country's financial situation and hydrocarbon potential, generated an explosive reaction. The credibility of the problems raised by the government and the need for the measure to apply to the entire population were by no means established. Given that the absence of legitimacy was one of the elements that shaped the debates and conflicts on energy policy, it is very important to acknowledge the argument used by the government to explain the causes of the problems, and then compare it with the interpretations and views against it.

Analysis of the socio-economic indicators shows that the problems highlighted by the government to justify the measure have verifiable quantitative grounds. Below, we consider the empirical evidence for the three problems justifying the measure.

The need to stimulate production The continuous increase in subsidies and the increasing importation of petrol resulted in a series of unsustainable economic features, i.e. the continuous fall of capital flows into exploration and exploitation activities from 2003, the static levels of natural gas production (current natural gas production capacity in Bolivia stands at around 43 million cubic metres/day, the level reached in 2005) and the fall in proven gas reserves: 19.9 trillion cubic feet (TCF) in 2005 to 9.94 TCF in 2010. Added to these problems Bolivia also witnessed the decline in petroleum productive capacity owing to the manufacture of liquefiable products associated with natural gas production, thereby making the productive levels of this form of energy impact negatively on the supply of petroleum condensate and natural petrol. These facts meant that it was impossible to sustain self-supply of final-consumption fuel for the internal market, which, along with growing internal fuel demand, results in the continuous increase in imports and government subsidies on diesel, petrol and LPG. In 2009, 48 per cent of the diesel oil consumed in the country was imported, a proportion that rose to 54 per cent in 2010. Until 2009, the country did not import gasoline and, in 2010, 6 per cent came from imports, while LPG imports increased from 1 per cent in 2009 to 7 per cent in 2010.[20]

According to Ministry of Hydrocarbons and Energy data,[21] the value of the actual transfers carried out by the National Treasury to cover its quota share[22] in subsidizing LPG and diesel significantly increased after 2007. In the case of LPG, the increase was 1,000 per cent, going from US$22 million in 2007 to US$250 million in 2008. In the case of diesel, this rose from US$123 million to US$180 million in the same years. In the opinion of the Ministry of Hydrocarbons and Energy,[23] by not achieving a productive reactivation, by maintaining the current liquefiable consumption structure[24] and considering the trend of increasing internal demand, the cost of government subsidies for the next ten years would reach a total of US$6,273 million just for diesel and petrol.

Analysis of sectoral performance shows serious problems in balancing an optimum economic revenue collection by the state and reasonable levels of investment incentives. The general scheme for subsidies along the sectoral chain places domestic prices for field products (natural gas, oil and LPG) and refined fuels at levels very far from those of the international market (in 2008, international prices exceeded prevailing prices for the domestic market by more than 80 per cent). According to rules that apply in the country, all increases in internal gas demand must be covered by companies before it can be exported. This situation is a disincentive for productive investment, which is the basis of the management model currently in force. This imbalance is not only a serious threat to the country's financial situation but also potentially risks loss of energy self-sufficiency.

Regressive fuel subsidies With regard to regressive subsidies on fuels, Medinaceli and Mokrani (2010a) carried out a redistributive analysis of the subsidy applied to petrol and diesel through an assessment of household spending on urban and inter-urban transport, based on the Household Survey of the National Statistics Institute. This is the only example of an assessment made of the redistributive effects of this policy owing to the fact that there are no specific surveys on the consumption of these forms of energy in both sectors. It reviews the average monthly spending on transport, expressed in current bolivianos, and the percentage of households that report positive spending in this category. It can be seen from the study's findings that households with a higher income and those who use the transport service most often spend more on this service.[25] This trend is sustained during the period analysed.

With regard to spending on inter-urban transport, the trend is similar to that for urban transport; higher-income households spend more on this sector. However, households belonging to the lowest income levels use this service more often. Thus, although these households spend less on this type of transport, they use it more intensively. This is due to the fact that they frequently live in medium-sized cities, or in the rural sector of Bolivia. With regard to average monthly household spending on car fuel for private use, the top-decile households spend more on this product, given that a greater proportion of them possess cars.

In order to deepen the analysis, the impact on household spending of the pricing of LPG, special petrol and diesel oil, starting from *export parity*,[26] was analysed for 2008[27] under certain criteria.[28] It can be seen that, if the price of petrol varied, all households together would spend an additional 48.6 million bolivianos, of which 2.5 million from decile 1 and 5.7 million from decile 10. There is, then, a moderate regression concerning LPG consumption; however, with regard to the impact of the subsidy on the other two derivatives, the regression is even greater, given that the costs that households would bear (which they do not owing to the subsidy) is proportionately higher. From these findings, it can be seen that the subsidy[29] on the price of LPG, special petrol and diesel oil shows a significant regressive impact. In particular, the impact observed in the last two derivatives is significantly higher than in the first. In fact, overall, it can be seen that of the total saved by households on account of having domestic market prices lower than those in international markets, almost 40 per cent corresponds to the highest-income quintile and 8 per cent to the lowest quintile.

This information allows us to conclude that the subsidy on special petrol and diesel oil prices shows a significant regressive component and that the households which benefit the most are those belonging to the top deciles. However, the frequent use of inter-urban public transport by poor households is striking. In this regard, it may be concluded that although the government rightly asserted that this redistributive mechanism was regressive, it is also true that reducing the subsidy on these fuels would result in higher inter-urban transport fares and that the low-income households would be affected. Likewise, a non-gradual decrease in subsidies on these fuels had significant short-term inflationary effects.

The need to eliminate smuggling With regard to smuggling, the government has indicated that out of the subsidy of US$380 million annually on hydrocarbons, US$150 million benefited smugglers and neighbouring countries. On this point, Ødegaard's chapter in this book presents us with an excellent analysis of fuel smuggling on the Bolivia–Peru border.

The violent social reaction to the reduction in fuel subsidies

The increase in fuel prices was a measure that sought to reorientate hydrocarbon policy towards a new model of prices, tariffs, taxes, royalties and company profits, the main beneficiaries of which were to be the National Treasury and the oil companies, in this order of priority. Despite the fact that there were serious justifications for the measure, it was not possible to sustain the measure in the face of the violent demonstrations against it. This was a process that showed that the 'objective economic rationality' upholding political reform did not grant it social viability.

The capacity for swift and widespread mobilization throughout the country must be understood as part of a long history of social organization, and the mobilization of groups and categories that have continuously called into question the government's power of action. This militancy of Bolivian society, composed of indigenous communities, trade unions, guilds and civic committees with the capacity for collective reaction to policies, programmes and actions, has frequently manifested itself through confrontation, road and street blocks, strikes and demonstrations that very often result in setbacks for government decisions.[30]

This model of state and society interaction can also be traced back to the consolidation of a deep-rooted political culture dating back to the colonial and republican periods.[31] The explosive reaction following the *gasolinazo* was therefore not totally surprising considering the history of Bolivian social mobilization. However, the new context of prominent legitimacy of the government of Evo Morales seemed to suggest that society could accept the surprise move and without prior debate on fuel subsidy reduction. This expectation was shared by decision-makers. Contrary to this belief, the *gasolinazo* and its protests were followed by a series of other major confrontations, including the Isiboro Secure National Park and Indigenous Territory (TIPNIS) crisis in 2011/12, and the clash between cooperative miners and salaried miners at the Colquiri mine in 2012.

The model of bilateral interaction between social and government actors, on the fringe of a formal institutional framework that brings together and manages tensions and disputes between specific interests, reproduces a kind of individualist and corporatist coordination between state and society that condemns decision-making to pragmatic logic and short-term 'fire-fighting', which most of the time means the government taking a position in favour of the groups that show more strength and capacity for pressure on the streets. This is a model that reinforces the rentier political culture.

In the case of the *gasolinazo* of 2010, the protests were not sparked in order to defend or attack hydrocarbon nationalization. The main motivation was the defence of popular sector purchasing power against the inflationary effects of this measure. Public discussion on the fuel subsidy reduction measure revolved around four main issues: (i) whether or not this was a neoliberal measure that contradicted the non-liberal nature of the current government; (ii) whether the MAS government was, in the end, rightist or leftist; (iii) whether the problems justifying the measure were certain; and (iv) whether this was the most appropriate measure to overcome them. The three first issues were part of the wider public debate, while the fourth was limited to specific circles of specialists in hydrocarbon issues.

The most concrete topics in the wider public debate were who would win and who would lose from the change in energy policy and 'the real reasons' for it. In the more selective debate among specialists, the topics of discussion were the lack of transparency in managing information on the situation of the hydrocarbon and financial sectors, consideration of the problems presented by the government, and the possible and desirable alternatives for confronting these problems.

Neoliberal or patriotic measures: right-wing or left-wing government
With its direct effects on the cost of living the reform of the hydrocarbons sector under Morales touched a raw nerve in the collective nationalist imagination and its close link with non-renewable natural resources. Mass support for the decree to nationalize hydrocarbons in 2006 was directly associated with the recovery of YPFB's role and

the national memory of the tragic loss of territorial sovereignty in the 'heroic defence' of the most important form of national wealth – oil – at the beginning of the twentieth century. Nationalization thus became the symbol for all previous losses to the territorial and economic sovereignty of the Bolivian state, including the policy of capitalization (privatization) in 1996.[32]

The significance of 'reclaiming the gas for Bolivians' was also derived from widespread social discontent with the neoliberal-model years. Social actors now posed the question: *Has nationalization not recovered national and collective heritage for the country and reversed the economic and political power acquired by foreign companies?* Nationalization and the independence of the country are closely associated in the collective imagination, and clash in the present as in the past with the government's position of acting according to foreign interests (generating incentives for foreign investment) against popular interests (increasing the prices of basic products and services). With the *gasolinazo* the emotive charge that characterized social mobilization against the neoliberal model returned to the public arena, thereby igniting popular indignation against the government of Evo Morales for the first time.

The adjectives shouted by protesters in the mobilization against President Evo Morales in the city of El Alto were the same as those used against President Gonzalo Sánchez de Lozada in 2003, when he had to flee the country. MAS was accused of adopting a measure that was in the spirit of the tenets of the Washington Consensus which had been embraced by previous political elites. For example:

> The *gasolinazo* is a structural adjustment measure recommended by the International Monetary Fund. Foreign investment feels 'encouraged' to invest, in this case, in further exploitation and exploration of oil fields. The Bolivian people, the poorest, the final consumers, pay for the adjustment, even if followed by 'compensatory' measures; social policies such as those recommended and even financed by the World Bank in these kinds of decisions. This policy is a clear sign of government subjugation to imperialist companies that are clearly blackmailing semi-colonial countries such as Bolivia and reveal that the MAS discourse of 'we are building a new state with sovereignty' is pure demagoguery and melts away in the face of the reality of semi-colonial subjugation.[33]

In the same way, the decision-making process of the MAS government was questioned, with accusations made of continuing 'neoliberal mechanisms for decision-making' in which the decision on the destiny of natural resources was still centralized and, therefore, the autonomy of indigenous peoples, regions and municipalities was not being defended despite the contents of the country's new Constitution. The debate on the meanings of decentralization, autonomy and democracy became important in contesting the limits of the central state's authority in defining energy policy.

The former consul of Bolivia in Chile and former representative of the MAS government at the Organization of American States (OAS), José Enrique Pinelo, stated:

> In the MAS there is no chance to discuss. There is no space for criticism and self-criticism. There is no organised instrument to assess the direction of adopted or still-to-be-adopted policies. The leaders have no information and obviously neither do the rank-and-file [...] The 'neoliberal' mechanisms for decision-making remain intact. Centrality in decision-making is the same. The power to decide the destiny of resources continues in very few hands. There is no real decentralisation; the rank-and-file do not decide the fate of local resources [...][34]

The perception that the government had betrayed the Bolivian people was widespread during the three days that the measure lasted, mainly from the sectors that constituted its political support bases. The rise in fuel prices (of up to 82 per cent) was perceived as an abuse of power and therefore illegitimate. This resulted in the direct questioning of state authority to define energy policies. The following is one expression of this sentiment: 'Evo Morales has the leaders of social organisations in his pocket, and with them inside he can batter the popular sectors as much as he wants.'[35]

Against this popular assault, the government tried to neutralize the meanings of the term '*gasolinazo*', arguing that it did not have the neoliberal intentions of tackling the fiscal deficit, reducing salary purchasing power or preserving the privileges of a political and economic elite that was living off favours from the state. It was argued that this new measure was a 'levelling of prices' because there was currently an economic surplus and debt capacity. Therefore, as Iván Canelas, the then spokesperson for the Palace of Government, commented, Evo

Morales' measure was not neoliberal but 'patriotic' and 'courageous', an argument that the population did not find convincing.[36]

The fight against smuggling versus measures against the grassroots economy The government of Evo Morales justified the *gasolinazo* as necessary by providing staggering statistics: US$150 million out of US$380 million annually benefits smugglers and neighbouring countries. The reaction against this justification centred on questioning the capacity of the state to assert its territorial sovereignty and to manage to control the flow of goods at territorial borders. Were not the armed forces preventing smuggling? Wasn't Juan Ramón Quintana, ex-Minister of the Presidency and current head of the strange Agency for the Development of Macro-regions and Border Areas, responsible for preventing smuggling? In the same way it was asked who, in the end, the smugglers were. Were they home-grown and small-scale smugglers or were we facing bigger agents closer to the state bureaucracy? It was stated: 'Evo comes out saying that the military control the large tanks, but not the home-grown or "small-scale" smugglers. Will these be the beneficiaries of the USD150 million in subsidies?'[37]

The problems of corruption of public officials, inefficiency in state regulation and control systems returned to the public discourse. Furthermore, mention was made of the continuing poverty and social exclusion at the roots of smuggling and the negative effects of the reduction in subsidies for the grassroots economy. In a profoundly angry tone, the official argument that the 'economic blow' would affect only the comfortable middle classes was counteracted. With the immediate rise in urban and interprovincial transport costs (up to 100 per cent) this argument was discredited in the eyes of the public. The government was accused of forgetting that the price increase would not only affect car owners but also public transport users. In the same way, the price escalation of other products due to the effects on transport was swiftly felt and denounced: 'Besides the farmers, prior to the "gasolinazo", we already had to pay a huge amount for a gas cylinder (22.5 bolivianos in the city and 45 bolivianos in the rural areas!!!) due to the effects on transport.'[38]

The existence of sectoral problems and how to face them Another aspect of political contestation focused on how to address the problems in the hydrocarbon sector. The government had completely dismissed the problem of low private investment in the sector. In associating this measure with the generation of privileges for oil companies and, therefore, for the economic elites, specifically transnational capital and imperialism, this measure clashed with popular understandings of the 2006 nationalization and the country's new hydrocarbon management model. For those sectors, nationalization meant a way for Bolivia to unshackle itself from the yoke of the international capital system and it was, therefore, unacceptable as a matter of principle to be subjected to the market game in which the generation of incentives (where benefits and privileges are synonymous) is an integral part. With a single stroke of the pen, the discussion on the reasonableness of the incentives to attract capital flows was dismissed. The reform became the most candid negation of the MAS attempt to retain the image of Bolivia as a self-sufficient and philanthropic state that had been reasserted through the last hydrocarbon nationalization.

Despite these efforts, analysts and many of the public recognized that the *gasolinazo* and the following crisis confirmed the sectoral and fiscal problems of the new hydrocarbon management model. According to specialists,[39] the crisis revealed the government's lack of sincerity and the failure of their hydrocarbon policies. Moreover, it revealed the non-transparent management of the country's financial situation, the inefficiencies of government management and the absence of alternative proposals to resolve structural problems that the *gasolinazo* sought to correct. For critics,[40] these problems were also linked to wider problems in the economy – the difficulties in providing food and construction materials, the rise in inflation, the increase in the internal debt and the unsustainability of government spending. Many saw these problems as having their root in the absence of a social, territorial and financial pact that would give the hydrocarbon policy legitimacy and rationality.

Critics questioned the government's model of bilateral negotiation with specific sectors in order to allocate resources in the form of favours in the short term, and proposed that it be replaced by a financial pact based on a long-term national agreement that defined investment and spending priorities, thereby determining the origin, composition and distribution of financial resources from an overall perspective on the problem.[41] It was argued that only with this public management instrument could an institutional structure be established that democratically mediated ideological disputes regarding energy policy and conflicts of interests on non-renewable natural resource revenue.

Conclusions: power, sovereignty and epistemology

The analysis of a singular event – social mobilization against hydrocarbon policy reform in 2010 – exemplifies the challenges faced by nation-states in ensuring sovereignty over control of energy resources. This event condenses different dimensions whereby, in order to define its energy policy, the authority of the state requires ongoing negotiation and adjustment. On the one hand, there are the global economic dynamics that resist new rules which do not conform to their interests, and on the other, there are the national and local dynamics that equally resist being subject to the power of the state to define national policies when these are not in line with their views, values and interests.

In the Bolivian case, there is clear resistance to the dynamics of global markets as the source of new rules for the hydrocarbon sector. Despite the widespread support for and social acceptance of nationalization by Bolivian society, international investors demonstrated their resistance through the reduction of capital flows. Owing to the fact that the new management model combined, on the one hand, state monopoly control over hydrocarbon resources and, on the other, substantial foreign private investment, its feasibility was threatened by investment stagnation. The *gasolinazo* reduction in fuel subsidies in 2010 was an attempt to respond to these global dynamics. However, this attempt faced powerful internal resistance that obliged the government to step down and suspend the measure within three days.

One of the novelties of the case studied is that the response was led by popular sectors against a government widely recognized as having political legitimacy. It is important to remember that Evo Morales won the election of 2006 with 54 per cent of the vote and won again in 2009 with 62 per cent of the vote. It is surprising that a year after this victory, unprecedented in Bolivian history, the government was faced by militant rejection of a measure that had reasonable technical justification.

In spite of its appropriate causal model, the measure was shown to be improvised, non-consensual and, finally, inefficient. The consequences went beyond the hydrocarbon sector. The failed attempt to change one of the most sensitive variables of the economy triggered a price imbalance in the Bolivian economy with serious consequences mainly for the most vulnerable social strata. Inflation shot up in the country, mainly in basic goods and services. The surprising nature of the measure in an environment of high expectations about government fiscal leeway, added to the contradictions between the official justifications for the measure and the discourse pronounced in recent years, resulted in people losing confidence in the rules of the political and economic game. Since then, sectoral and fiscal problems have been worsening and it has not been possible to further reform the hydrocarbon industry and use the revenue to sustainably overcome the mining model, the social result of which is inadequate economic growth, a more profound dependency on non-renewable natural resources and the deterioration of other productive sectors.

This singular event furthermore helps us to historically substantiate the paradox of states that export non-renewable natural resources in a context of international price increases, i.e. we see here the expansion of state jurisdiction without equivalent strengthening of state authority over energy resources. It is a paradox that can only be understood in terms of the theoretical approaches that recognize the role and significance of social actors in state formation, to non-renewable natural resources and to the appropriate form of public decision-making. It is a perspective that starts from the definition of state and government actions as being interwoven with historically constructed social and political relations.

Analysis of the continuous processes of negotiation, dispute and interpretation within specific institutional frameworks is required to understand the extent and limits of state actions. Anthropology and sociology provide some inroads into this because they have analytical tools to understand the ideas, perceptions, meanings and interests behind the formulation, response or acceptance of a public policy. It is therefore important to develop conceptual tools to understand the specific institutional frameworks within which the relations between society and state are established and to understand the political and social dynamics that act to maintain them.

From this perspective, it is possible to understand the dynamics of political economy that unfold within an institutional framework characterized by models of bilateral interaction between social actors and government. It is clear that this does not always favour unifying processes, but rather involves tension and dispute management among interests and views on natural resources, on the role of the state and the legitimate form of decision-making. These kinds of regularized practices that serve to guide stakeholders' actions do not contribute to strengthening the state's legitimate authority in order to manage revenue from non-renewable natural resources for the greater public benefit. In other words, the isolated events analysed above illustrate the social dynamics of a kind of institutionality that depends on individualist and corporatist coordination between state and society, which condemns decision-making to a short-term, highly conflicting logic. This is a feature shared by many states that export non-renewable natural resources.

Notes

1 Karl (1997); Roed (2004); Rosser (2006); Robinson et al. (2006); Mehlum et al. (2006); Ferrufino (2007); Cumbers (2010); McNeish and Logan (2012).

2 Evans (1995); Hall and Thelen (2009); Schneider (2009); Dobbin (1993); Deeg and Jackson (2007); Aguilar Villanueva (2012).

3 Grindle and Thomas (1991); Allison (2000).

4 Williamson (1981); Becker (1991 [1981]); Coleman (1994).

5 Weber (1978); Bourdieu (2001, 2005).

6 Ibid.

7 Hall and Thelen (2009).

8 Aguilar Villanueva (2012: 17).

9 Ibid.

10 Ibid.: 15.

11 Ibid.: 17.

12 Karl (1997).

13 Ibid.

14 Ibid.

15 This section is based on Wanderley et al. (2012).

16 Source: Ministry of Hydrocarbons and Energy and National Tax Services.

17 Dry gas obtained from the separation process in liquid extraction plants.

18 Wanderley et al. (2012).

19 www.jornadanet.com/ Hemeroteca /n.php?a=57469-1&f=20101227.

20 For more detail, see Wanderley and Mokrani (2011).

21 Data published in the *Bolivian Hydrocarbon Strategy*, 2009.

22 We refer to the term 'quota share' whenever the final prices of these fuels are also subsidized with frozen prices on inputs that are borne by the hydrocarbon field economy.

23 Presentation at the Oil and Gas Conference held in Bolivia in August 2010.

24 Natural gas as fuel for final consumption accounts for only 18 per cent of the national energy matrix, while 50 per cent of this consumption is constituted by derivatives from oil and liquefiable products associated with natural gas (LPG, diesel, petrol and others).

25 This outcome is due to the fact that the poorest families are found in the rural areas of Bolivia.

26 An international oil price of US$90.80 per barrel and for the LPG standard the Mont Belvieu price of US$50.59 per barrel are assumed; both prices are measured at the Custody Transfer Point.

27 The last year for which the survey was available.

28 The criteria of analysis are as follows: (i) the change in family expenditure is estimated vis-à-vis a change in

the price of LPG to its export parity price, assuming a price elasticity of the demand equal to zero; (ii) the change in household spending on urban and rural transport is estimated, assuming that the fuels represent 50 per cent of the total costs in this sector and also that the increase in costs is transferred in full to the tariff paid by the consumer; (iii) the change in household spending is estimated vis-à-vis a change in the price of special petrol and diesel oil to its export parity price, assuming a price elasticity of the demand equal to zero; and (iv) in *export parity* scenarios, the new price of LPG would be 41.08 bolivianos/container and the price of petrol and diesel oil 6.65 bolivianos/litre.

29 Whether by direct subsidies or by not setting the domestic market price at its international opportunity (indirect subsidy).

30 For a more detailed analysis of the pattern of interaction between Bolivian state and society, see Wanderley (2009a).

31 UNDP (2007).

32 For more detail on the loss of legitimacy of the neoliberal model of hydrocarbon management in Bolivia, see Wanderley et al. (2012).

33 estudioyrealidad.blogspot.com/2010/12/las-repercusiones-del-gasolinazo.html.

34 www.bolpress.com/art.php?Cod:2010122703.

35 estudioyrealidad.blogspot.com/2010/12/las-repercusiones-del-gasolinazo.html.

36 www.lapatriaenlinea.com/?nota=53444.

37 old.kaosenlared.net/noticia/repercusiones-gasolinazo-neoliberal-evo-morales.

38 cadeca-catequistas.blogspot.com/2010/12/que-se-vaya-este-gobierno-neoliberal.html.

39 Carlos Miranda, Mauricio Medinaceli, Hugo del Granado, among others.

40 www.plural.bo/editorial/images/pdfnuevacronica/nc76.pdf.

41 www.cedla.org/blog/grupo politicafiscal/?p=28.

References

Aguilar Villanueva, L. F. (2012) *Política Pública: Una visión panorámica*, La Paz: UNDP, Strengthening Democracy Programme.

Allison, G. (2000) 'Modelos conceptuales. La crisis de los misiles cubanos', in L. F. Aguilar Villanueva (ed.), *La hechura de las políticas*, México: Miguel Angel Porrúa.

Becker, G. (1991 [1981]) *A Treatise on the Family*, Cambridge, MA: Harvard University Press.

Biggart, N. W. and G. Hamilton (1992) 'On the limits of a firm-based theory to explain business networks: Western bias of neoclassical economics', in N. Nohria and R. G. Eccles (eds), *Networks and Organizations*, Cambridge, MA: Harvard Business School Press.

Bourdieu, P. (2001) *Las estructuras sociales de la economía*, Buenos Aires: Manantial.

— (2005) 'Principles of an economic anthropology', in N. Smelser and R. Swedberg (eds), *The Handbook of Economic Sociology*, 2nd edn, New York: Russell Sage Foundation, Princeton University Press.

Chang, H.-J. (2011) 'Institutions and economic development: theory, policy and history', *Journal of Institutional Economics*, 7: 4.

Coleman, J. (1994) 'A rational choice perspective on economic sociology', in N. Smelser and R. Swedberg (eds), *The Handbook of Economic Sociology*, 1st edn, New York: Russell Sage Foundation, Princeton University Press.

Cumbers, A. (2010) 'Petroleo en el mar del Norte, el Estado y las trayectorias de desarrollo en el Reino Unido y Noruega', *UMBRALES*, 20, La Paz: CIDES-UMSA.

Deeg, R. and G. Jackson (2007) 'The state of the art: towards a more dynamic theory of capitalist variety', *Socio-economic Review*, 5.

Dobbin, F. (1993) *Forging Industrial Policy: United States, Britain and France in the Railway Age*, Cambridge: Cambridge University Press.

Evans, P. (1995) *Embedded Autonomy: States and Industrial Transformation*, Princeton, NJ: Princeton University Press.

Fayol, H. (1949) *General and Industrial Management*, London: Pitman.

Ferrufino, R. (2007) 'La maldición de los recursos naturales: enfoque, teorías y opciones', *Coloquio Económico*, 7, La Paz: Fundación Milenio.

Grindle, M. and J. Thomas (1991) *Public Choices and Policy Change: The political economy of reform in developing countries*, Baltimore, MD: Johns Hopkins University Press.

Hall, P. and K. Thelen (2009) 'Institutional change in varieties of capitalism', *Socio-economic Review*, 7.

Karl, T. L. (1997) *The Paradox of Plenty: Oil Boom and Petro-States*, Berkeley: University of California Press.

McNeish, J.-A. and O. Logan (2012) *Flammable Societies: Studies on the Socio-Economics of Oil and Gas*, London: Pluto Press.

Medinaceli, M. and L. Mokrani (2010a) 'Impacto de los bonos financiados por la renta petrolera', *UMBRALES*, 20, La Paz: CIDES-UMSA.

— (2010b) 'Políticas de asignación y distribución del impuesto directo de los hidrocarburos en Bolivia – impacto departamental y municipal', Working paper, CIDES-UMSA.

Mehlum, H., K. Moene and R. Torvik (2006) 'Institutions and the resource curse', *Economic Journal*, 116, January.

Ministerio de Desarrollo Productivo y Economía Plural and Ministerio de Trabajo del Estado Plurinacional de Bolivia (2009) *Plan Sectorial de Desarrollo Productivo con Empleo Digno*, La Paz.

Mokrani, L. and M. Medinaceli (2010) 'Subsidios a energéticos fósiles: bases del debate, evidencias empíricas y análisis de regresividad para el caso de GLP, gasolina especial y diesel oil en Bolivia', Working paper, La Paz: CIDES-UMSA.

Presidencia del Estado Plurinacional de Bolivia (2006) *Plan Nacional de Desarrollo: Bolivia digna, soberana, productiva y democrática para el Vivir Bien*, La Paz.

Robinson, J., R. Torvik and T. Verdier (2006) 'Political foundations of the resource curse', *Journal of Development Economics*, 79, January.

Roed, E. (2004) 'Escaping the resource curse and the Dutch disease? When and why Norway caught up with and forged ahead of its neighbors', Research Department Discussion Paper no. 377, Statistics Norway, May.

Rosser, A. (2006) 'The political economy of the resource curse: a literature survey', Working Paper no. 268, Institute of Development Studies, April.

Schneider, B. R. (2009) 'Hierarchical market economies and varieties of capitalism in Latin America', *Journal of Latin American Studies*, 41.

UNDP (2007) *Informe Nacional de Desarrollo Humano – el estado del Estado*, La Paz.

Urwick, L. (1943) *The Elements of Business Administration*, New York: Harper.

Wanderley, F. (2009a) 'Prácticas estatales y ciudadanía colectiva e individual en Bolivia', *ÍCONOS, Revista de Ciencias Sociales*, 34, Ecuador: Etnografías del Estado, FLACSO.

— (2009b) 'Ciudadanía colectiva e documentos de identidade', *DADOS – Revista de Ciências Sociais*, Rio de Janeiro: IUPERJ.

Wanderley, F. and L. Mokrani (2011) 'La economía del gas y las políticas de inclusión socio-económica en Bolivia', Research document, Fundación Carolina y CIDES-UMSA, www.fundacioncarolina.es/esES/publicaciones/.../Documents/AI56.pdf.

Wanderley, F., L. Mokrani and A. Guimaraes (2012) 'The economy of gas and the politics of social inclusion in Bolivia', in J.-A. McNeish and O. Logan, *Flammable Societies: Studies on the Socio-Economics of Oil and Gas*, London: Pluto Press.

Weber, M. (1978) *Economy and Society*, Berkeley: University of California Press.

Williamson, O. (1981) 'The economics of organization: the transaction cost approach', *American Journal of Sociology*, 87.

— (1993) 'Calculativeness, trust, and economic organization', *Journal of Law and Economics*, 36, April.

8 | PASSIVE REVOLUTION? SOCIAL AND POLITICAL STRUGGLES SURROUNDING BRAZIL'S NEW-FOUND OIL RESERVOIRS

Einar Braathen

Introduction

In 2007 President Luiz Inácio Lula da Silva announced the discovery of the largest oil reserves found on the planet for the last several decades, and in Brazil's history. They totalled almost fifty billion barrels. The reserves are contained in offshore pre-salt layers under very deep water, off the coast of south-east Brazil. They have been termed the *pré-sal* in Brazil's public debate. 'The pre-salt is our passport to the future,' President Lula declared, and national euphoria was unleashed. The president started an express legislation process to prepare for the start of production in the pre-salt fields.

However, it took six years for the first pre-salt field to be made ready for exploitation. By then the national euphoria had evaporated. On 21 October 2013 it was announced that the pre-salt field named Libra was to be handed over to an international consortium of oil companies. The announcement was made by Lula's successor, President Dilma Rousseff, in a luxury hotel on a beach in Rio de Janeiro. The ceremony was accompanied by a national strike of the petroleum workers, one of their demands being an end to simply auctioning oilfields. Lawsuits against the government had been filed by nationally acclaimed economists and oil experts. A local branch of the international Occupy movement camped outside the main national petroleum institutions. Angry demonstrators lined up in the street. The authorities deployed 1,100 soldiers from the National Guard plus 5,000 police to protect the ceremony at the Rio de Janeiro beach.[1] What had happened with Brazil?

After the centre-left forces took over the government in 2003, the country saw a large extent of consensus within the political class, and there was ample political demobilization of society (Montero 2005). Paradoxically, when a technocrat (Roussef) followed the charismatic leader (Lula) as president of the country in 2011, more political and social conflicts arose. In June 2013 there was an upsurge of national street demonstrations, the largest in Brazil's history. The auction of the first pre-salt field in October 2013 did not avoid this wave of social protests.

This chapter assesses the politics surrounding Brazil's new-found oil reservoirs from their discovery (2007) to their first auctioning (2013). The analysis is based on interviews and mass media observations made from 2010 to 2013 in Brasilia and Rio de Janeiro. Inspired by Antonio Gramsci and the Brazilian social scientist Carlos Nelson Coutinho's interpretations of Gramsci (Coutinho 2007, 2008), the study combines a from-below approach to social movements (hegemonic struggle) and a from-above approach to state responses (passive revolution). Two sets of assumptions and research questions will be examined.

On the one hand, there has been a process of increasing contestation and politicization of the new petroleum policies. The chapter assesses three different initiatives to oppose the dominant views and public policies. The first initiative emphasized resource sovereignty and national control of the oil. The second initiative was linked to regionalist interests in the distribution of the royalties emanating from the oil production in Brazil. The third initiative represents environmentalist critique of expansion of oil production. These initiatives correspond to 'industrialist', 'post-industrialist'[2] and 'anti-industrialist'[3] discourses, respectively. The analysis traces the connections between the social origins, the (social) mobilization capacity and the political impacts of each of the three initiatives. This conceptual framework is inspired by Tarrow (2007)[4] and by previous work on social movements in Brazil (see Hochstetler 2008). It also connects the issue of policy discourses with hegemonic struggles – the attempts by social movements to challenge an existing ideological hegemony in a policy field (Coutinho 2007). These elements help us undertake a political analysis to find out: (i) What have been the political impacts of each initiative? To what extent have they changed or reshaped the hegemonic discourse in the policy field they have tried to influence? (ii) To what extent can the difference in impact be associated with the difference in social origins and mobilization capacity? (iii) Have the politics of the movements themselves, e.g. in terms of promoting contact and alliance with other movements and with political parties, been changed and played any role?

On the other hand, a very small and closed circle of national policy-makers had the upper hand and maintained control of developments, in spite of the increasing contestation. The two key persons have been Lula and Dilma Roussef. The latter was Lula's minister of minerals and energy from 2003 until she was appointed his chief cabinet secretary in 2006. In this position she chaired the inter-ministerial committee which drafted the new petroleum policy. As a president, Dilma has overseen the implementation, and not least the reinterpretation, of this policy. The federal policy under Dilma seems to have become more insulated from social movements and society initiatives than under Lula. But does this mean that there has been a significant shift in policy orientation along with the change of personnel in power – a political regime shift? What have been the approaches of the Lula and Dilma governments to the main class forces in Brazil, and what have been their main policy tenets? To assess this question we draw on a seminal work of a famous Brazilian political scientist, André Vitor Singer, who served as the press spokesman of President Lula (Singer 2012) and who draws on Carlos Nelson Coutinho. We also use recent studies of state–society relations since 2003 (see Baiocchi et al. 2013). Analysts of Brazilian politics seem to agree that a new political regime, *lulismo*, emerged under Lula's presidency. Besides trying to grasp this concept, we are faced by the challenge of defining what kind of resource sovereignty was pursued by *lulismo*, and in particular of understanding how the pre-salt energy resources have influenced political power and its contestations.

In this regard, Antonio Gramsci's concept of passive revolution is applied to the Brazilian context in order to understand *lulismo*. *Passive revolution* is state-led capitalist modernization from above with selective concessions to popular demands, in response to or in fierce opposition to a popular revolution from below. Passive revolution contains two core elements: first, the element of 'restoration', in the sense that it is a reaction against the possibility of radical and effective transformation from the bottom to the top; second, the element of 'renovation', in the sense that many popular demands are assimilated and put into practice by the old dominant layers (Coutinho 2008: 91–6). For Gramsci the Italian Risorgimento in the nineteenth century was a passive revolution, and even more so Benito Mussolini's fascism in Italy after 1922 (ibid.). However, neo-Gramscians such as Christine Buci-Glucksmann and Göran Therborn (1981) have claimed that passive revolutions have also taken place in consolidated democracies, in terms of state reformism and the post-Second World War welfare state. Therefore, Coutinho claims that a passive revolution with 'the cooptation of the political and cultural leaderships of the subaltern classes' occurred in a current liberal-democratic country in the capitalist periphery – Brazil – during the governments of Fernando Henrique Cardoso (1995–2002) and Lula (2003–10) (Coutinho 2008: 104–5).

Before embarking on the three initiatives that challenged political power from below, we therefore look at *lulismo* to get a clearer picture of the context and current politics in Brazil.

Lulismo – the sub-proletariat and the president

Luis Inácio Lula da Silva was elected president in 2002 and re-elected four years later. However, in 2006 Lula and his Workers' Party (PT) lost almost twenty million votes from the better-off organized working and middle classes in the south-eastern and southern states, while they gained a similar number of votes among the poor sub-proletarian masses in the less industrialized north-east of the country. This was one of the most remarkable electoral realignments in Brazil's modern history (Singer 2012).

On the one hand, PT experienced a decline in its traditional supporters. The social movements and civil society organizations that had aided the party since its birth in 1980 became increasingly sceptical about the party and its leader (Hochstetler 2008). After Lula's first election in 2002 they hoped that the new president would take the '*petista* (PT) way of governing' familiar in cities like Porto Alegre and São Paulo to the national level – with redistribution, good government and participation as the basic characteristics. However, they experienced no real redistribution benefiting the organized working class. Pragmatism, building broad coalitions with the conservative political and financial elites, was the main 'ideology' and modus vivendi of Lula's administration (Kingstone and Ponce 2010). Instead of becoming an exemplar of good government, the Lula administration was caught in one of the biggest political corruption scandals in Brazil's history – the vote-buying scheme in the Federal Congress, the so-called *Mensalão*. This scandal alienated the left-leaning liberal parts of the middle class who had earlier voted for PT (Hunter 2011). Instead of rising influence in policy-making through increased popular participation, the civil society organizations and their leaders were all but co-opted by jobs in the government and 'bottom-up' policy conferences without real impacts on national decision-making (Baiocchi et al. 2013).

On the other hand, there was a remarkable rise in support for Lula and his party in the north-east of Brazil and among the poorest social layers. André Singer (2012) argues that this was due to a socio-political transformation taking place under the Lula government. The policies attributed to Lula ensured a material uplift and some degree of social inclusion of the poorest 10 per cent of the population, through federal cash transfers directly to the bank accounts (opened for the transfer purposes) of the female heads of poor families (e.g. the *Bolsa Família*) and an increase in the minimum salaries set by presidential decrees. Labour market reforms increased the number of formal employment contracts, reducing the importance of informal labour, thus ensuring more socio-economic rights to the lowest-paid segments of the proletariat. Singer points out that these relatively modest reforms had big joint effects and made significant changes to the class dynamics of Brazilian society. What has almost for a century constituted 'the permanently super-impoverished working surplus population', a statistical category which Singer prefers to call the 'subproletariat', has become a class in itself, a modern 'new proletariat'.[5]

Lulismo and *lulista* are the labels used by Singer to describe the new political regime connected to this social transformation. 'Lulismo is in my view the meeting between a [state] leadership, that of Lula, and a class fraction, the subproletariat, through a program with the main points delineated between 2003 and 2005. [...] A power *apparently* above the classes which creates the integration of the subproletariat to the proletarian condition, in the same way as the *varguismo* of Getúlio Vargas did when integrating rural migrants into an urban working class by means of industrialization' (ibid.: 45).[6]

Lulismo created a new historical block, with its own project. To understand this new phenomenon, Singer argues that the notions of mass politics (Marx) and passive revolution (Gramsci) are very useful 'when the local colours have been filtered' (ibid.: 45). Karl Marx in *The 18th Brumaire of Louis Bonaparte* showed that class fractions that have difficulties in organizing themselves and achieving a consciousness of themselves present themselves in politics as a mass. The mass identifies with a leader who, from above, mobilizes the state to benefit them without risky social disorder (ibid.: 37). Unlike the

Bonapartism of nineteenth-century France, the *lulismo* of Brazil was not preceded by violent revolution or counter-revolution, and it was not headed by military figures from the elite.

It is therefore perhaps even more appropriate to use Antonio Gramsci's concept *passive revolution*, which is the opposite of a popular revolution from below. For Coutinho, the concept of passive revolution helps us to understand 'the whole transition process of Brazil towards a capitalist modernity' (Coutinho 2007: 198). For another political writer, Brazil 'can be characterized as the place par excellence of passive revolution' (Vianna 2004: 43). But in the case of *lulismo*, Singer claims that the ideological construction is the inverse of previous types of passive revolution: conservative elements are co-opted by leaders with progressive origins, not vice versa (Singer 2012: 37–8). Lula actively supported capitalist accumulation and secured the privileges of the ruling classes. In this way, he obtained their acceptance of gradual and cautious social reforms, financed by improved tax collection and economic growth rather than by zero-sum-type redistribution from the rich to the poor. In other words, *lulismo* combines 'gradual reforms' for the poor and 'conservative pacts' with the rich. Montero (2005) and other political scientists have highlighted 'reforms under oligarchic-conservative control' as the main characteristic of Brazilian politics after the new democratic constitution was in place in 1988. Although Singer agrees that there is a lot of policy continuity between the governments of Fernando Henrique Cardoso (1995–2002) and Lula (2003–10), illustrated by his notion of a 'conservative pact', his Marxian point is that Lula's presidency created social *mobility* as well as new conditions for social and political *mobilization* among the popular classes. 'The *lulismo* makes an ideological re-articulation and pulls out the centrality of the conflict between left and right and reconstructs an ideology on the basis of the conflict between the rich and poor' (Singer 2012: 32). In a comment on Brazil's largest-ever street demonstrations in June 2013, Singer claimed that the protests were part of 'the ascension of the new proletariat'. These people have gained employment and higher income, but their lives are still precarious, particularly in the larger cities (Singer 2014). 'The demonstrators want higher public expenditures, while the market forces demand austerity. This will put the current Dilma government at the crossroads' (Singer 2013).

However, a theme not treated in Singer's analysis of *lulismo* and the new 'social energy' of the 'new proletariat' is the role of energy policies and resource sovereignty. In Lula's first term the *pré-sal* discoveries had not yet been made. Still, there was a successful campaign to 'bring light to everybody', *Luz para todos*. And biofuel production was actively promoted and offered lower fuel costs to the working young men who now could afford a motorbike. These issues will not be focused on in this chapter.[7] Rather we are interested in Lula's second term and thereafter. Our assumptions are that (i) the new petroleum reserves brought more steam and self-confidence to Lula's pro-poor and pro-employment policies; (ii) they secured the election of his former minister of energy and then cabinet secretary, Dilma Rousseff, as the new president in 2010; and (iii) an emerging *pré-sal* gospel 'empowered' and fuelled the new proletarian demonstrations in June 2013.[8] At the same time, *lulismo* played an important role in dealing with (co-opting or rejecting) the three types of contestation we are discussing: nationalist, regionalist and environmentalist movements.

'The petroleum has to be ours'

O petróleo tem que ser nosso ('The petroleum has to be ours') was the slogan of the nationalist and labour-based movement rallying behind Getúlio Vargas, who was the head of the country from 1930 to 1945 (after a *coup de état* or 'revolution' from above) and from 1951 to 1954 (after democratic elections). The Vargas governments declared nationalization of the oil resources in 1939 and established a state monopoly company, Petrobras, in 1953. A few months later President Vargas committed suicide. Apparently this was because of the Petrobras law, which was considered to be a 'communist' measure and created hysterical reactions from international and local capitalist groups (Ribeiro 2001).

Nevertheless, Petrobras survived and was an important instrument in the nationalist and anti-communist '*desenvolvimentismo*' (developmentalism) policy of the military rulers from 1964 to 1985. In addition to expansion into offshore production, a big national petrochemical industry was developed in this period. The workers in the oil industries organized themselves in free and independent trade unions once the military rulers started to liberalize labour legislation and prepared for their own exit, largely due to the labour protests and strikes all over Brazil in 1979 and 1980. The figurehead of this new labour movement was Luis Inácio da Silva, nicknamed 'The Squid' (*Lula*).

Most of the oil sector unions joined the new labour confederation promoted by Lula: Central Unica dos Trabalhadores (CUT). And most of the leaders of these oil worker unions joined Lula in founding a new party, Partido dos Trabalhadores (PT). However, after a new democratic constitution was in place in 1988, Lula and the left lost the subsequent elections to the neoliberal right wing.

The sociologist-turned-economist Fernando Henrique Cardoso, elected president in 1994, made it a priority to liberalize the oil sector and to attack Petrobras. In response to that, the movement *O petróleo tem que ser nosso* was created. It was initiated by the various and often competing organizations of employees in the oil sector – basically CUT's oil workers' federation (FUP), the independent union of the Petrobras workers in the state of Rio de Janeiro (SINDIPETRO) and the association of the engineers of Petrobras. The movement was backed by the entire trade union movement. Not only the left-wing parties, notably the Workers' Party (PT), but also two more moderate labour parties (PDT and PTB), building on the legacy of Getúlio Vargas, supported the movement. There were several street demonstrations, but no strike actions against Cardoso's petroleum bills. He controlled the National Congress, so by 1998 new laws were in place that dissolved the monopoly of Petrobras and introduced the 'regime of concessions'. This meant that an international auction was announced for every field that was cleared for petroleum production, and a concession was handed over to the winning consortium (headed by an operator company). Although Cardoso promised that Petrobras was not to be privatized, it was given full corporate autonomy and 'freedom to compete' not only nationally, but also internationally.

When Lula became president in January 2003, he had promised before the elections, in a 'Letter to Brazil', that he would not attack the free markets, the fortunes of the richest families or the privileges of the largest capitalist groups. The implication was that any business agreement based on laws made by the previous president's administration would be respected by Lula's administration. The left forces and the trade unions, including *O petróleo tem que ser nosso*, despaired.

However, new opportunities beckoned. A major event in 2007 made radical policy initiatives possible. Early that year, Lula informed the public that Petrobras had made huge petroleum discoveries in the so-called *pré-sal* layers, offshore in the south and south-east of Rio de Janeiro. This event demanded some new and innovative initiatives.

Besides, Lula and his Workers' Party were in deep trouble – main cadres of the party and the government were investigated for illegal financial operations and buying of votes in the National Congress, the so-called *Mensalão* scandal. Although Lula managed to be re-elected in October 2006, large parts of PT's social base had lost faith and were considering other political options. In this context, Lula and PT declared that the new oil reservoirs were 'a gift from God to the Brazilian people'. PT insisted that the *pré-sal* had to be skilfully managed so that the whole population benefited from it, and so that Brazil could 'eradicate poverty'. PT did not employ electoral populism alone – it went back to its roots and encouraged the trade union movement to give *O petróleo tem que ser nosso* the opportunity for a comeback.

The organizers of the campaign say they felt that the ideas and initiatives from the organized working class had again entered centre stage in the country – not only in PT, but also in national public life. The main demand of the campaign was *renationalization of the petroleum reserves and petroleum industries*. Petrobras should again be the monopolist producer of petroleum in Brazil, but under full democratic and public control. Although Lula and PT did not back these demands, the *pré-sal* issue helped to mobilize general opinion and not least the organized working class for Lula and PT. Consequently it helped to regalvanize the unity of centre-left parties under the hegemony of PT and the leadership of Lula, in a situation where these allies questioned the moral authority of Lula and his party.

By the end of 2007, Lula had started a process of creating a new legal framework for the oil industry. He used pragmatic arguments accepted even by the right-wing opposition: the pre-salt reserves had created a new situation not foreseen when the concession regime was installed in 1998. The law had to adapt to new realities. The redrafting process was in the hands of a committee consisting of representatives of various ministries and the oil industry, including the CEO of Petrobras (Sérgio Gabrielli). Although there were no representatives of the trade unions or *O petróleo tem que ser nosso*, this committee worked in both the organizational and ideological spirit of Vargas and his corporatism. Representatives of the trade unions took indirect part, through links with the top management of Petrobras (Sérgio Gabrielli was a former militant of PT and adviser to the oil worker unions) and some of the ministries. However, the independent mobilization capacity of *O petróleo tem que ser nosso* was proved when it criticized many of the proposals of the committee and started to lobby for its own alternatives. This work started in March 2009. The campaign aligned with federal senators and deputies and organized public hearings which presented the views of trade unions, other social movements and critical experts. They also mobilized mass support on May Day rallies in 2009 and 2010 all over the country, and in special demonstrations in the city of Rio de Janeiro and in the federal capital Brasilia. In particular they contested the use of international auctions and the presence of multinational petroleum companies in Brazilian waters.

How much impact did this nationalist and anti-imperialist campaign have at the end of the day? It imparted a clear nationalist stamp on what the government called 'the package' (of oil-related bills) presented to the two chambers of the National Congress by the end of 2009. The Brazilian state was again to be the majority shareholder of Petrobras. One of the laws suggested that only Petrobras was allowed to be the lead 'operating company' of the pre-salt fields. The oil and gas fields were to be owned by the Brazilian state, cashing in from direct shares of the petroleum produced and sold. The revenues from these direct shares would be administered by a new federal agency. In this way, a regime of 'production sharing' was to replace the old regime of 'concession', whereby concessionaries expropriated the oil resources and paid only a few per cent of the revenues back to the public by way of royalties.

However, the logics of global competition and capitalist relations of production were to remain. An enormous number of Petrobras shares were issued for sale on the international stock markets, particularly in New York. The exclusive right to extract the oil of each field would be issued to consortiums selected through international competition. They would keep the lion's share of the large profits, paying a small percentage of the value of the production in royalties and a small percentage of tax on profits. The state agency taking charge of the revenues from the public direct shares of *pré-sal* would direct them all to profit-maximizing investments inside and outside Brazil, and only the return on capital would be allocated to social and public spending in Brazil.[9]

In hindsight, the main work of *O petróleo tem que ser nosso* was to build a trade-union-based bridge between the oil policy of Getúlio

Vargas and that of Luis Inácio Lula, respectively. There is a continuity from *getulismo* to *lulismo* in the attempts to reconcile the interests of both capitalists and workers, and in insisting that it is up to the national president rather than the working class itself to define what is in the workers' interest. The campaign raised an independent voice of working-class memory, concern and knowledge. It did not dispute, however, the closed technocratic character of the initial policy-making of the government. And it did not dispute a political party monopolizing representation of its interests.

Still, the campaign had a significant political impact in Brazil. Although there were politically opportunistic reasons why Lula and PT allowed it to have an impact, the campaign managed to enlarge the historical bloc behind the new petroleum policy by attracting political, intellectual and social circles that were inspired by Gétulio Vargas and the 'developmentalism' from the 1950s to the 1980s as much as by socialist and left-wing radicalism.

'The royalties have to be ours'

A huge conflict has emerged about the distribution of the around 20 per cent of the value of annual oil production paid to the Brazilian state – the royalties.[10] After Cardoso's new petroleum laws were enacted in 1997, the so-called 'non-producing' states and municipalities have received 8.75 per cent and the very few 'producing' states and municipalities 61.25 per cent of the royalties. The Federal Congress has declared that this is a gross injustice leading to increased inequality between the regions and states of the federation of Brazil. Since 2010 Congress has on several occasions voted overwhelmingly for a new formula: the non-producing states and municipalities should receive 54 per cent, and the producing ones only 26 per cent of the royalties.

The main beneficiary of the royalties has since 1997 been the state of Rio de Janeiro and some of its municipalities. Together they receive more than 50 per cent of the royalties. The royalties fund between 20 and 40 per cent of the total spending of the state and the receiving municipalities. There has been much critique locally of the way the municipalities and states have spent these royalty revenues. Nevertheless, the issue has motivated the elected leaders of the producing states and municipalities – particularly in Rio de Janeiro – to fight the new bills passed by the Federal Congress. Since 2010 there have been repeated one-day strikes when state and municipal employees

TABLE 8.1 Congress's decision on the distribution of royalties between the different territorial categories (percentage of the distribution)

	Old petroleum regime (1997–)		New petroleum regime (2012–)	
	Royalties	Special participation	Royalties	Special participation
Federal government	30.00	50.0	20.0	46.0
'Oil-producing' states and municipalities	61.25	50.0	26.0	24.0
'Non-producing' states and municipalities	8.75	0.0	54.0	30.0
	100.00	100.00	100.00	100.00

Source: Marcelo Castro, Federal Deputy (PMDB Piauí), interview, Brasilia, 18 August 2011

were ordered to take part in mass demonstrations in defence of the current royalty formula. This struggle has been fuelled by regionalist feelings, following removal of the capital from Rio to Brasilia in 1960, that the federal government has constantly been 'discriminating' against Rio de Janeiro. These feelings have been exploited by local politicians to mobilize against the '*paulista*' presidents – those coming from the rival metropolis of São Paulo – Fernando Henrique Cardoso and Lula (Natal 2007).

The leader of this regionalist bloc has been the governor of the state of Rio de Janeiro, Sérgio Cabral from the Brazilian Democratic Movement Party (Partido do Movimento Democrático Brasileiro – PMDB). The presidents – first Lula, thereafter Dilma – have vetoed the Congress decisions, mainly to ensure smooth implementation of the new policy and legal framework for the oil production, but also to maintain an important political alliance between the federal president, the governor of the state of Rio de Janeiro and the mayor of the city of Rio de Janeiro.[11] This has upset the Congress majority, which has therefore opposed the obvious compromise: old fields applying the old formula, while the new (*pré-sal*) fields apply the new formula.

A slightly revised bill was approved by Congress in November 2012. It established a formula in which the share of producing states and municipalities would drop gradually from the current 61.25 per cent to 38 per cent in 2013 and, finally, to 26 per cent in 2020. Although some kind of reformulation/redistribution was expected (and accepted as fair), the non-producing states and municipalities had pushed for changes that infuriated producing states and municipalities. The governors of Rio de Janeiro, São Paulo and Espírito Santo went directly to President Dilma and asked her to veto the changes for existing production fields. Trying to play reasonably and minimize political losses, President Dilma vetoed the law some weeks later. The story, however, was far from over. On 7 March 2013, Congress overruled President Dilma's veto and reinstated its own version of the bill.

Politically, that represented a big defeat for President Dilma and one more episode in her turbulent relationship with Congress. In contrast to former president Lula, who was able to manage demands from governors and mayors through carefully listening to their demands and seeking a financial compromise, President Dilma has a distant relationship with representatives from other levels of the federation. Governors and mayors are not only displeased by the fact that President Dilma demonstrates very little appetite for personally engaging in political negotiations; her recent economic policies had also been straining the purses of local governments, as fiscal incentives awarded to various industries (e.g. automobiles, construction materials) reduced the revenues that the federal government should transfer to states and municipalities. Moreover, the intent of the federal government to create a unified interstate tax raised the opposition of governors from poorer states who are afraid of losing revenues and tools to attract investment. The increased participation of non-producing states and municipalities, despite arguments about fair distribution, was considered by governors and mayors as clear compensation for declining revenues.

One could expect that the recent 'tour de force' of the non-producer states and municipalities in Congress would at least settle the debate and the federal government would be able to finally start the new concession rounds for the exploration of the *pré-sal*. That would be naive. The governor of Rio de Janeiro, Sérgio Cabral, announced that he wanted to impose new state taxes on the oil companies in order to compensate for the lost revenues, and threatened to suspend the environmental licences of the oil companies operating on the state's shores. Also, at city level, the mayor of Rio de Janeiro stated that without compensation the 2016 Olympic Games would be at risk. At the end of the day, the governors of the producing states contested the constitutionality of the new law in the Federal Supreme Court (STF). The STF was surprisingly fast in its proceedings and decided in April 2013 that the new law could not be applied to those oilfields already producing.[12]

That created, at last, peace within the ruling political coalition. A reformed regulatory and legal framework for petroleum activities could now be implemented. In June 2013 the federal government announced the first auctions for a pre-salt field to take place in October that year.

'The petroleum is a threat to sustainability'

The social character and origin of the environmental(ist) critique of the oil industry in Brazil in general, and of the *pre-sal* in particular, may reveal why the environmentalists have managed to mobilize only small fragments of society and have had little impact on petroleum policy in Brazil. Let us depart from the main environmental critique of the petroleum industry:

> (i) 'Expanding the oil production, at a time when every country should be committed internationally to "clean (renewable) energy" and reduced CO_2 emissions, is not a sound policy for sustainable development'.[13]

This critique of the petroleum industry has been aired by environmentalists all over the world for some decades. The critique faces three types of problems in Brazil.

First, there is no movement grounded in the country which puts this critique forward. There are some academics who raise questions, but they are not active in networks such as 'concerned scientists', etc. The main organization dealing with this issue is Greenpeace, with foreign Anglo-Saxon origins. Although it can attract mass media through a few spectacular shows of action by a few men and women, it moves very few people in society.

Secondly, the main official answer to this question is technological optimism: 'Carbon Capture and Storage (CSS)!', in line with the official doctrine for sustainable development: 'Growth with environmental protection'. This type of optimistic modernism is popular and is supported by a strong political coalition in a growing BRICS economy such as Brazil. There are many scientists and intellectuals who believe in CSS.[14]

Last but not least, for the money- and power-holders in Brazil there are very few economic incentives for not expanding oil production. There are no international sanctions or internal revolts at play if the

CO_2 emissions of Brazil continue to increase rapidly – at least, that is the current dominant perception within the Brazilian elite.

(ii) 'Expanding the oil production puts the environment at a high risk'.

This is the old and more conventional environmental critique of the oil industry. After the Deepwater Horizon catastrophe in the Gulf of Mexico in mid-2010, this critique found a relatively strong echo in Brazil when a field operated by Chevron made a significant oil spill in the Campos basin in November 2011. The big corporate media in Brazil paid much attention to the accident. For people who enjoy the beaches in this part of the country, oil production all of a sudden looked like a big threat. The government and politicians both locally and federally spoke out more strongly than they had ever done before against the oil industry, probably because the company responsible for the spills was North American and not Brazilian. The national politicians became more silent, however, when some media such as *O Globo* exposed the fact that Petrobras had also 'leaked' frequently – sixty irregular cases in total in 2011, causing the spillage of many more barrels in total than Chevron. On 31 January 2012, the first major 'leakage' from a pre-salt field (operated by Petrobras) was reported.

The federal police, state attorney and internal auditing authorities started their investigations into Chevron. However, there was no prolonged social mobilization or attention around the issue. Instead of politicization, the case was the object of judicialization, and it was a high probability that the impacts would be very small. On 21 February 2013, the news was that 'Brazil drops criminal charges against Chevron over spill'. At the end of 2012, the internal auditing authority (TCU) released a report about the national regulatory agency in the petroleum sector (ANP) and the national environment agency (IBAMA). TCU was very critical about the role of these agencies in the recent accidents, and it questioned their general capacity to supervise and control the companies and petroleum activities. The report pointed out the complete lack of prevention and reparation capacity in relation to offshore 'leakages'/accidents. It also criticized the low ratio (of 1:10) between investments in environment-related technology and capacities and total offshore investments. It offered many recommendations for change. However, there were few signs of government initiatives to impose more regulations on the oil industry, especially not in a situation where (i) the development of the pre-salt fields was reported to be far behind schedule, and (ii) the international share value of Petrobras had dropped like a stone.

What some informants have emphasized is that the 'anti-pollution' approach to the oil industry lacks a real social basis, leaving the terrain to oil companies, legal experts and state bureaucrats. First, the workers operating in the fields are the first to die in case of accidents leading to large spills. The rate of accidental deaths in off-shore activities is probably very high – even according to the public records and statistics, which are considered to be based on gross under-reporting and thus not reliable. The workers' unions are engaged in the issue, but they have been disconnected from the wider environmental movement and discourse. Secondly, fishermen could potentially be an important social force in the environmentalist movement. They have been in several severe conflicts with the oil companies (Bronz 2009). One issue is compensation for damaged equipment and smaller catches due to exploration and production. In particular, the seismic surveys cause stress for and kill fish. Another issue is that the oil platforms attract the fish because of the lights installed on them, and the small continuous leakage of oil, which stimulates the growth of organisms that feed the fish. However, fishermen are not allowed to enter large waters declared by the navy to be 'zones of exclusion' (or zones exclusively for petroleum activities). In addition they are forbidden to fish within 500 metres of any oil platform anywhere (ibid.). Our informants in some of the fishing villages in the Campos area of the state of Rio de Janeiro tell of many problems, and many local protest actions. It is a big structural problem for them as 'artisanal fishermen' that the main fishermen's federation is linked to the trawling industry. There are some class-conscious unions representing artisanal fishermen but this segment lacks well-functioning state-wide unions (e.g. within the state of Rio de Janeiro).

The fate of the three initiatives and *lulismo* – after the June 2013 protests

We have assessed three initiatives which contest the dominant views and public policies regarding the giant new oil reservoirs in Brazil. The three are: 'The petroleum has to be ours' (nationalism), 'The royalties have to be ours' (regionalism) and 'The petroleum is a threat to sustainability' (environmentalism). From the angle of passive revolution, we have seen that the state leadership has agreed to some of the nationalist and regionalist demands, thus increasing popular support for the 'passive revolution' project of *lulismo*. On the other hand, the government saw no need to give in to any of the environmentalist demands.

We found that the first initiative ('nationalism') had support in the trade union movement, in some of the political circles inspired by Getúlio Vargas (president 1930–45 and 1951–54), and in the government apparatus dominated by the Workers' Party (PT) since 2003. The trade unions managed to raise some social mobilization behind the initiative. These factors persuaded President Lula to pay attention. However, the main reason why it had a strong political impact was that Lula could use it as an instrument, on the one hand, to regain credibility in the trade union and social movements and, on the other, to pressure for an overwhelming majority in Congress for the government reform proposals. The result was a national political consensus around a stronger claim for resource sovereignty, with direct federal ownership of the oil resources, Petrobras as the mandatory operator of every pre-salt field, and the oil revenues to benefit ordinary people by improving public services and eradicating poverty. Jointly, the pre-salt reserves and the nationalist campaign they unleashed strengthened the power block and social-ideological project of Lula, giving shape to *lulismo*.

The second initiative (regionalism), which originated in the state of Rio de Janeiro, enjoyed large social support within the oil-producing regions. It displayed an impressive mobilization capacity. Hundreds of thousands of people were mobilized on the streets several times, and there was sustained support from the regions' representatives in the National Congress. Still, its political impact has not been as big as the mobilizations would suggest. Its main social basis was the political class controlling the royalties and other rents flowing through a very exclusive group of states and municipalities. This class mobilized mainly its public employees and clienteles. While its leaders managed to negotiate some sort of alliance with the federal presidents (Lula and Dilma), the regionalist game they pursued earned them nothing but defeats in the National Congress. Thanks probably to support from the oil companies, which always contest changes in regulatory and fiscal conditions, they partially won their case in the Federal Supreme Court. Still, it is clear that the oil-producing regions keep a privileged portion of the royalties only for a transition period. While the nationalist initiative demonstrated the enormous potential power of *lulismo*, the regionalist initiatives around the royalty issue showed its inherent contradictions and weaknesses. It depends on popular electoral strength in certain poor regions (north-east) on the one hand, and on conservative pacts with the political-economic elites in the south-east (e.g. Rio de Janeiro and São Paulo) on the other. Hence, the royalty conflict along territorial lines has been one of the most serious threats to the survival of *lulismo* with Dilma as president.

The third initiative (environmentalism) was the weakest in terms of societal support, mobilization capacity and political impact. This is surprising given the fast organizational growth and political development of the environmentalist movement in Brazil in the second half of the twentieth century. It changed from focusing on traditional conservation of habitats to a distinctive socio-environmentalism meant to address ecological destruction and social injustice simultaneously. The alliance with the indigenous movement and international campaigns to save the Amazonian rainforest from the 1970s was instrumental for this growth. It also changed focus from pollution control to more sustainable cities (Hochstetler and Keck 2007). However, from the mid-1990s on the environmentalist organizations underwent a certain transformative process 'from protest to projects' (ibid.): activist organizations have become 'NGO-ized' and bureaucratized, depending increasingly on foreign finances (Braathen et al. 2007). In this way they seem to have institutionalized their focus on rainforest and indigenous issues, expressed by the struggle against the giga-hydropower plant Belo Monte, and city sustainability problems. The only environmental organization actively monitoring petroleum and offshore activities in Brazil has been Greenpeace, an international organization. Although it has a significant constituency in Brazil, the government can ignore it and dismiss it as a foreign pressure group, and there is scepticism in civil society because Greenpeace is seen to be a highly centralized organization governed from abroad. Hence, fishing communities, oil workers and environmentalists with interests in enhancing control of the oil industry have not managed to attain a viable 'socio-environmental' unity of action. Consequently, they have not been able to alter national power relations, which favour the 'growth coalition'. This bloc has as its first priority economic growth and increased public revenues, focused on fast expansion of the national oil industry.

Nevertheless, the campaign to stop the auctioning of the Libra field in October 2013 opened a new chapter in the history of social movements in Brazil. Perhaps for the first time, the trade unionists, environmentalists and street activists were united in a vital energy policy issue. They agreed on a slower pace for and higher national control of the exploration of the pre-salt resources. In this way they aspired to counter-hegemonize the 'nationalist' resource sovereignty discourse, which so far had been skilfully controlled by Lula.

The mass protests in June 2013 may have affected popularity and even power relationships within the broad *lulista* coalition itself. President Dilma found citizens' approval of her performance undermined in light of the protests, but her popularity was partially regained in the following months. One of the main reasons may have been her initiative to earmark future oil revenues for education and health spending, in response to the 'voice of the street'. Her proposal was adopted almost unanimously by Congress. A much more unpleasant fate seemed to await the head of the regionalist bloc of oil-producing territories, Governor Sérgio Cabral of Rio de Janeiro. Being responsible for a considerable amount of police violence against street demonstrators, he became the most unpopular politician in living memory in the state of Rio de Janeiro. Also, the petroleum capital of Rio de Janeiro became the centre of the new urban activism in Brazil, with the FIFA World Cup and Olympic Games triggering new social protests in spite of petroleum money flowing in. The taste of pre-salt may not derail the struggles of the urban poor for an improved quality of life.

These developments may indicate that Brazil's *passive revolution*, closely connected to Lula's charisma, is coming to an end. President Dilma has limited room for manoeuvre to concede to popular demands. In a speech on 21 June 2013, she promised to listen to 'the voice of the street' and deliver new programmes to enhance public transport, education and health systems. Apart from the Congress decision to earmark future oil revenues for health and education, no real improvements were observed the following year. She also promised a thorough political reform, to deepen citizens' influence on decision-making and to end the power of money in electoral politics, calling for a constituent assembly and a national referendum. However, most of her promises to the people were blocked in the government and in the National Congress by the conservative government partners, in particular the PMDB. Brazil's economic growth has decelerated remarkably in spite of the increased petroleum production activities. President Dilma was re-elected in October 26, 2014, for new four-year period with only 51.6 per cent of the votes. The conservative candidate Aécio Neves got 48.4 per cent. A climate of social and ideological polarization was created in these elections. Fifty years after Brazil saw radical social mobilizations being halted by a military coup and counter-revolution in 1964, some social segments want a popular revolution. Time will show whether the new-found oil reservoirs can help the current national leadership to unite the country with a renewed passive revolution.

Notes

1 'Protestos contra leilão de Libra são duramente reprimidos por forças militarizadas', *Brasil do Fato*, 22 October 2013, www.brasildefato.com.br/node/26407.

2 'Post-industrialist' in the sense that they are oriented towards modern financial and other service sectors fuelled by the land rents extracted from the oil industry.

3 'Anti-industrialist' in the sense that they are critical of the oil industry, at least if it grows freely and operates without strong environmental and social control.

4 Presented in the introduction to the book.

5 Although usually unemployed or underemployed, the sub-proletariat in industrializing Brazil is not entirely excluded from the labour market. This distinguishes them from the lumpen-proletariat and 'the permanently super-impoverished working surplus population'. The sub-proletariat is typically organized in female-headed families. They often move from rural to urban areas, or from cities in the periphery in the north-east to the faster-growing parts of Brazil, in order to provide better job opportunities for their offspring. Hence, effectively there is a continuum and no sharp differences between the 'sub-proletariat' and the 'proletariat'.

6 Singer points out that 'lines of continuity between *varguismo* and *lulismo* have to be the object of careful research' (2012: 45). We suggest some potential lines in the next section, 'The petroleum has to be ours'.

7 On biofuel in Brazil, see Meyer et al. (2012).

8 These assumptions will not be put to empirical test in this work.

9 This mixed economy model is largely inspired by the system set up by Norway. Interview with the project management for the new petroleum laws, Ministry of Mining and Energy, Brasilia, 1 April 2011.

10 We subsume to royalties a technically different but similar category of taxation called 'Special Participation'.

11 This to some extent has also been an alliance between PT (the Workers' Party) and the pro-business party PMDB.

12 This account of the disputes between the president, Congress and the oil-producing regions is based on an article written by the author in cooperation with Yuri Kasahara and Henrik Wiig, 'Brasiliansk oljebrems', published in the Norwegian business daily *Dagens Næringsliv*, on 22 March 2013.

13 Interview with a Brazilian representative of Greenpeace, Rio de Janeiro, 11 June 2012.

14 Although one of the main scholars conducting research and development on CCS admitted in an

interview with us that CSS was all but science fiction, particularly when facing the extreme deep waters and peculiar geology of the pre-salt.

References

Baiocchi, G., E. Braathen and A. C. Teixeira (2013) 'Transformation institutionalized? Making sense of participatory democracy in the Lula era', in K. Stokke and O. Törnquist (eds), *Democratization in the Global South: The Importance of Transformative Politics*, Basingstoke: Palgrave Macmillan, pp. 217–39.

Braathen, E., H. Wiig, M. Haug and H. Lundeberg (2007) *Development Cooperation through Norwegian NGOs in South America*, Study 2/2007, Oslo: Norad.

Bronz, D. (2009) *Pescadores do Petróleo. Politicas ambientais e conflitos territoriais na Bacia de Campos, RJ*, Rio de Janeiro: LACED.

Buci-Glucksmann, C. and G. Therborn (1981) *Le défi social-democrate*, Paris: Maspero.

Coutinho, C. N. (2007) *Gramsci, um estudo sobre seu pensamento político*, Rio de Janeiro: Civilização Brasileira.

— (2008) *Contra a corrente. Ensaios sobre democracia e socialismo*, São Paulo: Cortez.

Hochstetler, K. (2008) 'Organized civil society in Lula's Brazil', in P. Kingstone and T. Power (eds), *Democratic Brazil Revisited*, Pittsburgh, PA: University of Pittsburgh Press, pp. 33–53.

Hochstetler, K. and M. E. Keck (2007) *Greening Brazil: Environmental Activism in State and Society*, Durham, NC: Duke University Press.

Hunter, W. (2011) 'Brazil: the PT in power', in S. Levitsky and K. M. Roberts (eds), *The Resurgence of the Latin American Left*, Baltimore, MD: Johns Hopkins University Press.

Kasahara, Y., H. Wiig and E. Braathen (2013) 'Brasiliansk oljebrems', *Dagens Næringsliv*, 22 March.

Kingstone, P. R. and A. F. Ponce (2010) 'From Cardoso to Lula: the triumph of pragmatism in Brazil', in K. Weyland, R. L. Madrid and W. Hunter (eds), *Leftist Governments in Latin America: Successes and shortcomings*, New York: Cambridge University Press.

Meyer, D. et al. (2012) 'Brazilian ethanol: unpacking a success story of energy technology innovation', in A. Grubler et al., *The Global Energy Assessment*, Cambridge: Cambridge University Press.

Montero, A. P. (2005) *Brazilian Politics*, Cambridge: Polity Press.

Natal, J. (2007) *O Rio discriminado? (pelo Governo Federal)*, Rio de Janeiro: Armazem das Letras.

Ribeiro, J. A. (2001) *A Era Vargas*, Rio de Janeiro: Casa Jorge.

Ricci, R. (2010) *Lulismo. Da era dos Movimentos Sociais à Ascensão da Nova Classe Média Brasileira*, Brasilia: Contraponto.

Singer, A. V. (2012) *Os sentidos do Lulismo. Reforma Gradual e o Pacto Conservador*, São Paulo: Companhia das Letras.

— (2013) 'A energia social não voltará atrás' [The social energy cannot be turned back], Interview, *Epoca*, 21 June, revistaepoca.globo.com/tempo/noticia/2013/06/andre-singer-energia-social-nao-voltara-atras.html.

— (2014) 'Rebellion in Brazil. Social and political complexion of the June events', *New Left Review*, 85, January/February.

Tarrow, S. (2007) *Contentious Politics*, Boulder, CO: Paradigm.

Vianna, L. W. (2004) *A revolução passiva: iberismo e americanismo no Brasil*, Rio de Janeiro: Revan.

Weffort, F. (1978) *O populismo na política brasileira*, Rio de Janeiro: Paz e Terra.

9 | DOING WELL IN THE EYES OF CAPITAL: CULTURAL TRANSFORMATION FROM VENEZUELA TO SCOTLAND

Owen Logan

In Los Angeles [...] we had a large cathedral completely full of mothers and fathers of very low income families, and I could see mothers and fathers prostrated in front of their children with their uniforms and their [musical] instruments, completely overwhelmed by the dignity that the children bear. We have to realise that the moment a child receives an instrument he stops being a poor child. A child with an instrument is no longer poor. A child with an instrument and a teacher is no longer excluded. So our project is seen and perceived as an instrument of social inclusion and this represents for us an extraordinary mission, and a mission that goes beyond Venezuelan borders. (José Antonio Abreu, founder of El Sistema, speaking at 'Reaching for the Stars', a forum on music education held at the University of California, Berkeley, 28 November 2012)[1]

[...] if they [young people] say, 'I want to be like him – I want to have his passion, and joy and being' – you can be like him; but you'll have to practise eight hours a day, six, seven days a week, and you'll have to get up in the morning and think about doing this day in and day out and then you will be, as it were, your own [Gustavo] Dudamel. So it's a wonderful, wonderful model to give to the world because it's based on passion, it's based on effort, it's based on goodness, it's one of the most productive icons the world has produced recently, I thank God for Venezuela that it's produced not only El Sistema but it's now producing individuals that the world is reckoning on and learning from. (Film voice-over by Richard Holloway, chair of Sistema Scotland, and former bishop of the Scottish Episcopal Church, speaking about El Sistema's celebrated 'graduate', the conductor Gustavo Dudamel. Shown at the 'Big Concert', Stirling, Scotland, 21 June 2012)

Introduction

In 2008 a children's orchestra was founded on the outskirts of Stirling in Scotland. Within only four years local children were performing under the baton of the world-famous conductor Gustavo Dudamel, and their performance was broadcast internationally by the BBC. This would be a much more remarkable story were it not part of a systematic effort to perform the same 'miracle' time and time again. Using discourse analysis and drawing from fieldwork interviews, this chapter explores the role of Sistema youth orchestras operating among deprived communities.[2] Founded in Venezuela during the oil boom of the 1970s, in recent years the Sistema model of social action has been adopted and independently developed in well over fifty countries. Playing in an orchestra is the key teaching method, not simply the goal offered to children. Taken together the orchestras inspired by the Venezuelan model may be regarded as the largest social inclusion project in the world.

Governments in Venezuela and Scotland appear to be at opposite ends of the neoliberal belief system, yet in both these oil-producing countries the mainstream political left has moved away from theories of exploitation to those of exclusion and inclusion. The orchestral training examined here exemplifies the latter school of thought, as the above statements from José Antonio Abreu and Richard Holloway illustrate. They regard playing in an orchestra as a virtuous means of individual and social advancement for the disadvantaged. In what follows I look at how the political economy of oil boosted this relatively new discourse of socio-economic transformation, despite the evidence of its failings.

The labour theory of value was once at the heart of socialist and labour movement politics internationally. In the British Isles throughout the nineteenth and twentieth centuries the mantra of socio-economic transformation, 'labour is the source of all wealth', was emblazoned on numerous trade union banners and was the subject of popular analysis in countless political tracts about capitalist exploitation. Two insights were central to the political praxis that sought to confront exploitation. One was the realization that capitalist production not only creates unemployment but is actually dependent on a reserve army of labour made up of landless populations destined to spend long periods in virtual destitution. Technology cannot offer beneficial free time for workers or for societies at large because technological development is geared to the capitalist need to control and intensify labour in order to extract profit. The second and closely related insight is derived more particularly from Karl Marx's (1818–83) analysis that the ultimate measure of wealth is the free time workers seek to seize back from capitalist production. This was articulated most clearly by collective struggles for shorter working hours. According to Marxian thinkers this struggle over the production and use of free time would eventually become the pivotal point in capitalist development. Either the system implodes because of the denial of the potential benefits of technology or it creates a sufficient number of new areas of labour for some form of capitalist exploitation to continue (see Cleaver 2000 [1979]: 81ff.).

The discourse of social exclusion arose in France in the 1980s (see Boltanski and Chiapello 2005). In a relatively short time it has become as influential as the labour theory of value. No doubt for those not acquainted with Marx's writings, classical theory may appear anachronistic now that so much productive labour is technologically transformed or outsourced. As may be gleaned from Richard Holloway's commentary above, time-consuming and laborious cultural participation is envisaged as a remedy for the social ills now thought to be the results of exclusion from an otherwise functional capitalist economy. Not only is social policy justified on the basis of the potential for greater inclusion in this apparently healthy economic system, but classical understandings of social justice and the combined exploitation of labour and nature have been dropped. In Britain New Labour leaders may even be heard describing capitalists as *the* wealth creators. These ideas have far-reaching implications for class politics and the public understanding of key political issues such as progressive taxation and what might constitute sustainable development. There are knock-on effects to be seen in every area of policy.

My main focus is the shift in cultural policy whereby issues of cultural democracy highlighted in this chapter are overtaken by a concern for the morale of the poor, who are said to suffer from insufficient motivation and aspiration. They are now seen to be spiritually not politically impoverished. Consequently the idea that was so prevalent in community arts in the 1970s about people changing art has been turned on its head. Participating in the given structures of the arts is now supposed to change people, enriching them spiritually and making them upwardly mobile. The implication of this discourse, which echoes nineteenth-century attitudes about class and culture, is that it is no longer possible to pursue socialist policies to restructure education, welfare and culture on egalitarian grounds. The survival of egalitarian policy elsewhere, seen for example in the highly praised Finnish education system,[3] would suggest that this is a pessimistic ideology. However, it fits with an economic system that appears to be set on dismantling the ethical frame and the material gains of socialism in the twentieth century.

As I go on to discuss, the overarching claims made for the benefits of the Sistema model in tackling problems of education and crime are as misleading as Abreu's passionate account of a concert in Los Angeles which is transcribed above. Are we really to believe that a child receiving a musical education is no longer poor, as Abreu says? A plethora of historical and contemporary photographs of beggars could show us that poor children playing musical instruments are still poor children and they can grow up to be destitute adults (Figures 9.1 and 9.2). Realist photographers have gone further to hint at the complex ways music can turn poverty and suffering into a spectacle (Figure 9.3). At the level of political analysis I argue in a similar vein that El Sistema effectively impoverishes policy discussions concerning culture, inequality and education. In doing so the organization may be seen as part of a general trend which is now obscuring the public interest in many areas of education by blurring the boundary between public and private interests and diminishing the 'publicness' of public policy (see also Haque 2002). In strictly monetary terms, one of the results is that El Sistema commands more financial resources than the entire Ministry for Culture in Venezuela. Like many non-governmental organizations (NGOs) and charities, the organization is careful to manage a message; in the case of El Sistema it is that the arts can do no harm at the level of social policy and by and large they are a progressive factor. Against this false impression I argue that El Sistema is a threat to egalitarian education and cultural policies and its rise is an example of political opportunism driven by the financial and cultural exploitation of the poor and their worsening social conditions.

Among other things in this chapter I contrast 'Maestro Abreu's' paternalistic leadership – which prompts people to describe themselves as disciples and followers – with more genuinely egalitarian cultural organizations as well as traditional conservative youth organizations which have fallen out of favour. As other writers have argued,

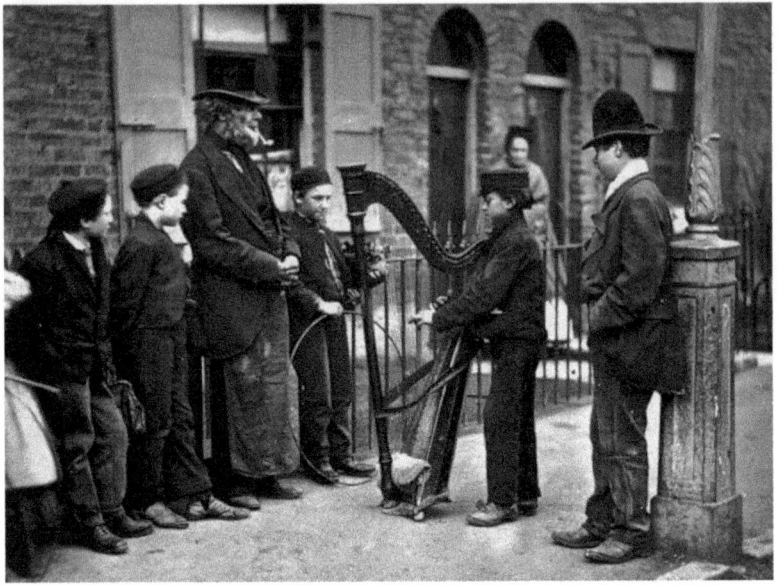

In the first photo-essay on poverty in London (published in 1877), the socialist writer Adolphe Smith wrote: 'there is an element of romance about the swarthy Italian youth to which the English poor cannot aspire'. Considering the labour conditions of children working as street musicians, and their public appeal at the time, Smith concluded: 'The traffic in very young children should be checked; the older children will continue to come over here of their own accord. The well-founded hope of gain, and the love of travel and change will suffice to attract a constant supply of street musicians. These will be obtained readily without any resort to kidnapping, or the creation of white slave trade. The present evils, whether exaggerated or not, would soon disappear if the Elementary Education Act could be made to extend to foreign children, and if the *padroni* were brought under the Factory and Sanitary Acts which govern other employers of labour. In the meanwhile it is hoped that the English public will continue to welcome with their pence all who cheer with good music our dull streets.'

9.1 Italian street musicians photographed by John Thomson. From J. Thomson and A. Smith (1877), *Street Life in London* (London: Sampson Low, Marston, Searle & Rivington)

Venezuela's revolution is still replicating capitalist inequalities (Dangl 2010; Ellner 2009; Martinez et al. 2010; Guerrero 2010). None of these left-wing analysts belittles the progress made in Venezuela under the banner of twenty-first-century socialism. And according to UNESCO's 2010 Education for All Development Index (EDI), the country has seen significant improvements in education. El Sistema is only one of several projects and reforms that can take part of the credit. However,

In March 2013 Lois Beath wrote in her blog: 'Unemployment is quite high in Spain right now, so in some areas we saw quite a number of beggars. [...] In Palma de Mallorca [...] the beggars in the old part of town all had their own style. Some sold art or trinkets, but that's nothing new. Others though, they dressed up in costumes that got weirder and weirder as we worked our way through town. [...] Other beggars perform as street musicians in hopes people toss coins their way in appreciation of the entertainment.' Photograph courtesy of Lois Beath. See mycruisestories.com/2013/03/09/beggars-of-spain/.

9.2 Palma de Mallorca. The creativity of 'beggars' who perform in Spain is celebrated in an award-winning blog, *My Cruise Stories*, by Lois Beath

to put the EDI measurement in perspective, Venezuela ranks 59th and the United Kingdom 9th. Approximately 8 per cent of pupils in the United Kingdom go to private schools.[4] In Venezuela the 15 per cent of pupils found in private schools rises to 26 per cent at high school level (see Pearson 2010). At the other end of the spectrum are the estimated 195,000 Venezuelan children who are still not receiving primary school education (ibid.). Yet Venezuela's El Sistema is now widely regarded as an international model for improving educational equalities. Major structural inequality in Venezuela and the different history and political challenges in a country like the United Kingdom have been ignored. Undoubtedly this form of depoliticization makes El Sistema increasingly attractive to neoliberal politicians in a time of

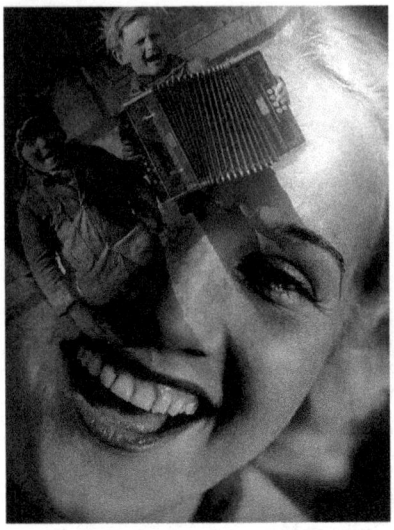

Realist photography influenced by the rise of workers' illustrated publications and avant-garde aesthetics in the 1920s and 1930s took a radical anti-naturalist turn. The realist intention to reveal more about class relations in this period can be seen in Karel Hajek's 1934 photomontage using a picture of child musicians. In making music the children of the poor become a charming spectacle for the disturbingly maternal gaze of a brightly lit fashionable young woman.

9.3 Photomontage by Karel Hajek (1934). From V. Birgus (2002), *The Czech Photographic Avant-Garde 1918–1948* (Cambridge, MA: MIT)

capitalist crisis and austerity. The same accounts which propagate El Sistema as a way of improving the education system are seen by some as left-wing simply because they acknowledge inequality as a problem that should still be addressed by the state. As the theoretical means of doing so 'inclusion' goes largely uncontested, allowing egalitarian educational policies and ideas of cultural autonomy to be demoted.

The dramatic move from the classical analysis of exploitation to the orthodox criticism of exclusion and the drive for inclusion is not, of course, confined to the countries considered here. These ideas are spreading everywhere. Nevertheless the politics of resource sovereignty in Venezuela and in the dis-United Kingdom have accelerated the implementation of inclusion as the transformative discourse advanced by contemporary social policy. In both Scotland and Venezuela oil revenue more than taxation is seen as a consensual means of protecting and supporting 'the left hand of the state', and this creates a degree of ideological uncertainty for the left.[5] The exploitation of energy resources is very broadly contested, but if there is such a thing as a short cut to socialism the redistribution of resource revenue does not appear to be it.

In Scotland the right-wing policies of different Westminster governments on the one hand, and the prospects of a much greater share of North Sea oil revenue on the other, combined to make nationalism more attractive to socialists of various hues. In the long run-up to

Scotland's 2014 independence referendum on 18 September, when almost 45 per cent of voters supported independence, key moral arguments about the political economy were swamped by more technical debates about the prospects for increased oil revenues for Scotland. From the monarchy, nuclear weapons and land reform to schooling, welfare and economic 'regeneration', many radical policy questions seemed to wither rather than grow in importance. Arguably the political conjuncture that may be described as resource sovereignty is particularly amenable to corporatist ideological contracts which may be difficult to undo once in place. El Sistema exemplifies these dealings, whereby the political pressure for social transformation is *financially* converted into the politics of social resilience. In the opinion of one bank director, all Venezuelan citizens must be united as in an orchestra.[6] Notwithstanding the anti-imperialist credentials of 'socialism for the 21st century' in Venezuela, the high-sounding ambition obscures what I see as a more experimental and often faltering process still based on the influential idea of *'sembrar el petroleo'* (sowing the oil). We will return to the critical issues we associate with resource sovereignty in Chapter 10, where we dicuss its characteristics in light of all the case studies in this volume.

The discourse of social inclusion is very much a language in action. The praxis of inclusion advanced by organizations like El Sistema is usually taken at face value as an unquestionable good. Therefore the main issue often appears to be whether or not the model works as a mechanism of socio-economic integration for the young who would otherwise fall by the wayside were it not for Sistema. Even if the answer to this question was a resounding yes – for which there is no evidence because comparable countries without the history of Sistema have done no worse and sometimes have fared rather better – there are still vital issues to consider about the differences between cultural instrumentalism coming from above and from below. Some of the key issues of instrumentalism from above are captured by what the sociologist Pierre Bourdieu (1930–2002) (1977) called 'the good faith economy', which he saw mediating the historic expansion of commerce and the rationality of the market. For Bourdieu the relations, loyalties and beliefs maintained by the good faith economy, like the notion of 'art for art's sake' in the art market, veil an economism which knows no other interests than those of capital. From Bourdieu's perspective it is precisely because economism really seeks to transform every aspect of life according to market logic that supporters of a subtle form of plutocracy mystify culture and ritualize art, creating networks of obligation and dependency in the process. These aspects of the good faith economy are articulated in cultural traditions which may, on the surface, appear to be the domain of spiritual and moral conduct rather than socio-economic mediations.

However, the interesting thing about El Sistema is that in general little or no pretence is made about its paternalism or its socio-economic role – quite the opposite. Proponents of the organization stress that it is a social project much more than an arts project and on this basis it receives funding directly from governments and banks as well as arm's-length arts agencies and private foundations and patrons. It is argued that music-making in orchestras alleviates poverty by promoting community cohesion and encouraging the virtues of cooperation, dedication and mutual respect. In the words of Abreu, the petroleum economist, conductor and devout Catholic who founded the organization: 'I realized one of the most efficient ways to fight poverty was to introduce excluded children and young people to musical education [...] and turn their families and communities into our allies.'[7]

The victory of El Sistema and its claims

It is estimated that Abreu has received some sixty prizes and awards internationally (Borchert 2012: 11). In Venezuela his project has grown to roughly 285 '*núcleos*' or ensembles around the country. These provide free access to classical instruments and music tuition for somewhere between 310,000 to 370,000 children.[8] To many people El Sistema, not the public school system, appears to be the means of delivering a musical education to children. Critics of the organization such as the well-known Venezuelan folk singer Cecilia Todd regard the organization as an aesthetically homogenizing project that encourages the development of a robotic musical culture.[9] Following on from this sort of criticism a project Sistema Alma Llanera devoted to folk music was set up. Its status is in doubt and it presently appears to be a tokenistic project. So the question of what ought to be the appropriate structure for musical education and how to educate children about musical diversity remains. However, it is not socio-aesthetic issues which figure in major funding assessments of El Sistema. Rather, it is argued that the organization combats crime because regular music tuition keeps the youth off the streets during after-school hours when they would otherwise be at risk of becoming victims of violent crime or becoming involved in deviancy. This rationale has often been promoted by Abreu. For example, he claims that 'a child who is given a violin will not pick up a gun because music produces an irreversible transformation in a child [...] he may become a doctor, study law, or teach literature, but what music gives him remains indelibly part of who he is forever'.[10] Abreu also declares that 'the most holy of human rights is the right to art'.[11] On the back of such authoritative-sounding statements it is frequently implied that El Sistema is an integral part of the battle against many varieties of deviancy among the poor and that it advances their cultural rights.

These socially instrumental claims have given El Sistema an important place in the repertoire of contemporary inclusion policy and allowed its proponents to lever increases in funding and to gain political prestige unlike that accorded to any other artistic or cultural venture. Indeed, the conductor Gustavo Dudamel compares the national youth orchestra at the peak of El Sistema to the Venezuelan national flag. There are powerful reasons to think of the organization in such terms. Not least that the results of all the other inclusion programmes supported by Venezuela's revolutionary governments, including the eradication of illiteracy, are harshly criticized by the opponents of *Chavismo* at home and abroad for their superficial nature and misleading use of statistics.[12] Although Abreu was occasionally criticized for associating too closely with the Chávez governments, the twin claims made for El Sistema, namely that it is an effective means of combating poverty and crime and that it supports human rights in the field of culture, have not been subjected to the same sort of scrutiny by the Venezuelan government's increasingly vociferous opponents.

The welfare economics philosopher Amartya Sen reminds us that human rights do not only have an instrumental significance, they have a constitutive importance too (Sen 1999: 17ff.). People have the right to participate in culture for its own sake. In itself this has instrumental significance because human rights at the level of culture entail the right to *change* culture rather than to merely reproduce its consecrated forms. Looked at from this perspective, Abreu's idea that 'anything good, noble and praiseworthy must be reproducible [...] so what's good for one underprivileged child has to be good for all underprivileged children' risks reducing children to the status of performing monkeys.¹³ Despite the frequent references to human rights issues concerning freedom of expression in Venezuela, usually raised by the political right, increased official support for El Sistema has escaped such scrutiny. To understand the reasons for this double standard, and the real depth of the issues that the organization claims to address, we need to appreciate how the organization's rationale has interacted with political power.

Venezuelan politics was determined by military strongmen throughout the first six decades of the twentieth century. The political field thereafter is marked by the connected vices of corruption and what is called excessive presidentialism. The latter reduces politics to the power of presidents and makes for a poor weapon against corruption. No matter how honestly they are used, the office and the power of the president cannot impose a meritocracy on society, which in many instances prefers personal bonds and clientelism to capitalist hopes for relative transparency and an open market. Even writers, who stress consensus about the liberal democratization process dating back to the 1958 *Punto Fijo* pact, also emphasize the crisis of legitimacy which was to destabilize the political class, the business and banking elites and organized labour (Crisp and Levine 1998; Gates 2010; Logan 2007). All these established arenas of power have been tainted by corruption. Indeed, the historical context for Abreu's project is this growing crisis of political legitimacy, which turned into a major crisis of governance in the mid-1990s.

In the midst of this highly polarized polity – represented by profoundly distrusted elites – the classical orchestral form made inclusive by El Sistema looks like an invitation to a near-perfect meritocracy. This is voiced by a young woman violinist filmed for a promotional documentary about the organization who says: 'There is no difference between classes here, nor white or black, [or] if you have money or not. Simply, if you are talented, if you have the vocation and the will to be here, you'll get in, and you'll share with us and make music.'¹⁴ In an insightful dissertation about the organization, Gustavo Borchert (2012) notes that its advocates have identified El Sistema with meritocracy and identify meritocracy as the country's path to progress.¹⁵ However, contrary to the suggestions that El Sistema has successfully institutionalized meritocracy, research by the ethnomusicologist Geoff Baker (2014) reveals more of the murky patronage system described as *palanca* (string-pulling) in the organization. As one former Sistema musician told Baker: 'Just like there's betrayal at work, there's betrayal in the orchestra. Just like there's *palanca* at work, there's *palanca* in the orchestra.'[16]

During his tenure as Venezuela's President, Hugo Chávez (1954–2013) increased support for Abreu's project using the presidential budget derived from the nation's oil revenues. This support for what was seen as an aesthetically conservative project was not uncontroversial. Given the objections raised by artists such as Cecilia Todd, mentioned above, it is worth asking how Chávez's unexpected support for El Sistema fitted the wider political field that brought him to power. From his first election campaign until his death, Chávez retained political support from within Venezuela's business cadre. Leslie Gates' (2010) study of Chávez's 1998 election gives evidence of the financial support his candidacy received from members of the business elite, and according to Gates' interviews some business leaders discounted Chávez's socialist messages and lent their support in the hope of gaining access to state contracts (ibid.: 108). Whether support for Abreu's project could in some way deflect crude demands to 'collect on [...] financing and secure a favour' from the government in return – as one of the business leaders described the logic in his sector – or to what extent financing from business people was actually reciprocated, is an open question. However, there is no suggestion that increased support for El Sistema, which was solidified only in 2007, was part of a corrupt quid pro quo. Rather, as Gates shows, early business support for Chávez did not just reflect cynical positioning but also internal political conflicts within the business sector which stemmed from the way the sector was ideologically discredited by Venezuela's 1994 banking crisis (ibid.: 114–31).

What can be said is that against this background of intra-elite conflict Chávez's support for Abreu's project was a signal that the government and the business elites could still find common ground. This was increasingly significant because in other areas of the arts, government policy could be heavy handed. For example, private sponsorship at the state-owned Teresa Carreño theatre was forbidden when what really needed to be demonstrated, from the government's socialist position, was that private patronage did not determine or influence the publicly funded repertoire.[17] Several people involved in the cultural sector in Caracas interviewed for this study see Chávez's support for Abreu's project merely as the result of skilful lobbying at the elite level and nothing revealed here would contradict that interpretation. Nevertheless, in Venezuela and elsewhere there are few means for politicians to legally reciprocate support received from business, but, as I go on to discuss, support for the pyramid-like form of the classical orchestra is one of them.

From the European conquest of South America in the sixteenth century and the introduction of African slaves, the blending of musical traditions on the American continent has been the norm. What is unusual about El Sistema is the revival of the cultural conquest which attempted to discipline this musical production through the combination of spirituality and commerce. For the early Jesuit missionaries who set up the 'reductions', semi-autonomous trading settlements, in the sixteenth century, the baroque orchestra provided a powerful means to convert the continent to Catholicism and European notions of civilization (see Sadie 2001: 474–9). For Voltaire (1694–1778) this appeared as atonement for the brutality of imperial conquest and was nothing less than 'the spectacle of the world' (Voltaire 1819: 351). Abreu's lifelong project to turn the poor into the allies of the ruling belief system today does not require the creation of such protected economic communities. Rather, the orchestra has become an idealized symbol of the business community's belief in meritocracy and its belief in itself.

Listening to Abreu's narratives, the historical lineage and the contemporary business of classical music seem unquestionably 'noble'. In the twentieth century the horrendously close relationship between classical music and Nazi genocide is sordid (see Lasker-Wallfisch 2000), but just as unmentionable in this context is the eighteenth-century craze for castrati singers; this phenomenon involved castrating thousands of boys annually in the early decades of the century with lethal consequences for up to 80 per cent of them (see Barber 1996: 11). Today the skilled labour of entire orchestras is turned into the cultural and financial capital of individual conductors, and in this sense the visible performance of classical music is analogous to capitalist enterprise. It is no less an outgrowth of baroque socio-economics for that. Musicians are stakeholders, and although industrial actions are not unheard of they are rare and tend to take place *in extremis*. The tension of the organizational form sometimes surfaces; for example, the virtuoso violinist Nigel Kennedy condemns conductors for greed and opportunism.[18] Such public outbursts and the voices of dissent brought to light by Baker's ethnography, mentioned above, belie a more deeply felt sense of vocation and apprenticeship within an avowedly meritocratic system, and of course without such a work ethic the hierarchical structuring of orchestras could not be reproduced. That it is reproduced so vividly in Venezuela is perfectly in keeping with Latin America's colonial heritage. To 'make America' the youth of the Spanish and Portuguese merchant class in the nineteenth century worked very long days under close supervision of the *patrón*. Salaries were also given over to invest in the business as he chose, all in hope of eventually receiving greater rewards and status in return for total dedication to the enterprise (see Stein and Stein 1970: 152–5).

The same sense of dedication finds a latter-day echo in El Sistema's motto, *tocar y luchar* (to play and to fight). The first obvious victory for Abreu's project took place on 30 April 1975 in Venezuela's Ministry of Foreign Affairs. Domingo Sánchez Bor, who played in Abreu's orchestra performing at the ministry that night, recalls: 'we knew it was for real, all that about the scholarships, the chances for work, our careers as musicians, it was all possible, it had come true. That evening we had found the purpose we needed and which we had searched for so hard' (Borzacchini 2005: 44).

That El Sistema propagates the idea of meritocracy in a country bedevilled by corruption and crime is not in doubt. But does the organization also foster solidarity and function like a sort of social glue as its proponents insist? There is a barrage of similarly anecdotal bits of evidence coming from institutional studies and official reports as well as journalists' articles, books and, perhaps most influentially, from documentary films, which are used to suggest that it does. The uncomfortable fact consistently brushed aside by these flattering assessments is that crime has steadily risen alongside the rise of El Sistema in Venezuela. Moreover the increased support given to the organization by Chávez did nothing to abate a steep rise in violent crime between 1999 and 2011. According to recent data sets for this period the murder rate climbed from 25 per cent to 45.1 per cent.[19] In schools, where one might expect the organization to have the greatest positive impact, the occurrence of serious violence and killings is equally alarming.[20] If El Sistema appears unable to stem the tide, then it ought to be asked whether the organization is actually part of the social problems that have gone hand in hand with its growth in Venezuela.

Both supporters and opponents of the Bolivarian Revolution

recognize 'a culture of criminality' as a real social issue in Venezuela.[21] It is seen as a weak point in the revolutionary process, the effects of which are exploited by the opposition. Apart from supporting Sistema there is hardly any agreement about the policies which would be appropriate to meet this challenge. Over the same period of increased violence, data sets show that extreme poverty, unemployment and inequality have fallen. Many criminologists would expect these indicators to be accompanied by a fall in crime. The fact that the opposite is the case when it comes to violent crimes such as murder and kidnapping is a paradox which can be seen creating a discursive vacuum at the level of policy debates.[22] Consequently the message that the sort of social cohesion fostered by El Sistema is a means to reduce crime goes largely unquestioned. This is a striking lacuna in the discourse which has grown up in support of Abreu's project given that mainstream sociology, especially Bourdieu's now classic study on *Distinction*, cautions us that inequality is reproduced most effectively and most deeply in taken-for-granted cultural processes.

If we follow Bourdieu and look at culture as communication and process then the complex problems exacerbated by the organization's meritocratic discourse are more readily seen. We may get a brief glimpse of this first in the testimony of two young trumpeters filmed for *El Sistema*, one of the typically anti-realist documentary films which promote Abreu's worldview.[23] In this film the two boys who are firm friends reflect proudly on the development of their local orchestra but one pauses unexpectedly to comment on the problem of percussionists. Percussionists, as the boys seem to agree, are *bochincheros* or rowdies. Unlike the young trumpeter who has spoken of his career ambitions, these 'slackers' and 'troublemakers' as they are also described are too disorganized to get on in life. At this discordant point in the film's message of orchestras delivering social cohesion in a crime-ridden society, the film-makers cut to shots of a bird of prey flying over Caracas. Following this menacing symbolism, Abreu is brought back into the film to make the point that El Sistema is incomplete and the number of Venezuelan children touched by the project will rise to a million in the next decade. In this way what viewers have just witnessed is brushed aside. It is as if the answer to the social fractures created by using music to instil the belief in meritocracy and transforming play into labour is to proceed with more of the same with even greater gusto.[24]

What's wrong with this picture? Without recourse to Bourdieu's sociology, why is it so easy to dismiss the social divisions which El Sistema can be seen to deepen even within a single orchestra? Although revolutionary leftists recognize that the Bolivarian Revolution has traditional social democratic aims they argue that in the face of neoliberal hegemony these become revolutionary aims.[25] In Britain the relationship between the erosion of social democracy and the rise of meritocracy was first examined by the sociologist Michael Young (1915–2002). Young coined the term meritocracy in his 1958 book *The Rise of the Meritocracy*, where he formulated its core principle as 'IQ + Effort = Merit'. As a social democrat who had served in Britain's 1945 Labour government, Young was also able to recognize that the meritocratic ethos was a ruthless one. Not only does meritocracy blur the boundaries between labour and play for the sake of success but it reinforces inequality. Because people are made to feel they deserve the advancement and privileges that are in fact delivered by capitalism, their toleration of structural inequality may know no bounds. For example, in very novel satirical terms for a sociologist, Young envisaged Britain in 2034 as a country where the most promising children of the poor would be purchased by the wealthy to boost the prospects and opportunities of their own families. In this way he illustrated the creation of the neo-feudal system he saw taking shape under the auspices of social democracy.

More recently the French Marxist philosopher Étienne Balibar (Balibar and Wallerstein 1991) has argued that the dilution of socialism into excessively nationalist forms of social democracy means that the issue of racist attitudes – often laid at the door of the petite bourgeoisie by leftist historians – must be fundamentally rethought in terms of 'class racism'. Instead of looking at which social class is most supportive of racist ideas, Balibar argues we should be alert to the spread of petit bourgeois thought about class and culture. The same issue can be seen from a more ethnographic angle in the English writer Owen Jones' (2012) book *Chavs – the Demonisation of the Working Class*. In great detail Jones unpicks the stereotyping and blame for the victim mentality which underpins the British economy today. The context for Young's pioneering critique was the emergence of 'the opportunity state' in the 1950s and 1960s and the rise of correspondingly technocratic attitudes, especially among upwardly mobile members of the working class. Since there can be no real meritocracy without equality the perennial moral quandary for those who believe in meritocracy is how to reconcile personal advancement with political, economic and institutional reproduction of inequality. As Young saw it, people look to the state to equalize opportunities while at the same time looking for advantages for themselves and their children. Although these critical insights about the corrupting aspects of meritocracy emerged from Britain's social democratic project they clearly have a much wider relevance.[26]

The basic formula essential to meritocracy – intelligence plus effort – is not confined to legal or virtuous transactions in society; as Young realized, the ideology of meritocracy promotes a dubious public morality. Murder rate statistics in Venezuela rather blankly describe a social reality where armed robbery and kidnappings are the routine facts of life and death. Few people, especially in Caracas, have not been touched by these crimes, the successful execution of which may well demand intelligence plus effort plus daring. The life of an orchestra is about as far removed from physical daring as it is possible to be and, as I go on to discuss in the next section, there are good reasons to see El Sistema hardening such cultural distinctions and therefore as being imbricated with the rise of violent crime, or in other words to see the organization as part of the problem and not part of the solution. However, Venezuela's oil economy was an important factor driving this so-called inclusion project forward and spreading its message internationally despite the evidence that crime was getting much worse in Venezuela, a trend ruthlessly exploited by right-wing ideology.

Abreu has offered a fairly candid account of the combination of religious, musical and political beliefs underpinning his mission (in Borzacchini 2005: 16–29). Politically he identifies himself most with the writer and politician Arturo Uslar Pietri (1906–2001), regarded by some to be the best president Venezuela never had. Seeing the dangers of an extractive mono-economy in the 1930s, Pietri coined the very influential phrase '*sembrar el petroleo*', and although he moved from a statist vision to a much more market-friendly one, he argued throughout his life that Venezuela's vast oil wealth must be used for all-round development.[27] Pietri ran unsuccessfully for president in 1961, and it was his support for Pietri's movement which Abreu says led him into being elected to the Venezuelan parliament and to various political roles, including his appointment as minister of

9.4 Onlookers at a street performance (below) in 2013 about crime and sinfulness by an evangelical church in Caracas. Photographs by Owen Logan

culture during the second presidency of Carlos Andrés Pérez from 1989 to 1993. Of all the country's presidents Pérez (1922–2010) is the one most associated with Venezuela's boom-and-bust oil economy.

Like other resource-rich nations Venezuela has long been a hive of debt creation for lenders from North America and Europe. Thanks to greater public awareness concerning the dynamics of usury and speculation, since its partial exposure by the global financial crisis, and partly too because of popular published revelations about white-elephant development schemes that were sold to countries like Venezuela as 'must have' projects, the age of obscurity concerning the demi-monde of finance capitalism may be drawing to a close. There is certainly much less emphasis put on grandiose development projects, which are viewed with greater suspicion now. However, calls for the systemic reform of central banking and the international financial institutions and a raft of other initiatives such as the localization of currency systems – initiatives all designed to democratize socio-economic development – have been muted rather than amplified by the divisive politics of austerity. In the wake of repeated debt crises lending is increasingly justified on the basis that the end results reach the poor and directly transform their lives. In an often-quoted evaluation of their 2007 loan of US$150 million to the Venezuelan government to build regional centres for El Sistema, the Inter-American Development Bank (IDB)[28] estimated that in the long term every dollar loaned would save the Venezuelan public purse 36 cents by increasing employment opportunities and lowering crime levels.

The IDB's (2007) projection is based on the stated results of El Sistema in reducing the school dropout rate from 26.4 to 6.9 per cent. This positive effect is underpinned by parental need for safe and free childcare after school. El Sistema provides this only by default. This basic effect on school attendance may of course be replicated and improved on by other less regimented means, not all of which require funding. This includes the need for democratization and responsiveness in the school system as various educational reforms in Venezuela demonstrate (see Hazel 2007). The IDB report does not discuss such educational options in what is an opaque comparative framework. From the basic effect on school attendance ascribed to El Sistema, the bank's cost–benefit evaluation instead goes on to make sweeping projections based on a slur – namely that unless children join an orchestra they will be much more likely to become criminals. As the IDB report states, El Sistema's main objective is 'the development of human capital, training in civic values and good behaviour, the creation of future employment opportunities, and offering alternatives for the non-criminal use of free time […]' (see IDB 2007: Project Objective and p. 5). No consideration is given to the all too obvious risk that involving large numbers of children in the same cultural activity so defined might not be a deterrent to delinquency, and indeed it could even increase feelings of alienation, resentment and 'exclusion' among others.

Since 2007 other credit lines have been opened to Venezuela to bolster El Sistema's infrastructure, giving the organization a degree of permanence that could outlive attempts to reform educational and cultural provision. In line with IDB's human capital rationale for tackling problems of exclusion, the Development Bank of Latin America (Corporación Andina de Fomento – CAF) loaned US$350 million for the construction of El Sistema's new headquarters in Caracas, due to open in 2017.[29] Interestingly, the authors of a 2012 CAF report on *Public Finance for Development* recognize the historical linkage between increased taxation, increased state action and greater political debate and accountability – a nexus which is identified with socio-economic development in Europe and North America in the twentieth century (see Sanguinetti and Villar 2012). On this comparative basis the report highlights the deficiency of taxation systems in Latin America (especially where states rely on resource revenues) and the consequential deficits in civic consciousness and state accountability in the provision of public goods. However, the picture the CAF report paints of historical development of these issues in the countries of advanced capitalism is all but denuded of the traces of ideological conflict over the social contract and issues of socio-economic justice, despite the obvious effects of two world wars on policy and public attitudes. By neglecting the role of ideological struggle over the role of the state in developing education, welfare and public enterprise in Europe and North America, not to mention the history of 'military Keynesianism' in the United States, the CAF report blithely envisages a very different process of consensus-building for progressive taxation in Latin America. Unsurprisingly, for a major lender that promotes external investment and free trade, this is not seen in the light of the varieties of anti-capitalism but rather in terms of investing in human capital.

Although El Sistema is acclaimed as a social project by major funding assessments, these assessments fail to evaluate the organization in the same broad terms. Thus El Sistema has not been looked at critically as an essentially private inroad into public education; one that substitutes a necessarily diverse musical education in children's early years for what in many cases is an intensive training which borders on indoctrination. Given the range of socio-economic benefits which bankers promise from the continued development of this project – which has in fact been the sonorous accompaniment to the social problems they identify – I will now turn to holistic academic and grassroots perspectives on the related issues of crime and cultural democracy in Venezuela. I argue that the notion of social action advanced by El Sistema is profoundly invalid both as a means to combat crime and as an articulation of cultural rights. It is detrimental to the understanding of both and hinders genuine public policy.

Critical perspectives on culture and crime

The IDB's rationale for its US$150 million loan represents the most worrying victory for Abreu's ideas about orchestral music constituting an efficient means to combat poverty and crime. The fact that most poor people are not actually criminals appears to be overridden by the old fear – so prevalent until the twentieth century – that potentially most are. Given the history of banditry the fear of 'underclass' violence has never been totally irrational. However, as the historian Eric Hobsbawm (2000) shows in his influential book *Bandits*, their numbers have always been a very small proportion of the populations effected. In Colombia, during a particularly murderous episode of *la violencia* in the early 1960s, only one person in a thousand was a member of an armed group. Even in such extreme situations statistical evidence shows how unusual it is for poor people to pick up guns (ibid.: 24).

The motivations among the minority who do pursue violence are complicated. As Hobsbawm and others recognize, pathological crime is not devoid of meritocratic and ethical evaluation. Such reasoning supports a powerful Robin Hood myth and real violent crimes are measured against it, and therefore relativized by it. Moreover the same reasoning may be shared by violent criminals and law-abiding citizens alike. Whatever social injustices might be used to explain or justify brutality tend also to be reinforced by punishment in corrupt criminal justice systems. For instance, an ex-convict in Caracas who was interviewed for the present study remembered sewing up his own mouth with a needle and thread in protest. Now a poet with a public sector job thanks to recent government policy, he recalled his self-inflicted pain as a necessary and almost unremarkable part of the hunger strike he pursued with some of his fellow inmates.

In the last analysis it is quite absurd to think that a subcultural elite – one made up of the very small number of people who are prepared to secure their advancement through physical daring and the use of violence – could be diverted from such a course of action by the expansion of musical training among the children of the poor. Moreover, as I go on to discuss here, the spread of a project that falsely promises upward mobility as a solution for criminal deviancy is a fatal diversion from other cultural activities capable of tackling the real roots of crime.

Despite El Sistema's rhetoric of transformation the organization focuses on children, not families or communities as a whole or the need for structural egalitarian reform in education. Indeed, the goal of *saving* children from their social circumstances is blatant, and although it stigmatizes communities, it makes El Sistema ideologically amenable to different politicians and governments. The idea that children need to be introduced to an entirely different culture to the one they know is both implicit and explicit in the organization's mission. When it comes to criminality the message is equally clear. Popular culture and music do not merely reflect, reiterate or glamorize social malaise; rather they are aspects of a criminal pathology and an obstruction to social mobility. Therefore the potentially deviant outlook of disadvantaged children must be sanitized and remade according to more respectable and law-abiding mores. This project of sanitizing culture is no novelty, of course. It articulates a 'class racism' (see Balibar and Wallerstein 1991: 153–216) which has crypto-fascist and outright fascist roots that are entwined with those of economic liberalism, but it is not my purpose to trace these linkages here.[30] Nor will I deal with El Sistema's work in prisons, the rigidity of which is critiqued by Borchert (2012). What I stress here is that the ideas El Sistema promotes are damaging, simply because they detract from genuinely effective attempts to combat cultural inequality and the social relations underpinning crime and violence in society.

Critical realist criminologists today regard crime as a refraction of the structures of power and inequality in society. In examining the causes of crime such research regards inequality within disadvantaged groups as just as important to analyse as the more obvious disparities of class, race and gender. Indeed, critical realists have accused the political left and right of not taking crime seriously as something which is part of the neoliberal package, namely the pro-market and anti-welfare, anti-planning and anti-regulation policies being articulated globally. William Wilson's (1987) book *The Truly Disadvantaged* was a major influence on this school of thought. Wilson starts off by examining the paradoxical deterioration of Afro-American communities in US cities after the passing of the Civil Rights Act in 1964, and finds that they were made weaker by new opportunities for upward mobility which had the effect of dividing and weakening communities increasingly identified in terms of colour not class. How do these realist insights compare with the experience of Venezuelans?

Ricardo Romero was born in Caracas in 1968. Romero has conducted writing workshops in prisons over several years and, influenced by anarchist political thought, he has become an ardent critic of the criminal justice system. In conversation he illustrates the failure of the system not so much through its many documented abuses but with the hypothetical assertion that if every prisoner in Venezuela was released at once the effect on the crime rate would be minimal. The reason Romero gives is that crime is organized in Venezuela vertically and the powerful organizers and their corrupt collaborators are hardly ever to be found in prison. Romero's view that widespread corruption and organized crime go hand in hand is shared by many Venezuelans and is one of the reasons grassroots activists put so much trust in Hugo Chávez but did not extend this trust to his cabinet or his government officials.[31]

The political complexity of *Chavismo* as a social movement is evident in the various community media organizations that have benefited from government support and which are essential organs of revolutionary consciousness. Petare is a well-known poor barrio to the east of Caracas and is home to Petare TV, one of Venezuela's community media organizations run on egalitarian principles with the aim of overcoming divisions of labour. One of the station's founders is Charles Mendez, who was born in Petare in 1963. Mendez says it has made very little difference to the organization whether the local mayor is a *Chavista* or not, the station has been in conflict with both. One of Petare TV's community journalists was murdered in 2007 during the course of an investigation into corruption. As a result of threats at the time some local programme makers gave up their involvement with community broadcasting. However, such experiences have also hardened the resolve of Petare TV, particularly in their resistance to the idea that 'culture' needs to be brought to the

9.5 Onlookers during the news digest on *Petare al Dia*, broadcast by Petare TV in 2013, in an open-air studio. Photographs by Owen Logan

people of the barrio. Although people at Petare TV are tolerant of El Sistema, largely because of Chávez's support for the latter, Mendez says that if Petare TV is to survive and develop it will be because the community gains a better critical understanding and 'intelligence' about its own life and cultural affairs, partly through local programme

9.6 Musicians and production crew on *Petare al Dia*. Photographs by Owen Logan

making and partly through civic-social networks that can defend the genuine interests of the community.[32] Along with defeats there have been significant local victories in confronting the rise of violence.[33]

Petare TV is not a parochial cultural project. One of the station's innovations has been to compare and contrast reporting from the world's media, creating a more media-savvy environment where there is greater awareness of the politics of broadcasting and the issues of media manipulation. There has been a considerable amount of academic and activist writing looking at the development of community media in Venezuela. Naomi Schiller (2011) and Sujatha Fernandes (2011) describe the rather fraught relations of organizations like Catia TV and Radio Bemba with the commercial right-leaning media on the one hand and the 'official' revolution on the other. As the tragic murder of one of Petare TV journalists would also suggest, the situation and autonomy of community media projects is politically precarious. Much of their funding is project based or comes via government advertising.

Among the factors that Mendez sees influencing crime in Petare is the overspill of the long-standing conflict in Colombia, with the result that paramilitary narco-gangs have effective control in some areas and paid hit-men operate with impunity more broadly. This is a politically volatile subject, but cultural organizations like Petare TV are in the position to offer important insights on crime and violence. They are forums for communities to discuss the interconnected nature of social problems that tend to be taken in isolation and become overly charged by ideology, as the rationale of El Sistema shows. According to Mendez and many other voices on the left, crime has been used opportunistically by the right-wing opposition, who seek to benefit electorally from the social malaise. There is more to consider here than political opportunism at election times. Simplistic accounts which collapse poverty, crime and 'bad behaviour' into one phenomenon make matters much worse because once such a discursive environment is created and sustained by organizations like El Sistema and in the media more generally, the real dynamics of political power, local authority and criminality are harder to confront. Areas like Petare are naturalized in the public consciousness as breeding grounds for criminality in terminal decline, instead of being regarded as spaces of commercial, political and cultural exploitation. The latter may come in many forms from respectable international lenders to narco-commerce to the proponents of cultural and educational transformation from above. In different ways each contributes to the cheapening of life in places like Petare.

What is the potential for the development community media as a countervailing force? For community media to prosper in Venezuela in any significant democratic form its position as a public good and its independence from both business and the government of the day would need to be strengthened. If the development of human capital was an area of potential consensus-building for public finance, as it appears in the CAF bank report discussed above, then the funding prospects for independent community media projects would be good. However, the logic of human capital is at one with capitalist economism and is anything but neutral. Whether support for community media was raised from credit lines, resource revenues or taxation (possibly on corporate advertising campaigns), in their present form people's organizations like Petare TV represent an obvious challenge to the socio-political status quo. If there is to be redistribution of public finance through culture, then the sort of social action promoted by El Sistema is infinitely more attractive in the eyes of capital. The meritocratic discourse of equal opportunities for social climbing is usefully devoid of actually existing egalitarianism. While democratically run community media projects with complex educational and cultural functions articulate the constitutive and therefore unpredictable aspects of human rights in culture, El Sistema focuses instead on narrower economistic goals. It should be no surprise, then, that lenders want to make the organization more concrete – quite literally – and distort evidence to make the case for investment.

Using the example of Petare TV, I have argued here that important and courageous cultural projects run along egalitarian lines are less likely to prosper as a result of El Sistema and the economism articulated by its powerful backers. It may be objected that this is a gloomy and entirely speculative projection. Given the history of the fracturing and atomization of communities in the United States and elsewhere, it is much less speculative than the sweeping cost–benefit projections made by bankers to support their lending to Abreu's organization. As I discuss in the conclusion, it is necessary to sound an alarm about the somewhat deceptive functions of the good-faith economy. The egalitarian and autonomous character of organizations like Petare TV is typically precarious and is always at risk. Such radically minded community organizations are subject to financial and institutional interventions which negate social solidarity by turning civil society into the sphere of liberal competition for top-down funding and patronage. In the competition for funds, politically and economically unquestioning organizations like El Sistema have an obvious fighting advantage, but the terms of competition for relatively small amounts of funding for less well-established projects can easily distort or artificially mould their functions (see Folorunso et al. 2012).

Oscar Sotillo Meneses is a founder of *La Mancha* (The Stain), a respected left-wing and somewhat avant-garde cultural journal published in Caracas since 2002. The publishing organization also broadcasts a radio show and maintains good relations with a variety of leftist cultural projects and groupings, including Petare TV. Sotillo thinks the political phenomenon of *Chavismo* now being carried forward by President Nicolás Maduro can be roughly divided into two halves: those who think El Sistema is wonderful and those who think it is awful. Describing the institutions and networks of high culture he says:

> the artists and support workers in those institutions have the advantage of knowing what culture is. It is things that they've been doing for decades and centuries and that's also the advantage which the political right possess. Whereas, we on the left are never sure what culture is because we're in the process of trying to remake it.[34]

In terms of social milieu *La Mancha* may be regarded as more of a bohemian organization than a bourgeois one. Typically there is a high degree of scepticism about the aesthetics of consecrated culture and conventional upward mobility among its network of poets, writers, artists and musicians. In these circles El Sistema is regarded as a 'mafia' operating in the cultural field. However, the left project of remaking culture, as Sotillo puts it, is a lot more attractive when people have the intellectual and personal respect of their peers in the cultural arena and some key administrators or politicians. It is a much more arduous process in the face of professional snobbery, political violence and poverty. It is this risky moral space between political virtue and aesthetic virtue which El Sistema keeps apart, while organizational networks like *La Mancha* attempt to draw them together.

Conclusion: what harm can it do?

According to Abreu the discipline of the classical orchestra is 'the most beautiful expression of national unity' (quoted in Borzacchini 2010: 60). Speaking in a similar vein he says 'art ought to be "a flag bearer of citizenship"'.[35] If it can be said that El Sistema also demotes the significance of a broad musical and therefore a broader cultural education, this is focused on those who are least able to resist – the poor. Sistema vigorously promotes a ready-made culture, it does not promote public accountability in education or culture. Yet Abreu and other proponents of the organization are also explicit about their desire to mould good citizens. Proponents of the Scouting movement and other youth organizations with similar social engineering ideas about building harmonious group identity, national unity and so forth may fairly claim to perform a valuable role despite their ideological orthodoxy. However, this orthodoxy also prevents these organizations, now in decline in Britain and elsewhere, from calling on much support from the public purse because the public interest in supporting 'culture' at all depends first and foremost on the recognition that it is infinitely more nuanced and radical than the discursive spectrum of representative party politics. This is equally true in Venezuela and the UK. Therefore organizations which promote cultural orthodoxy in the name of social unity are still open to conscientious official questioning when they look for public money. Notwithstanding the elitism and barely accountable patronage system run by the Arts Councils in Britain (now abolished in Scotland under nationalist government) as arm's-length bodies, the Arts Councils were intended to be the guardians of this cultural gap between representative politics and everyday life.[36] El Sistema is an arts project for all to see and hear. However, by playing down this obvious fact and proclaiming its social role instead the organization appears to escape many key questions of the public interest in culture.

The idea of 'sowing the oil' and latterly resource-driven nationalism has boosted Abreu's project immeasurably. Indeed, to observers concerned about cultural democracy or 'cultural eco-systems' El Sistema may look like the equivalent of an oil slick. Given the flaws in the Sistema rationale it is worth asking why governments at the opposite ends of neoliberalism have gone out of their way to support the organization. Here critical aesthetics may expand on socio-political analysis. In his book *O Som e o Sentido*, the Brazilian composer and writer José Miguel Wisnik describes the following characteristics of orchestral musical performance:

> The inviolability of the written score, the horror of making
> mistakes, the exclusive use of melodically tuned instruments, the
> silence demanded from the audience, all makes one hear traditional

erudite music as representative of a sonorous drama of melodic-harmonic tones within a chamber of silence, wherein noise would ideally be excluded (the bourgeois concert theatre turned out to be this chamber of representation). Such representation depends on the possibility of enclosing a universe of sense within a visible frame, a box of verisimilitude that must be, in the case of music, separated from the paying audience, and ringed in silence.[37]

Politics is also about the dramatic and highly ritualized control of sound and voices. More so because strategic political debates and decision-making increasingly take place in international political arenas where politicians and members of unelected power elites congregate behind closed doors. While they are not always 'ringed in silence', ordinary people have little or no access.

Arguably the G8 and the offices of the European Union are Scotland's delimiting political structures more than UK government. From an avowedly anti-imperialist position, Venezuela has been at the forefront of the development of regional power structures such as the Community of Latin American and Caribbean States (CELAC), founded in Caracas in 2011. Part of the novelty of the Sistema model is that the poor were symbolically present when Gustavo Dudamel and the Simón Bolívar Youth Symphony Orchestra gave a concert for the 2011 CELAC summit. Not even the now notoriously misleading 'Make Poverty History' campaign led in 2005 by rock stars like Bob Geldof and Bono during the G8 Gleneagles summit in Scotland could engineer this level of inclusive symbolism.[38] Orlando Chirino, the Trotskyist trade union leader and former Chávez supporter turned opponent, observes that these events exclude labour and social movements from meaningful participation. With the history of the European Union in mind, it is easy to see that the apparently progressive socio-economic aims of regional economic blocs can ossify institutionally and be reversed. They give way to a blatant capitalist technocracy, which in the case of the EU does not merely encourage an orchestrated democracy; periodically it has been seen to demand it.[39] As vice-minister for the promotion of the cultural economy, Humberto González was interviewed in this study before the death of Hugo Chávez. Symptomatic, perhaps, of the politics of excessive presidentialism, González saw Chávez's leadership role as the main defence against CELAC going the same way as the EU.

Perhaps reflecting more of his own Trotskyist background, González also spoke critically about the competition for government money coming from oil rents and the difficulty of fostering independence and pluralism in such a mono-economy. In general terms González was candid about 'mistakes' made by the government.

The development of a homogenizing force in education and culture like El Sistema ought to be seen in the historical context of Venezuela's oil economy. If looked at as a subconscious model for political participation and representation then 'the inviolability of the written score' that Wisnik describes closing a wider 'universe of sense within a visible frame' fits very well with ritualized political events of inter-elite horse trading mentioned above. Consciously or not, the attempt to create model citizens who do not question the terms of inclusion doesn't stop there, though. Globally El Sistema also fits into the everyday 'grassroots' political phenomenon known as NGOism or NGOization, namely 'the professionalization and institutionalization of social action' (see Choudry and Kapoor 2013; Opoku-Mensah et al. 2007). For all the talk of rights which emanates from non-governmental organizations they are not generally conducive to active trade unionism among their workforce and sometimes virulently opposed to it (see Sharma 2006). Because NGOs are advancing neoliberal government policies by creating bridges between state and private business interests, critics see their non-governmental tag as something of a misnomer which conceals various levels of dependency. The development of El Sistema and the state umbrella organization FundaMusical Bolívar is an example of the bewildering variant known as a GONGO – a Government Organized Non-Governmental Organization. Several studies have shown that when NGO-type organizations do side with labour activists and social movements their involvement tends to weaken rather than strengthen popular power because of their chronic external dependencies and professionalized outlook (see Folorunso et al. 2012).

Conservative GONGOs like El Sistema operating in the good-faith economy and reproducing notions of the deserving poor and treating children as human capital are not merely an alternative to people's organizations and authentic democratic action. I have argued the Sistema model ought to be seen as a serious *threat* to both. This is the case because unless El Sistema changes beyond recognition this type of organization cheapens the language of solidarity and unity it often deploys; it is already implicated in the atomization of politics and the 'colonization' of social space that typify NGOization. The most important weapon the poor have in the fight against poverty, inequality and the socio-economic precarity which is spreading upwards as a result of capitalism is anything but meritocratic notions of upward mobility or the development of income-generating human capital. If the history of socialism in the twentieth century is a guide, the crucial weapon the poor have to hand is the political capital gained from self-organization and class solidarity – the very things Abreu's model of cultural mobilization counters aesthetically and socially.

Coda

Critical realists point out that when studying crime and related social ills inequality and fragmentation within groups are as important to consider as the wider structural inequalities in society. For example, William Wilson, cited above, sees the black consciousness movement in the United States in problematic terms. He argues that the growth of racial identity politics in the late 1960s and 1970s obscured the economic fragmentation of black communities along class lines. This began to take place before theories of exploitation were displaced by theories of exclusion. Therefore democratic rights and the struggles of the labour movement were still entwined in many cases. On 3 April 1968, the day before he was assassinated, the civil rights leader Dr Martin Luther King made his last and most famous 'mountain-top speech' before the striking sanitation workers in Memphis, Tennessee. He stressed that direct action always needs to be 'anchored with the power of economic withdrawal'.[40] However, the assassination of black leaders like King and Malcolm X (1925–65) eroded class and economic thinking and hardened racial thinking, just as Malcolm X (its most celebrated black proponent) had abandoned it. Although blackness was then to mask an emerging middle-class politics based on commerce and individual self-advancement, it also articulated a certain understanding of exploitation rooted in the history of slavery. For these reasons it is worth comparing Abreu's rhetorical address at the outset of this chapter with the rhetoric of that earlier movement in the United States, also in Los Angeles, also in the context of music and also under the influence of religious idealism.

In 1972 the young Reverend Jesse Jackson gave the invocation at the Wattstax Music Festival to an audience of around 100,000 people in the Los Angeles Memorial Coliseum. Towards the end of Jackson's invocation transcribed below, the people rose to recite the black national litany. The event marks a historic moment in the emergence of the black culture industry and the commercialization of blackness. Yet it owes everything to confronting exploitation and promoting self-organization.

> This is a beautiful day, it is new day, it is a day of black awareness, it is a day of black people taking care of black people's business, today we are together, we are unified and on one accord, but when we are together we've got power, and we can make decisions. Today on this programme you will hear gospel and rhythm and blues, and jazz. All those are just labels, we know that music is music, all of our people have got a soul, our experience determines the texture, the taste and the sound of our soul. We say that we may be in the slums but the slums are not in us, we may be in prison but the prison is not in us. In Watts we have shifted from burn baby burn to learn baby learn. We have shifted from having a seizure about what the man got to seizing what we need. We have shifted from bed bugs and dog ticks to community control and politics. That is why we've gathered today to celebrate our homecoming and our own sense of somebodyness. That is why I challenge you now to stand together and raise your fists together and engage in our national black litany. Do it with courage and determination:
>
> I am somebody! I may be poor but I am somebody! I may be on welfare but I am somebody! I may be unskilled but I am somebody! I am black, beautiful, wild and must be respected, I must be protected. When we stand together what time is it? Nation Time! When we stand together what time is it? [...]⁴¹

From Jackson's invocation we can witness one understanding of power. It is based on large numbers of people acting in concert. As we discuss in the concluding chapter, power is much more complex than a numbers game, and among other things it involves economic leverage. Historically, relatively small numbers of workers have counteracted the power of even smaller numbers of capitalists. It is such challenges which contemporary organizations like El Sistema negate, devoted as they are to the ideological unmaking of the working class from the earliest years. Our times call for the cultural and political renewal of the labour movement and the recovery of class-based understandings of exploitation. This appears to be *the* imperative for self-organization and genuinely public policy today. Lessons from the recent past, not least the massacre of striking workers in post-apartheid South Africa and the rise of the far right internationally, suggest that we have no other option.

Notes

1 Translation by Rodrigo Guerrero. See www.youtube.com/watch?v=eEYSfn2-P9k, accessed July 2013.

2 The Simón Bolívar Music Foundation (FundaMusical Bolívar) is the state state governing body of the National System of Youth and Children's Orchestras and Choirs of Venezuela, widely known as El Sistema. The State Foundation for the National System of Youth and Children's Orchestras of Venezuela (FESNOJIV) was founded in 1996 and renamed FundaMusical Bolívar in 2011. This 'umbrella organization' is attached to the Ministry of People's Power for the President's Office of the Bolivarian Republic of Venezuela. In this chapter I also refer to the sister organization Sistema Scotland and the international organizational model which has less formalized connections to the state simply as Sistema. For the official description of El Sistema see www.fesnojiv.gob.ve/en/el-sistema.html, accessed July 2013.

3 See, for example, *The Finland Phenomenon* (*El Sistema Educatiu Finés*), documentary by John A. Compton, New School Films, www.youtube.com/watch?v=GhVpAuz9ccA, accessed December 2013.

4 UK Government statistics quoted by the BBC, www.bbc.co.uk/schools/studentlife/debate/2008/42_state_vs_private_school.shtml, accessed July 2013.

5 See Bourdieu (1992), also available online at www.variant.org.uk/issue32.html, accessed July 2013.

6 Alberto Dao in Borzacchini (2010: 213). See also Govias (2011) on the five principles.

7 From an interview with Abreu by Alan Yentob, in 'How music saved Venezuela's children', BBC *Imagine* series, programme produced by Alan Yentob and Janet Lee. Transmitted 18 November 2008. See www.youtube.com/watch?v=43tqQhOTCgQ, accessed July 2013.

8 Tunstall (2012: 36). There is some variance in the numbers given for El Sistema. See 'El Sistema' entry on Wikipedia, en.wikipedia.org/wiki/El_Sistema.

9 See 'La música popular merece tanto apoyo como el sistema de orquestas', interview with Cecilia Todd, *Ciudad CCS* (Caracas: Fundación para la Comunicación Popular CCs de la Alcaldia de Caracas), 12 December 2011, www.ciudadccs.info/?p=240320, accessed July 2013.

10 From an interview on the CBS *60 Minutes* programme, presented by Bob Simon, 13 April 2008, www.cbsnews.com/video/watch/?id=4011959n, accessed July 2013. Another spokesman (Raphael) for El Sistema interviewed on the same programme responds to the question why popular music would not 'work' for poor children. He answers that 'what they have at home on the radio is popular music all the time, their father who drinks every day gets drunk with that music. So you have to give them something different. And when they sit in one of these churches in the orchestra they think they're in another country, another planet and they start changing.'

11 Quoted in 'Venerated high priest and humble servant of music education', by Daniel Wakin, *New York Times*, 1 March 2012. Here Abreu also states

that 'Orchestras and choirs are incredibly effective instruments against violence'. www.nytimes.com/2012/03/04/arts/music/jose-antonio-abreu-leads-el-sistema-in-venezuela.html?pagewanted=all&_r=0, accessed July 2013.

12 See, for example, Francisco Toro, 'Finding George Orwell in Venezuela', *International Herald Tribune*, 16 August 2012, latitude.blogs.nytimes.com/2012/08/16/finding-george-orwell-in-venezuela/, accessed July 2013.

13 From *El Sistema*, film by Paul Smaczny and Maria Stodtmeier, produced by EuroArts Music (2008–11).

14 From *Jose Antonio Abreu: The El Sistema music revolution*, filmed for the Ted Prize, www.ted.com/talks/jose_abreu_on_kids_transformed_by_music.html, accessed July 2013.

15 Borchert (2012: 51). See also Govias (2011), whose account is endorsed by Abreu and which proclaims the meritocratic idea of El Sistema.

16 See Baker (2014). I extend my thanks to Baker for sharing insights from his ethnography in advance of publication.

17 From author's interviews.

18 Nigel Reynolds, 'Nigel Kennedy criticises "greedy" conductors', *Daily Telegraph*, 12 March 2008. See www.telegraph.co.uk/news/uknews/1581463/Nigel-Kennedy-criticises-greedy-conductors.html, accessed July 2013.

19 From 'How did Venezuela change under Hugo Chávez?' posted by Ami Sedghi, *Guardian*, data blog, 6 March 2013, www.guardian.co.uk/news/datablog/2012/oct/04/venezuela-hugo-chavez-election-data#zoomed-picture, accessed July 2013.

20 See Milagros Rodríguez, *Venezuela: la violencia salpica a las escuelas*, infosurhoy.com/es/articles/saii/features/main/2013/03/29/feature- 02?change_locale=true, accessed December 2013.

21 See Guerrero (2010) and Pearson (2009) for well-considered pro-Chávez assessments.

22 See, for example, Eva Golinger and Roberto Briceno-Leon discussion, www.aljazeera.com/programmes/insidestoryamericas/2012/06/20126554927373645.html.

23 See *El Sistema*, film by Paul Smaczny and Maria Stodtmeier, produced by EuroArts Music (2008–11).

24 See Baker (2014). I extend my thanks to Baker for sharing these insights from his ethnography in advance of publication.

25 Gates (2010), p. 108.

26 Young's book was translated into seven languages.

27 By 1992 Pietri regarded the state, not markets, as impediments to all-round development. See 'Not sowing the oil by Arturo Uslar Pietri', translated at the Devil's Excrement, devilsexcrement.com/2005/09/08/not-sowing-the-oil-by-arturo-uslar-pietri, accessed July 2013.

28 The lending members of the IDB are: Austria, Belgium, Canada, China, Croatia, Denmark, Finland, France, Germany, Israel, Italy, Japan, the Netherlands, Norway, Portugal, Republic of Korea, Slovenia, Spain, Sweden, Switzerland, the United Kingdom and the United States.

29 Geoff Baker enquires about the ongoing use of these funds. See geoffbakermusic.wordpress.com/el-sistema-the-system/el-sistema-blog/constructing-el-sistema/development, accessed July 2013. See also CAF news at www.caf.com/en/currently/news/2012/11/caf-approves-$310-million-in-loans-for-cultural-and-urban-development-in-venezuela, accessed July 2013.

30 For a clear examination of the historical overlap of liberal and fascist thought and practice, see Landa (2012).

31 See Lebowitz (2010) review of a book by Iain Bruce.

32 It is often said that Chávez's support for El Sistema was strategic and

for that reason a degree of toleration is evident among people involved in producing Petare TV. Yet nobody interviewed in this study offered an analysis of Chávez's strategy or its possible outcomes for organizations like Petare TV. In October 2013, in keeping with the station's goals of fostering participation and debate to strengthen the revolution, I was given the opportunity to critically discuss El Sistema on air and recall Chávez's early admiration for Tony Blair and the 'Third Way'. I am grateful to Petare TV for this and subsequent discussions which informed this chapter.

33 Mendez gives the example of identifying and closing down illegal bullet factories.

34 From author's conversation with Sotillo Meneses in Caracas.

35 Ibid. See 'Reaching for the Stars' presentation, note 1.

36 The replacement of the Scottish Arts Council by Creative Scotland, an organization widely perceived to lack cultural independence, has been fraught with problems, leading to the resignation of the directors.

37 Translation by Gustavo Borchert. Quoted in Borchert (2012: 50).

38 Hodkinson (2005), also available online at www.redpepper.org.uk/G8-Africa-nil/, accessed July 2013.

39 See, for example, the controversial Lisbon Treaty, en.wikipedia.org/wiki/Irish_European_Constitution_referendum, accessed December 2013.

40 See AFSCME union website at www.afscme.org/union/history, accessed July 2013.

41 See 'Wattstax' documentary by Mel Stuart (1973), www.youtube.com/watch?v=dwuAwSbxbNk.

References

Baker, G. (2014) *El Sistema: Orchestrating Venezuela's Youth*, Oxford: Oxford University Press.

Balibar, E. and I. Wallerstein (1991) *Race, Nation, Class – Ambiguous Identities*, London: Verso.

Barber, P. (1996) *The World of the Castrati, the History of an Extraordinary Operatic Phenomenon*, London: Souvenir Press.

Boltanski, L. and E. Chiapello (2005) *The New Spirit of Capitalism*, London: Verso.

Borchert, G. (2012) 'Sistema Scotland: a critical inquiry into the implementation of the El Sistema model in Raploch', MA dissertation, Department of Music, University of Glasgow.

Borzacchini, C. (ed.) (2005) *Venezuela Bursting with Orchestras*, Caracas: Banco del Caribe.

— (2010) *Venezuela en el cielo de los escenarios*, Caracas: Fundación Bancaribe.

Bourdieu, P. (1977) *Outline of a Theory of Practice*, Cambridge: Cambridge University Press.

— (1992) 'The left hand and the right hand of the state', Reprinted in *Variant*, 32, Summer 2008.

Choudry, A. and D. Kapoor (2013) *NGOization – Complicity, Contradictions and Prospects*, London: Zed Books.

Cleaver, H. (2000 [1979]) *Reading Capital Politically*, Leeds and Edinburgh: Anti/Theses and AK Press.

Crisp, B. and D. Levine (1998) 'Democratizing the democracy? Crisis and reform in Venezuela', *Journal of Interamerican Studies and World Affairs*, 40(2), Center for Latin American Studies, University of Miami.

Dangl, B. (2010) *Dancing with Dynamite: Social Movements and States in Latin America*, Oakland, CA: AK Press.

Ellner S. (2009) *Rethinking Venezuelan Politics: Class, Conflict, and the Chávez Phenomenon*, London: Lynne Rienner.

Fernandes, S. (2011) 'Radio Bemba in the

age of electronic media', in D. Smilde and D. Hellinger (eds), *Venezuela's Bolivarian Democracy – Participation, Politics and Culture under Chavez*, Durham, NC: Duke University Press.

Folorunso, F., P. Hall and O. Logan (2012) 'A country without a state? Governmentality, knowledge and labour in Nigeria', in J.-A. McNeish and O. Logan (eds), *Flammable Societies – Studies on the Socio-Economics of Oil and Gas*, London: Pluto Press.

Gates, L. (2010) *Electing Chavez – The Business of Anti-Neoliberal Politics in Venezuela*, Pittsburgh, PA: University of Pittsburgh Press.

Govias, J. (2011) 'The five fundamentals of El Sistema', *The Canadian Music Educator*, Fall, Toronto: Canadian Music Educators Association.

Guerrero, M. E. (2010) *12 dilemas de la Revolución Bolivariana*, Caracas: Fundación Editorial el Perro y la Rana.

Haque, M. S. (2002) 'The diminishing publicness of public service under the current mode of governance', *Public Administration Review*, American Society for Public Administration.

Hazel, L. (2007) 'Venezuela: overcoming historical oppression through education', Carlisle, PA: Dickinson College Student Papers, www2.dickinson.edu/departments/commstud/mosaic/destinations/venezuela/Venezuela%2007/studentprojects07.html, accessed July 2013.

Hobsbawm, E. (2000) *Bandits*, London: Weidenfeld & Nicolson.

Hodkinson, S. (2005) 'G8 – Africa nil', *Red Pepper*, November.

IDB (Inter-American Development Bank) (2007) *Venezuela. Proposal for a Loan for a Program to Support the Centro de Acción Social por la Música, Phase II*, Washington, DC: Inter-American Development Bank.

Jones, O. (2012) *Chavs – the Demonisation of the Working Class*, London: Verso.

Landa, I. (2012) *The Apprentice's Sorcerer – Liberal Tradition and Fascism*, Chicago, IL: Haymarket.

Lasker-Wallfisch, A. (2000) *Inherit the Truth: A Memoir of Survival and the Holocaust*, New York: St Martin's Press/Thomas Dunne.

Lebowitz, M. (2010) 'Exploring the dialectic of the Bolivarian Revolution', *Monthly Review*, 61(9), February.

Logan, O. (2007) 'Living on oil under democracy: from Texas to Patagonia and home', *Variant: The Oil Issue*, Spring.

Martinez, C., M. Fox and J. Farrell (eds) (2010) *Venezuela Speaks!: Voices from the Grassroots*, Oakland, CA: PM Press.

Opoku-Mensah, P., D. Lewis and T. Tvedt (eds) (2007) *Reconceptualising NGOs and Their Roles in Development: NGOs, Civil Society and the International Aid System*, Aalborg: Aalborg University Press.

Pearson, T. (2009) *Crime in Venezuela: Opposition Weapon or Serious Problem?*, Venezuelanalysis.com, 30 March.

— (2010) *UNESCO: Education in Venezuela Has Greatly Improved*, Venezuelanalysis.com, 27 January.

Sadie, S. (ed.) (2001) *The New Grove Dictionary of Music and Musicians*, 2nd edn, vol. 1, Oxford: Grove/Oxford University Press.

Sanguinetti, P. and L. Villar (eds) (2012) *Public Finance for Development: Strengthening the connection between income and expenditure*, Caracas: CAF Development Bank of Latin America.

Schiller, N. (2011) 'Catia sees you – community television, clientelism and the state in the Chávez era', in D. Smilde and D. Hellinger (eds), *Venezuela's Bolivarian Democracy – Participation, Politics and Culture under Chavez*, Durham, NC: Duke University Press.

Sen, A. (1999) *Development as Freedom*, Oxford: Oxford University Press.

Sharma, A. (2006) 'Crossbreeding institutions, breeding struggle: women's employment, neoliberal governmentality, and state (re)formation in India', Middletown, CN: Social Sciences Division II Faculty Publications, Wesleyan University.

Stein, S. and B. Stein (1970) *The Colonial Heritage of Latin America – Essays on Economic Dependence in Perspective*, New York: Oxford University Press.

Tunstall, T. (2012) *Changing Lives: Gustavo Dudamel, El Sistema, and the Transformative Power of Music*, New York: Norton.

Voltaire (1819) Œuvres complètes de Voltaire, vol. III: *Essai sur les moeurs et l'esprit des nations*, Paris: L'imprimerie Crapelet.

Watkins, K. et al. (Global Monitoring Report Team) (2010) *Reaching the Marginalised*, Paris: Unesco.

Wilson, W. (1987) *The Truly Disadvantaged – the Inner City, the Underclass and Public Policy*, Chicago, IL: University of Chicago Press.

Wisnik, J. M. (1989) *O Som e o Sentido: uma outra história das músicas*, São Paulo: Companhia das Letras.

Young, M. (1994 [1958]) *The Rise of the Meritocracy*, New Brunswick, NJ: Transaction Publishers.

10 | LATIN AMERICA *TRANSFORMED*?

John-Andrew McNeish

> Amerindians in the South and Latinos in the North, among others, are now telling the rest of the world that we can have globalization on our own terms, rather than on those of supposedly unstoppable forces. They remind us that we are freer than we think, to reinvent and reconstruct. (Gardiola-Rivera 2010: 4)

Introduction

In perhaps one of the most provocative contributions to recent decolonization literature, Gardiola-Rivera (2010) poses the following question and proposal in a recent book title: *What if Latin America Ruled the World? How the South will take the North into the 22nd century.* Tracing in his book a kaleidoscopic timeline from prehistory to the recent past of social movement activism, crisis to economic growth and Latino immigration to the North, Gardiola-Rivera concludes, as the quote above suggests, Latin America and its inhabitants are now ready to realize their own potential. For a non-Latin American the optimism and pride reflected in these words must be respected, even admired. Certainly the position taken by Gardiola-Rivera is one that has at least in part been shared by many other intellectuals in Europe and Latin America. Indeed, for the international political left Latin America's economic growth and political development over the last decade have been seen as providing the region and the world with guidance for the future. Industry leaders, foreign statesmen,[1] media coverage[2] and scholars alike flag examples from the region as emerging powers (Leira 2014; Reid 2014), innovators in progressive social and economic policy[3] and/or the vanguard of socialism for the twenty-first century.[4] Indeed, to the right and left of the political spectrum there is broad agreement that economic and political development has taken hold in Latin America. However, while there is broad recognition of significant change, it is also clear, with an analytic eye to the details of Latin America's *great transformation* over the last decade, that a somewhat more nuanced and more cautious reading of change should be considered.

With particular focus on the region's persisting economic dependence on natural resource extraction and the fundamentally energy-fuelled nature of recent developments, the chapter proposes that the region remains a captive of its colonial past. Current economic conditions, political tensions (evident in media coverage of current events in Venezuela, Bolivia and Brazil) and now myriad local, regional and national socio-environmental struggles focused on energy resources draw attention to the political dynamism of the region, but also reveal the fiction of a globalization made on any region's terms, and of any real harmony in Latin American efforts to rethink their relationships to the world. Recognizing these dynamics, I suggest in this chapter an approach that balances Latin America's developmental wins with a more critical consideration of the region's persisting challenges and reliance as an enclave economy.

While inspired by recent decolonization debates (to which Gardiola contributes) and in particular contributions on the discursive nature of development (Mignolo 2005, 2011; Moraña et al. 2008; Quijano 1992, 1998; Escobar 1999, 2010), I emphasize in this chapter the need to give greater emphasis to the resource-dependent nature of Latin American economy and society. In line with Mitchell (2013), I maintain here that in Latin America, as in the Middle East, energy extraction and the development of political formation are mutually constitutive. Hence it is not energy resources or related *rent-seeking* per se, but rather their role in competing projects to define power, justice and development which generates varying levels of conflict. An account of energy politics also reinforces the constructed nature of Latin America as a region and the globalized nature and reliance of its politics, economics and intellectual life. Rather than a heroic autonomous Latin America, the complexities of energy politics in the region reveal instead the dominance of Occidentalism, i.e counter-discursive reworkings of the Western imaginary in tension with persisting material and epistemological impositions (Coronil 1996). As such, I argue that to rethink Latin America as an energetic society it is important to draw together the dual political-economic and epistemological nature of contests over development. I also suggest that such an approach reveals with increased empirical clarity the relationships between natural resources and competing expressions of sovereignty, and their combination as the basis of contestation in the region.

Aiming to trace a course that takes both the political-economic and epistemological nature of development seriously, and resource sovereignty as an inroad into understanding regional energy politics, the chapter moves from an initial exploration of Latin America's recent decade of economic growth and development to a consideration of recent critical insights. Following an initial outline of the main features of energy and economic development in Latin America, I use my research on recent socio-environmental struggles in Bolivia as a springboard into a more nuanced exploration of the wider limitations to the use of energy as a means to cultivate sustained development and political power. As such, the chapter moves through multiple scales in order to analyse the linkages between energy and development, and to highlight the inherent tensions between competing versions of sovereignty and regionalism. The chapter concludes by arguing that while the region's energy politics reveals the flaws of recent expectations of a new post-neoliberal Latin American era, it nonetheless bolsters evidence of creative tendencies that test the limits of standard recipes for development.

Energetic development in Latin America

In a period when Northern economies have suffered a dramatic slowdown, Latin America has re-emerged as one of the world's powerhouses of economic and social development.[5] Driven by a steady increase in international commodity prices,[6] Latin American energetic politics over the last decade can furthermore be characterized by a continent-wide push to open new frontiers of resource exploration and extraction. In South America almost half of foreign investments (43 per cent) are now in natural resources (ECLAC 2013).[7] Given that Latin America is home to one of the largest reserves of oil and gas outside the Middle East, the hydrocarbons sector is particularly implicated in this boom.[8] Foreign direct investment (FDI) in many countries (e.g. Peru, Panama, Guatemala, Colombia) in the region has also increasingly focused on the extractive sector (Tissot 2012).

A parallel development has taken place in domestic energy consumption as the general level of wealth and wages in the region has grown. Indeed, the rising foreign and domestic demand for raw materials and energy stimulated in Latin America an average GDP growth rate of between 4 and 5 per cent in the years 2003–12 (ECLAC 2013: 7). With rising levels of wealth and middle-class consumption, the internal industrial and private demand for energy and in particular fossil fuels (diesel and petroleum) for transport has also rapidly expanded (Tissot 2012: 6). In the last decade, overall energy consumption has increased at a rate of 3 per cent per year. At the same time Latin America's primary energy matrix has not changed substantially, with fossil fuels[9] accounting for 74 per cent of energy needs. With a population of nearly 400 million and a large unsatisfied demand for public transport, Latin America has become the world's fastest-growing vehicle market (50 per cent in the last five years). Nearly 80 per cent of the region's population lives in urban centres, but infrastructure deficits leave most Latin American countries without viable mass transportation systems, making private motor vehicles the preferred mode of transport. While other non-fossil-fuel energy sources make up a modest 4 per cent of the region's energy mix, significant increases have been made in renewable energy production and consumption. In 2010, the world's ethanol production was close to 1.5 million bpd, with Latin America contributing around 31 per cent of that.[10] The International Energy Agency (ibid.: 18) forecasts that Latin America's ethanol production will also increase to 34 per cent by 2016. Significant increases in the number of hydroelectric power projects have also been seen in recent years in response to increases in electricity demand throughout the region. Indeed, it is as a result of the dominance of hydropower in regional electricity production that Latin America has the cleanest overall energy matrix in the world. Projects for solar power, wind and geothermal energy production are also rapidly expanding throughout the region as power consumption grows.

According to ECLAC, the positive economic and social outcomes of the recent extractive boom can partly be attributed to the macroeconomic management of resource revenue that 'created sufficient fiscal space to manage the effects of the global financial crisis without jeopardizing fiscal sustainability' (2013: 7). The introduction of macroeconomic policies aimed at lowering debt and increasing public finances combined with higher primary export prices strengthened economic stability and provided resources for implementing anti-poverty programmes and increasing access to basic public services. According to an ECLAC/OECD report (ibid.: 7), solid economic performance created the possibility for 'transforming the state, enabling the adoption of ambitious public policies that lock in the prospect of long-term development and mitigate short-term risks'. The linkage of state financing gained from expanded extractive activities to economic diversification and a range of social programmes and cash transfer schemes – of which Brazil's *Bolsa Família* programme might be considered the strongest representatives – has furthermore had significant results in reducing poverty levels and inequality (Soares 2013; Lustig et al. 2013; Gudynas 2011a: 195).[11]

While not without critical qualification, a number of authors link the reduction in inequality in the region with left-wing mobilization for more inclusive fiscal policies and institutions (Maxwell and Hershberg 2010). In contrast to previous neoliberal administrations, the governments of the 'left wing wave' or 'pink tide' (not quite red) are widely recognized by analysts and international institutions alike (including the World Bank, IMF and OAS) as having generated solid economic growth and reduced inequality in the region through their introduction of redistributive and participatory measures. *The Economist* has even attempted to dub the last ten years the Latin American Decade,[12] pointing to prudent management of commodity revenues, increased FDI and the increasing demands for Latin American resources in Asia (Reid 2010). Moderate governments in Brazil and Chile are seen to have generated solid economic growth, reduced poverty and inequality and created innovative and fiscally sound programmes, while respecting the basic principles of the free market and liberal democracy. More radical governments, such as those of Hugo Chávez in Venezuela and Evo Morales in Bolivia, while remaining rhetorically controversial, are also recognized for having had some success in cutting severe poverty levels and generating needed state financing through increased levels of state intervention, the promotion of national resource ownership and popular participation (Weyland et al. 2010).

What if Latin America did not rule the world?

It is on the crest of the wave of clear developmental and political change that Gardiola-Rivera and other intellectuals in Latin America and Europe have hailed the region as an example for the world. However, while recognizing that important economic and social transformations have taken place in Latin America, some analysts on both the right and left of the political spectrum are more sober in their evaluation of the development gains in the region over the last decade. Indeed, it is principally with reference to the manner in which development in the region has been founded on energetic resource extraction that caution is expressed. Instead of sustaining an image of Latin America as dynamic, these writers together rather warn of the manner in which the region remains mired in and captive to the principles of the global economic system and its own lack of economic diversification.

One set of writers warns that although high commodity prices have driven economic development in the region over the last decade, rent-seeking behaviour inherent in extractive economies and resource nationalism in particular are responsible for a cooling in growth rates (from 6.6 to around 4 per cent) over the last two to three years. Growing demand for energy resources both within and outside the region, coupled with the decline of conventional petroleum economies and reserves, encouraged governments in the region to renationalize natural resources and/or demand higher levels of rents and taxes from private companies involved in extractive activities. This coincided with the ascendancy of new left-wing regimes throughout the region which used natural resource wealth as a tool to stimulate national development, project geopolitical influence and replace market liberalization policies adopted in the previous decade. While new left-wing resource nationalist governments were able to significantly increase their political legitimacy and the availability of state funds through new models of state and mixed ownership and taxation, they have been less successful in attracting the financial and technological investment needed to sustain production and exploration levels.

In an apparent repeat of the conditions that ended Latin America's earlier experimentations with import-substitution industrialization (ISI), analysts argue that as a result of the failure of governments to attract sufficient levels of these investments the region's extractives and energy resource boom is now entering a period of stagnation (Talvi and Munyo 2013). According to researchers at the Brookings Institution, while commodity prices remain high the production possibilities in the region are being exhausted because 'improvements in physical and technological infrastructure and human capital may not have kept up with the strong output performance of the past few years'.[13] Indeed, substantial investment throughout the petroleum value chain is said to be required if Latin America is to meet its growing demand for energy and oil products in particular (Tissot 2012: 16–17). According to the International Energy Agency (IEA), Latin America needed to invest nearly US$3.5 trillion in total energy infrastructure between 2011 and 2013 and approximately US$1.9 trillion in the oil sector alone. The Organization of the Petroleum Exporting Countries (OPEC) forecasts that Latin America's refining sector alone needs more than US$100 billion in investment over the next twenty-five years. Recognizing the limitations to the further development of the hydrocarbons sector and the need to expand the total energy matrix to match demand, some analysts emphasize the possibilities of expanding the renewables sector (ibid.).

While some writers question the autonomy and sustained growth of the region's energetic productivity, others question the negative impacts of the region's capitalist extractive model on society and the environment. As Haarstad (2012: 2) has commented, amid the optimism and achievements of regional development 'it is critical to remain attentive to the enduring challenges that are involved when various social actors make claims to revenues, territorial sovereignty, and participation within expanding extractive economies'. The economic boom in Latin America has not only produced new developmental and political opportunities. It has produced an increase in rent-seeking activity and the number and intensity of social and environmental conflicts, primarily between extractive companies and local communities. As such the claims of developmental progress made in recent years are characterized as 'bad development' given their reliance on finite resources and an incapacity to really improve the quality of human life or protect the natural environment (Tortosa 2001). Here governments of all political colours and ideologies are implicated. Indeed, while Gudynas (2010) characterizes *neo-extractivism* as the basis of a new form of regional developmentalism practised by new left governments committed to the increased redistribution of power and wealth, they are also seen to maintain much of the earlier liberal economic model based on the extraction of natural resources.

In this light neo-extractivist economies can, according to some critics, be seen as just a recent update in the history of economic imperialism and capitalist primitive accumulation in Latin America. Latin American governments promote global capitalist integration and the fragmentation of national territories through the creation of extractive enclaves, while negative social and environmental impacts remain unaddressed. Here the effects of degradation, contamination and pollution caused by extractive activities on the people's health and the environment are highlighted. Gudynas (2011b) also points out that neo-extractivism offers very few real economic benefits in that 'the externalization of the social and environmental costs represents a severe economic cost, exaggerates the economic dependency, reducing the capabilities of the diversification of production', and creates limited employment (ibid.: 267; Acosta 2009). Bebbington (2009a; Bebbington and Bebbington 2011) similarly emphasizes a continent-wide push to open frontiers for extracting hydrocarbons, mining, producing biofuels, harvesting timber and investing in agro-industry. In this push, he argues, progressive governments are just as likely as neoliberal governments 'to tell activists and indigenous groups to get out of the way of national priorities, just as likely to allow extractive industry into fragile and protected ecologies, and just as determined to convince indigenous peoples that extractive industry is good for them too, without fulfilling their rights'. He also argues that as extractive activities have increased, 'the environment has become an increasingly important domain of contention and social mobilization, becoming both a vehicle and an objective of contentious politics, influencing the way in which politics is organized and performed' (Bebbington 2009a).

Responding to a picture of growing socio-environmental contestation in the region linked to extraction and infrastructural activity, Gudynas (2011b) and other political ecological and post-colonial thinkers (Mignolo 2011; Moraña et al. 2008; Escobar 2005) argue for the establishment of an alternative post-extractivist economic model. Recognizing that the alternatives offered by contemporary development thinking are insufficient in general – and particularly inadequate for dealing with extractivism – Gudynas argues that we should jettison conventional ideas of development. Drawing on the wider post-developmentalist critique that emphasizes its discursive and neo-imperialist formation (Escobar 2005; Quijano 1992, 1998), he suggests it is necessary to go farther and think about alternatives to the very idea of development. To specify an alternative, Gudynas (2011a) draws inspiration, as many Latin American scholars have been doing (Fatheuer 2011; Choquehuanca 2010), from the constellation of ideas referred to in contemporary pluri-legal and constitutional debates in Andean countries such as Bolivia and Ecuador regarding indigenous ideas of *buen vivir* (loosely translated as *living well*). He suggests (2011a) that *buen vivir* can be characterized by its critical approach to the ideology of progress and the expression of this in contemporary development as economic growth, the intense exploitation of nature and the corresponding material interventions. *Buen vivir* is characterized as an idea that aims to ensure people's quality of life, in a broad sense that goes beyond material well-being (to include spiritual well-being) and the individual (to include a sense of community), as well as beyond anthropocentrism (to include nature). Under *buen vivir*, the values inherent in nature are recognized, and therefore also the duty to maintain its integrity at both the local and the global level. This perspective aims to transcend the dualism that separates society from nature, as well as break with the linear idea of history that assumes our countries must imitate the lifestyles and culture of Northern industrialized nations (ibid.: 444).

Rage along the road to development

Despite the acknowledgement of general increases in prosperity in the region over the last decade, there are, then, two main critiques that are made of the extractive energetic model shared by most Latin American countries. On the one hand critics of the changes resulting from resource nationalism emphasize the economic investment and technical limitations of countries in the region, and on the other hand critics of the persisting extractive economic model emphasize the social and environmental damage it continues to cause. Here both the political-economic and epistemological difficulties of the region's energetic model are emphasized, but largely kept separate from one another. In the remaining pages of this chapter I want to demonstrate that while these sober evaluations of Latin America's energetic and extractive model are a necessary antidote to more heady claims of unbridled transformation, supporters of a more political-economic or epistemological ecological line would be well served to place their perspectives in tighter tension with one another. As such I echo the suggestion by Castree (2002: 111) that the seeming stand-off between political economy and political ecology is misleading. Indeed, it is my contention that such a balanced (dual) approach adds to an understanding of the current energetic context of Latin America by revealing more fully the dynamic workings of the socio-economics of energy, i.e. the operation of resource sovereignty. Allowing for a more comprehensive 'mining' of the details of recent cases of energetic politics in the region, such a perspective explodes the myth of a singular and pure Latin American rationality, just as it does ideas of Latin America as an autonomous economic and political region.

Although Bolivia might be viewed as an exceptional case for different reasons (topography, political history, ethnicity, the longevity of its resource extractive experience, etc.), the TIPNIS case – a socio-environmental protest – can be seen as sharing many similar features and dynamics to those witnessed following the extractives boom in other parts of Latin America in recent years. Although clearly the product of a particular history and social relations, common to other examples found elsewhere (see, in this volume, Braathen, Reyes, Borchgrevink, Ødegaard, Boyer et al.), the case demonstrates the linked material and epistemological complexities of energetic politics, i.e. the bracketing together of material and social resource claims and interests, or what we have called resource sovereignty (McNeish and Logan 2012 and Introduction). While centred on a particular territory of contestation, the TIPNIS case also shares a multilevel nature in common with many of the other contexts studied in this book. Here local socio-environmental struggles are tightly intertwined with national political processes, if not questions and discussions of the terms of wider regional and international economic development. It is also a case that I know with some degree of intimacy given my field research on the case (McNeish 2013a, b).

In short, the TIPNIS case refers to the political crisis resulting from a series of indigenous protests focused on a project to build a road through a national park and indigenous territory in the lowlands of Bolivia, i.e. the Territorio Indígena y Parque Nacional Isiboro Secure. Formally, integration and development were the two arguments made by the Morales government as the basis for the construction of Tramo II of the Villa Tunari–San Ignacio de Moxos road (300 kilometres). What was not initially made transparent by the government was that there were a number of other interests that seriously impinged on the park's territorial integrity and are recognized by local people as motivating the government's decision to build a road. From the 1970s onwards, Aymara and Quechua campesino colonizers from the Altiplano and high valleys began to settle in the area south of the park, close to the city of Cochabamba and adjacent to the coca-producing Chapare area – a product of structural adjustment. With time, their numbers increased, and together with large-scale cattle farming farther east placed increasing pressure on the lands of the natural reserve. To the west of the park territory several blocks of natural gas have been identified and under earlier governments contracts were signed with

Repsol for the exploitation of these fields. Under the current nationalized oil and gas industry these contracts have been transferred to the Bolivian–Venezuelan joint venture Petro-Andina (YPFB-PDVSA). According to a report produced by a La Paz-based research institute, a third of the park area was marked out in government development plans as blocks designated for oil and gas production.[14]

The government claimed that the road would be built respecting national legal norms and protections governing environmental social impacts. The 2009 Constitution lays down general requirements for the practice of prior consultation, making it obligatory for the state to carry out such consultation of indigenous peoples and their organizations where non-renewable natural resources on their lands are to be exploited. Still, the details of how to carry out such consultations had not been clarified. Moreover, in the case of TIPNIS the local indigenous populations pointed out that no effort was made by the government to carry out a prior consultation exercise until long after protest had taken place. Recognizing that government actions unmasked undesired impacts and contradicted official statements on promises of integration and development, the communities of the TIPNIS drew on their connection with the Confederation of Indigenous Peoples of Bolivia (Confederación de Pueblos Indígenas de Bolivia; formerly Confederación de Pueblos Indígenas de Oriente Boliviano or CIDOB) to organize opposition to the road project. The protest platform grew as the thirty-four other indigenous organizations in the lowlands of the country, the National Council of Ayllus and Markas of Qullasuyu (Consejo Nacional de Ayllus y Markas del Qullasuyu – CONAMAQ), the principal indigenous organization from the highlands, the national human rights ombudsman, and a group of environmental and human-rights-oriented non-governmental organizations agreed to join the inhabitants of the TIPNIS on the march. On 15 August 2011, around two thousand marchers left the city of Trinidad, the lowland regional capital of the Bolivian department of Beni, to follow a route that would take them 600 kilometres before reaching the capital city La Paz.

The march continued for over forty days before its free passage was blocked by a counter-protest of coca growers, *colonos* (colonizers), on the road between San Borja and Yucomo. Claiming fears of a violent confrontation between the march and the coca farmers, the government sent over five hundred police to the area to 'keep the peace'. On Sunday, 25 September 2011, Bolivian police raided an encampment of several hundred of the indigenous protesters encamped near the town of Yucumo on the border between the Bolivian departments of Beni and La Paz. Within minutes of arriving, the police poured into the midst of the camp. Present at the encampment that day, I personally witnessed that men, women and children were beaten to the ground, and had their mouths taped and their hands tied behind their backs before being hauled away into a fleet of waiting hired buses (McNeish 2013a). Dozens of television crews and newspaper reporters gathered images of bloodied protesters being arrested and dragged by their limbs or hair out of the field for the national and international media. Hundreds of protesters, myself included, avoided arrest by fleeing into the forest behind the camp.

Following the massive media coverage of images and reports from the police raid,[15] there was a massive public outcry in Bolivia in support of the TIPNIS march. In many of the main towns and cities of the country, the streets filled with protesters – marking individual associations' and many of the main union organizations' disgust with government action. From a high of 70 per cent popularity in January 2010, Morales plunged by mid-October 2011 to an average 35 per cent approval rating across the major cities of La Paz, El Alto, Cochabamba and Santa Cruz.[16] Counter-protests against the police raid, organized by indigenous communities in the highlands and lowlands of the country, blocked many of country's main roads. The human rights ombudsman and opposition parties tried to launch legal action against the government, and a series of international organizations, including the Organization of American States (OAS), made statements condemning the police raid. The hundreds of detained protesters from the camp where the police raid occurred were freed from captivity in the regional airport of Rurrenabaque because of the massive turnout of militant support. Following this, the protesters regrouped and continued their march to the capital. They arrived to a heros' welcome in the city of La Paz on 19 October 2011.

A magnified flashpoint

It is not my intention to elaborate much more in this chapter on the empirical details of the TIPNIS case. It is, however, important to emphasize that the TIPNIS case did not end with the entry of the protesters into La Paz in October 2011. Indeed, a series of controversial processes (new laws for territorial intangibility and prior consultation after the fact) and debates (for and against the road, and regarding political responsibility for the police raid) linked to the case have continued to influence Bolivian political life. It is also important to emphasize that the TIPNIS case not only sent shock waves through domestic political debates and policy, but resulted in a significant movement for the re-evaluation (including the defection of intellectual support from the government to the political opposition) of both the operation of the Bolivian state, and of the country as an energetic society. Indeed, importantly for the intentions of this chapter, the TIPNIS crisis has notably become a landmark case in recent academic research on resource and energy politics in the region (Haarstad 2012). In a recent book Kaup (2013: 7) posits: 'contemporary Bolivian struggles and conflicts surrounding processes of neoliberalism are not exceptions. Instead, they are magnified flashpoints of what has happened, is happening, and potentially will happen around the globe.' Indeed, social phenomena such as the TIPNIS case are comparable precisely because they are historically connected and mutually conditioning (McMichael 1990).

To date most of the analysis of the TIPNIS case has focused on the role and response of the government to the TIPNIS march. Indeed, coverage of events by the media, environmental organizations and other NGOs, and by Bolivian and foreign analysts, has largely been focused on discussing and criticizing the response of and action taken by the Morales government. The overwhelming message that has been repeated is that the government's muddled handling of the TIPNIS affair, both in terms of failing to guarantee the security of the protest marchers and in their apparent manipulation of law and corporate connections with coca growers and peasant farmers, is reminiscent of earlier right-wing governments.

Intellectuals within and without the country have widely broadcast the message that nothing has really changed in Bolivia from neoliberal times (Webber 2011). Although the MAS government has repeatedly spoken out against capitalism, attempting to form an alternative socialist state, according to recent critics the reality is that the Bolivian economy functions as part of a world capitalist system (Kohl and Bresnahan 2010). Levitsky and Roberts (2011) recently situated Bolivia's economic policies, alongside those of Argentina and Ecuador, in a 'heterodox' camp between the 'orthodox' free market policies of Brazil, Chile, Uruguay and Peru, and the 'statist' policies of Venezuela. In the same edited volume Raúl Madrid (ibid.: 240) points out that while the Bolivian government frequently engages in 'radical, even incendiary, rhetoric', its 'economic and social policies [...] have not represented a dramatic break with the past'. 'Despite its periodic criticisms of capitalism,' Madrid argues, 'the government has not sought to carry out a transition to socialism or change the existing pattern of development,' an argument substantiated by the fact that the economy remains 'focused largely on the export of natural resources', under the control of foreign capital, and that 'the government has largely respected private property and has sought to encourage private investment' (ibid.: 248).

It is argued that if the political context looks and feels the same, surely it must be the same. A number of Bolivian social scientists have noted that contrary to common expectations reports on Bolivia by the IMF over the last few years have been full of praise for Bolivia's 'solid macroeconomic performance in recent years', rooted in 'prudent macroeconomic policies', garnering 'record-high net international reserves' for the Bolivian state. Reacting to the media coverage surrounding these reports, a well-known analyst commented: 'Even the IMF is happy with Bolivia's economy, imagine the irony of that' (see Webber 2011: 177). Commonly the TIPNIS controversy is also not seen as an isolated event but connected to a chain of other governmental 'failures', including the *gasolinazo* (the government's failed attempt to end fuel subsidies) and confrontations with miners in the Bolivian highlands (see Wanderley, this volume). Kaup has, for example, recently called current policy in Bolivia a form of 'neoliberal nationalization' (2010), and as such emphasizes that this economic model continues to have some serious structural limitations. Ongoing talk of the need to 'increase reserves' now means that the government is, as Gustafson has recently commented, 'increasingly willing to make conditions more attractive for investors'.[17] Far from nationalization – or sovereignty – Bolivia is seeking new deals with Repsol (Spain), Total (France), Gazprom (Russia), British Gas (UK) and others. Reacting to the Morales government's recent efforts to create secret alliances with business interest, a writer for the Centre for Latin American Research and Development (CEDLA) also notes: 'The premises of the neoliberal model abound in this government just as they did previously: such as competition, rentism, commercial exchange and the exploitation of this model. The government of MAS despite continuously denying it is continuing to apply neoliberal politics, while claiming also to be going to great lengths to change the economic model.'[18]

As well as emphasis on the economic and political contradictions and constraints of the Bolivian development model, criticism is levelled at the Morales government's attempts to resolve the problem of exploitative privatization by sustaining an extractive economic model. Widely publicized efforts were made by the government in its first term to draw on indigenous cosmology to establish an alternative 'people's' model of development that recognizes the rights of Mother Earth. Drawing on the experience of the World People's Conference on Climate Change in 2010, a new Law of Mother Nature and Integral Development for Living Well (Ley 071) was introduced in 2012 aimed at creating the basis for balancing humans' needs with environmental protection. The law formally defines Mother Earth as 'a collective subject of public interest', and declares both Mother Earth and life-systems (which combine human communities and ecosystems) as titleholders of inherent rights specified in the law (Article 5).

Claiming that Law 071 is rife with vague and contradictory promises, and particularly the claim of being able to both extract natural resources and protect the rights of nature or Mother Earth, many intellectuals and the leaders of the country's two main indigenous organizations (CONAMAQ and CIDOB) have questioned when the practice of a truly alternative ecological model of development will start. National development policy was not only to recognize the rights of Mother Earth, but to express in practice the linked Aymaran philosophical moral value of *suma q'amaña*, as mentioned above in connection to regional post-developmentalist debates. Mirroring his comments on the regional extractivist model, Gudynas questions as an ecologist the continuing anthropocentrism of the Bolivian government's continuing leftist emphasis on progress and modernity in its development policies (2010). Similar characterizations of the TIPNIS confrontation and other cases of socio-environmental confrontation in the country have also been supported by a series of environmental organizations working in Bolivia and abroad.[19] In a recent article Friedman-Rudovsky writes for *Yale 360*:

> Morales has gained an international reputation as a modern day *Captain Planet*. As a staunch defender of pachamama, or Mother Earth, he has bounded from United Nations meetings to global summits

lambasting the developed world for wreaking environmental havoc with its insatiable appetite for oil and consumer goods [...] Unfortunately, Morales did not heed his own advice, expressing support for the TIPNIS road without consulting the park's indigenous people.[20]

In the 2005 electoral campaign Garcia-Linera became the main intellectual voice of the MAS economic development programme. While Morales continued to invoke many of the symbols that conjured up the radical past of MAS, Garcia-Linera placed emphasis on the impossibility of establishing socialism in Bolivia for at least fifty to one hundred years. Instead, Bolivia must build an industrial capitalist base. The capitalist model he envisions – Andean Amazonian Capitalism – projects a greater role for state intervention in the market (Garcia-Linera 2012). The formula essentially means capitalist development with a stronger state to support the transformation of peasant mercantilism so that it will eventually become a powerful national bourgeoisie and drive Bolivia into true successful capitalist development. It is argued by the vice-president that it is only after this long intermediary phase of capitalism has matured that the fulfilment of socialism will be materially plausible. In his conclusions Webber (2011: 225–6) comments on this thesis as being part of the 'nostalgic relapse of the current MAS administration', as it helps to explain the disappointing failure of the Morales government to break with neoliberalism in any sustained and serious manner.

Resource sovereignty

Interpretations of the energy politics surrounding the TIPNIS case, then, boil down to one of two separate insights, i.e. that resource nationalism is unrealistic about the pressures, institutional, financial and technical requirements of global capital, or that the government, despite its rhetoric, holds a *developmentalist* perspective on change and modernity that contrasts with its electoral support base. These interpretations have some value, but again, as at the regional level, while critiques are correctly made of political-economic and epistemological limitations these critiques are again kept largely separate from one another. Put succinctly, there is little consideration of the way in which political-economic path dependencies impact on epistemological developments, or of how epistemological path dependencies impact on political-economic choices and orientations.

Studying the actions and interests of all actors involved in the case, and not just the reflex actions of the government, reveals these interpretations to be overly schematic. Concern here has to be given to the interconnections between material and social claims: this is evident in the TIPNIS case, but I would also suggest evident in other examples of current socio-environmental contestation. In early work (McNeish and Logan 2012) and in this volume I have proposed with colleagues that a reconceptualization of sovereignty is a way to capture the complex material and social dynamics of resource claims. I suggest again in this chapter that the idea of resource sovereignty – which brackets together political-economic and epistemological concerns and claims – is relevant to revealing the intricacies of path dependencies and the significance of the energetic politics played out in the case of the TIPNIS and Bolivia, and perhaps suggestive of the same in other socio-environmental struggles in Latin America. Specifically it pushes us to consider the interlinkages and interplay between a wider set of actors, between multiple levels and expressions of sovereignty, as well as the foundations for their claims. It perhaps also requires us to hold several thoughts in the head at the same time, i.e. supporting the critique provided above, but balancing this with some explanation of government muddling.

It is evident from the public and intellectual debates surrounding the TIPNIS campaign that much of the commentary has been written by left-leaning analysts, many of whom are 'not so much against the [MAS] government as [they are] [...] for recovery of the "process of change"' and an effort to uphold its original ideals (Achtenberg 2011). Many of the people who have made fierce critiques of the government nonetheless defended Morales against a right-wing coup attempt in September 2008 and helped re-elect him in December 2009 with 64 per cent of the vote and in November 2014 with 61 per cent. They understand that Morales is better than his neoliberal predecessors, and better than the neoliberal opposition, but also that Bolivians still deserve better than what they're currently getting. Moreover, much of their anger derives from their perception that Morales' actions and inactions are in fact empowering the right by alienating popular sectors and failing to pursue genuinely revolutionary policies of massive redistribution, decentralization and respect for indigenous rights and the environment.

In her 26 September 2011 letter of resignation, former defence minister Cecilia Chacón cited this concern, saying that 'the measures taken [against the TIPNIS marchers], far from isolating the right, strengthen its power to act and manipulate the March with the aim of attacking the process of change for which Bolivians have sacrificed so much'.[21] On 28 September Pablo Solón, the former Bolivian ambassador to the UN and the coordinator of Bolivia's 2010 World Peoples' Conference on Climate Change and the Rights of Mother Earth, sent a letter of protest to Morales expressing a similar sentiment: 'To block the right, which wants to take advantage of the protest in order to return to the past, we must be more vigilant than ever in defense of human rights, the rights of indigenous peoples, and the rights of Mother Earth.'[22] Despite the intentions of these critiques, linkages are found between social movement mobilization and the political right in the country. The government's mismanagement of the TIPNIS protest led to rising support of both the environmental organizations involved in logistical support, and enabled oppositional political groups and parties to exploit the political atmosphere and build support for their 'alternatives'. There is also evidence to demonstrate the links between the right-wing opposition, foreign interests and TIPNIS community leaders in the run-up to the protests. These date back to before Morales' election and have continued since, as demonstrated in US embassy cables made public by Wikileaks.[23] Further evidence of ongoing meetings and communication can be found by simply visiting the websites of USAID and CIDOB.

None of this gets the government off the hook for the violent police intervention at Yucumo and the failed investigation of governmental responsibility. Nor does it remove any of the contradictions of a politics based on a rhetoric of ecological respect and an economic practice of expanded resource extraction and confrontation. However, it leads in the direction of recognizing that contradictions can be found on all sides, and that claims and positions are not always what they first appear.

Political and economic limitations on epistemological development

Although the governments of the recent left-wing wave (or pink tide) have had ambitions to change the rules of the political game in the region, material limitations stand in their way (investment, technology, infrastructural and institutional capacity, etc.). Indeed, the realistic limits of leftist government operating within the global capitalist system are under-emphasized in current critiques.

Recognizing that Bolivia remains one of the poorest countries in Latin America, it was inevitable that under a left-wing government efforts would be made to reduce the nation's high level of economic inequality through new cash transfer programmes.[24] However, it would also be inevitable that to create an effective social safety net there would also be a need to sustain or even increase economic growth. This is, however, difficult, now as in the past, because, as Robinson has asserted, Latin America is still 'deeply tied – and subordinated – to the larger world capitalist system that has shaped its economic and political development from the conquest in 1492 right up to the present period of globalization'.[25] Although its commodities have attracted significant investment through price increases in recent years, Latin America has been largely unable to change its peripheral role of supplying non-value-added commodities such as natural resources to wealthier core countries within the capitalist system, leaving these states without many viable options for independent development. Commenting on Latin American countries' reliance on foreign finance, Coronil (2002) has also suggested that

> Even when these nations try to break free from their colonial heritage, that is, their dependence on the export of primary products, through the implementation of development plans directed at diversifying their economies, they generally need foreign currency to achieve this. Although the stated objectives of Bolivia's ruling party include socialism, the peripheral political-economic position that Bolivia shares with much of Latin America also necessitates building infrastructure, such as highways, to facilitate the trade of metals, timber, agricultural products, and other resources. In these processes, the environmental and social costs of development can be limited more than they have been, but are nonetheless unavoidable.

Following the nationalization of its energy resources, the changes made by the Bolivian government, including the introduction of a hydrocarbons tax, allowed it to utilize increased profits from the hydrocarbon sector to both stabilize the economy and pursue an array of redistributive social programmes. To protect against external shocks and increase internal economic stability, the government increased foreign reserves from around US$2 billion to US$8 billion. However, the Morales government inherited a hydrocarbon sector in which very little investment had been made in exploration activities since the late 1990s. While the Morales government attempted to direct more funds towards YPFB and exploration activities, the push to decentralize state funding over the past twenty years by both Morales supporters and opponents has resulted in a distribution of hydrocarbon rents that directs profits towards departmental and municipal governments. Within this context, the Morales administration has been constrained in its efforts to properly fund YPFB, and this has made it difficult for the state company to appear a viable investment partner in potential joint ventures with foreign firms. For transnational corporations, nationalization inevitably led to lower returns. Whereas they were willing to accept impacts on existing investments, they became less willing to promise future investment. As a result, by 2010 industry estimates of reserves plummeted and Bolivia lost nearly half its prospective gas reserves, or approximately 4 per cent of the country's GDP.[26]

Frequent critique of the Morales administration has highlighted that as a result of nationalization and government subsidies on petrol (i.e. keeping the domestic price at US$27) the conditions for foreign investment in Bolivia have been negatively affected. It is also highlighted that deficiencies in existing infrastructure (pipeline networks) and the technical capacity and numbers of educated personnel have made it difficult for the government to meet contractual agreements with Argentina and Brazil for the sale of gas. Critique is made of the failures to expand oil extraction levels and the need of the government in the last two years to import petroleum from Argentina (Norwegian Ministry of Foreign Affairs 2013). The national media have evidenced the government's failures to meet the growing demand for both gas and electricity in the country. Domestic demand for natural gas has risen sharply in recent years and has bumped up against the country's export obligations, the main source of export revenues and foreign currency reserves. Bolivia needs between 8 million and 8.5 million cubic metres of natural gas a day to satisfy its needs at home. The country's current electricity capacity is around 1,300 MW, where demand is 1,090 MW. It is estimated that there will be an 8 per cent increase annually in coming years. News articles and opposition commentary frequently highlight the queues and frustrations of people trying to get hold of gas canisters for domestic cooking, and the blackouts that occur as a result of the poor administration of the electricity supply.[27] These difficulties are widely recognized, even by government officials. Government officials I personally interviewed in the autumn of 2011 repeated the opposition's concerns about the insufficient levels of foreign investment and state funding to meet the heightening levels of domestic energy consumption. However, while these detractors are difficult to circumvent, current discussions have largely told only a fraction of what is a complex energy story.

First, energy consumption increases in the country can be understood only by recognizing the rising material development and prosperity levels in the country.[28] Secondly, to meet this demand in a country which is roughly two-thirds forest and one third agricultural land, it is impossible to avoid environmental and social impact. It is perhaps worth noting in relation to this point that currently only 39 per cent of rural Bolivian households (in contrast to 89 per cent of urban households) have been linked to the national electricity supply.[29] This currently forces rural families into a costly and largely unregulated and frequently illegal informal economy of accessing batteries, firewood and gas canisters. Thirdly, consideration should be given to the difficulty the government experiences in withstanding the destabilizing forces of elite factions in the country (the highland mining and lowland agricultural lobbies). These elite factions have deep-seated domestic historical rooting, international networks and clientelistic relationships with different parts of civil society (Kaup 2013). Bolivia's earlier extractivist social relations therefore affect the possibilities for new actors to alter the country's socio-economic trajectory, and force reconciliation with those they originally appeared to oppose.[30] Fourthly, while foreign investment has undoubtedly been affected by state regulation, this does not mean that foreign companies have abandoned Bolivia and the possibility of increased investment in the future. Indeed, foreign capital is clearly active in a range of areas within Bolivia's diverse energy economy.[31] Fifthly, since its election the Morales administration has not only used nationalization to add to the public purse available for social development, but has used the increased levels of finances staying in the state (encouraged by high commodity price levels in the international markets) to invest in the country's extractive technical capacity, build its capacity for the industrialization of hydrocarbon products, i.e. the construction of liquid separation plants for gas to liquid gasoline (GLT), and aim at the expansion of its entire energy matrix.[32] Lastly, while questions remain regarding their impacts on the environment and local communities' health and economy, the government has also taken seriously the idea of increasing national investment in renewable technologies. A great deal of attention has been given in recent years to the Bolivian government's efforts to mine the world's largest reserve of lithium from the salt flats of Uyuni.[33] Less attention has been given to its efforts to develop power generation in areas of natural thermal activity and encourage community usage of solar panels through microcredit programmes. Recognition also needs to be given to the fact that all of these efforts to expand and diversify the country's energy matrix away from non-renewables inevitably confront material dilemmas of economic costs, technological capacity and environmental impacts.[34]

Epistemological limitations of political economic development

As much as political and economic change have been affected by material limitations, it is also important to highlight their linkage to epistemological limitations. Here emphasis needs to be placed on the dynamic and sometimes confusing nature of the relationship between economic dependencies, social history and identity formation.

While the contradictions between the rhetoric and actions of the Bolivian government should be a cause of serious concern, much of existing analysis of the TIPNIS controversy relies on a series of assumptions about the class and identity interests of those involved in the protests. Most crucial to reconsidering these relationships and to exploding renewed stereotypes, attention needs to be paid to the historic linkages of Bolivian indigenous communities to extractive industries and global commodity markets.[35] Indigenous groups in the country share a relationship to extractive economies that has been emphasized from conquest to the current day. Indeed, many of the earlier ethnographic texts on Andean communities highlight and discuss the consequences extractive capital would have in determining indigenous identity and culture (Nash 1993 [1979]; Taussig 1983). It might be argued that these earlier texts refer only to the Andean area. However, it would be a mistake to ignore both the cultural and political links resulting from migration from the highlands to the lowlands, or the lowlands involvement in other parts of the political economy of resource extraction.[36] I agree with Fabricant and Gustafson (2011: 8) that there is a need to recognize the articulations across spaces that have occurred in Bolivia and which, through their economic and social pressures, left both *lo andino* and *el oriente* embedded both politically and culturally in each other.

Also of note is the particular manner in which national economic and political transformations have squeezed ethnic and class identities together. Indeed, it should be recognized that far from being simply marginalized subjects, indigenous peoples have had a particularly important role in the modern history of the country (Gotkowitz 2008). In 1991 the communities of the TIPNIS organized the first march for 'territory and dignity' which not only won legal recognition from the government of their status as an indigenous territory, but brought about a series of legal and constitutional changes recognizing the rights and diverse interests of indigenous communities throughout the country. The country's ratification of ILO169 is tightly related to this experience. Moreover, the sustained militancy of the TIPNIS is widely cited within the country as an event marking the start of the gradual transformation of the country into its current form as a plurinational state (República del Estado Pluri-Nacional de Bolivia 2009).

By complicating classic stereotypes of indigenous peoples, Bolivia complies with a recent reconsideration of the meaning and expression of indigeneity. Despite earlier anthropological efforts to isolate and contain the singularity of indigenous identity and interests through ethnographic description, new studies suggest that indigenism has never been a singular ideology, programme or movement, and its politics resists closure. Citing Stuart Hall's (1996) influential conception of black cultural politics, De la Cadena and Starn (2007) have proposed that indigenous activism is 'without guarantees'. Local indigenous demands 'tend to disturb political agendas and conceptual settlements, progressive and conservative alike' (De la Cadena 2010: 335). Drawing on ideas of a 'cosmo-politics' initially developed by Latour (1993) and Stenger (2005), De la Cadena argues that 'participating in more than one and less than two socio-natural worlds, indigenous politicians are inevitably hybrid, usually shamelessly so' (2010: 353). Recognizing that a cosmo-politics opens up an understanding of politics to one based on hybrid positions demands a recognition of power disputes that not only take place within a singular world, but where there is the possibility of *pluri-versal* (Ranciere 1999) adversarial relations, i.e. where both humans and nature (the cosmos) interact.

In the TIPNIS case we can see the coming together of connected, but also conflicting, 'cosmo-politics', or resource sovereignties. Here blood was being spilled because of differences of opinion, understandings and relationships with the natural environment. However, if we look closer at the demands of the TIPNIS protesters and consider the range of interests that surround the march, we end up with a different picture to that commonly assumed by external analysts or environmental interest groups and the political opposition. The fragmentation within the ranks of the march, between the communities of the TIPNIS, and the face-off with the coca growers and government, are indicative of the way in which expressions of indigeneity are marked by contrasting historical experiences of state formation and the operation of the international economy. Historical experiences have meant, as I have suggested elsewhere (McNeish 2013a: 235), that different sectors of the population refer to different points on a sliding scale between class and ethnic identification.

Recognition of such epistemological diversity and historical path dependency is important because it should give us cause to question not only current readings of Bolivia's 'energetic' politics, but also some of the proposals for alternatives. As has been mentioned above, both at a regional level and at a national level, a series of regional analysts and thinkers – including many of the foremost decolonialist scholars (Gardiola-Rivera 2010; Gudynas 2011a; Mignolo 2011; Escobar;[37] Moraña et al. 2008) – have backed the idea of *living well*, or *el buen vivir/sumaq qamana*, as a radical and truly Latin American basis for an alternative development model. It has become the basis for a model that rejects the damaging effects of capitalist growth premised on primitive accumulation, or accumulation by dispossession, through its eco-sophic emphasis on the post-extractive establishment of harmony and inclusion between society and nature. In its most general sense, *buen vivir* denotes, organizes and constructs a system of knowledge and living based on the communion of humans and nature and on the spatial-temporal-harmonious totality of existence. That is, on the necessary interrelation of beings, knowledges, logics and rationalities of thought, action, existence and living (Walsh 2010). The eco-sophic goals here are inspiring, but recognizing that indigenous peoples do not comply with simple stereotypes of defenders of the forests, but rather like us struggle with the terms of modernity and change, there should be cause for a critical pause in considering indigenous ideas such as *buen vivir*. If this idea is a solution, why are indigenous people – who have supposedly applied and institutionalized it – still at odds with one another, themselves and with nature? Indeed, it should draw attention to the difficulties indigenous intellectuals themselves experience in trying to fill the idea with meaning and application at the national level.

It might be worth here remembering Rappaport's (2008) critique of the Barthian (1969) notion of ethnicity based on her co-productive research with indigenous leadership. The Barthian model is seen here as problematic because it makes little sense of the multiple and contradictory processes of identification that have been harnessed by indigenous political actors to contend with both their organizational needs and their own subjectivities as cosmopolitan intellectuals. Anthropological treatments of ethnicity focus on how individuals negotiate ethnic boundaries, not how political organizations – which are themselves palimpsests of multiple ethnic boundaries that are continually negotiated and renegotiated – create and maintain them. According to Rappaport (2008), what is needed is a new look at who participates in identity politics, at how intercultural organizations create new forms of identification and negotiate the fluid boundaries of their constituencies.

Constitutional debates in Bolivia and Ecuador in recent years are suggestive of different readings of the significance and meaning of the idea. Indeed, as Walsh (2010: 20) has highlighted, when the idea of *buen vivir* has become integrated in national constitutional debates, questions have to be asked about it 'becoming another discursive tool and co-opted term, functional to the State and its structures and with little significance for real inter-cultural, inter-epistemic and plurinational formation'. Rather than representing the positioning of indigenous eco-sophic values, in its transference into national debates the term the 'good life' has become more akin to 'alternative visions of development emerging in the Western world'. Walsh concludes by asking to what extent this new binary *buen vivir*–development enables a de-envelopment of the developmentalisms present and past. There is a suggestion here that, as with other counterculture proposals, what is being marketed by academics and activists alike has in reality little impact on dominant capitalist development models (Heath and Potter 2006). Indeed, it is perhaps worth considering the links between *buen vivir* and other concepts already active in neoliberal developmentalism, i.e. multiculturalism, development with identity, cultural liberty

(Hale 2006; UNDP 2004), if not in Catholic theology (Orta 2004). Certainly we hit here an epistemological cul-de-sac, where as a result of history and political process '*otros saberes*' (other knowledges) inevitably collide and, even if uncomfortably and unequally, merge with those that are dominant.

As such I argue here that rather than a pluri-verse, we see more of a intra/inter- or hybrid epistemology in formation. If ideas of customary and state legal norms do not exist as separate entities, but because of history and globalization as inter-legalities (De Sousa Santos 1987), it surely should follow that epistemologies (knowledge) are the same. Indeed, this hybridity can clearly be related to Coronil's ideas of Occidentalism. From Marx (1981) to Fanon (1967) and Coronil (1996), we are reminded that while recognizing historical formation of the moment a social revolution must strip itself of the past. Coronil (ibid.) moreover urged us to distract our attention from the problematic of 'Orientalism', which focuses on the deficiencies of the West's representations of the Orient, to that of 'Occidentalism', which refers to the conceptions of the West animating these representations. Occidentalism is thus the expression of a constitutive relationship, both positive and negative, between Western representations of cultural difference and worldwide Western dominance. Occidentalism can then be recognition of the impact of the West – of its material and epistemological impositions, but also an inversion of the Western imaginary, the world turned upside down, or a counter-discourse. Such 'critical border thinking' underlines a response to modernity that does not reject it and retreat to 'fundamentalist absolutism' but rather raises the possibility of its subaltern reworking (Grosfoguel 2011).

Latin America transformed?

It would be unrealistic to suggest that Bolivia is not an exceptional context. There is no doubt that the country has a distinct political economy and political ecology. However, I argue here as above that it might be an exceptional example that to some extent proves the rule of wider complications in energy politics and socio-environmental conflicts in the region. It is not that the specific details of the TIPNIS case and the national debate surrounding its larger economic and energetic significance should match with others – in some elements it will, in others it will not. Rather, it is the observation of the operation of competing resource sovereignties and the difficulties caused by both material and epistemological path dependencies which are of significance to wider analysis. With their *flammable* nature, resource and energy politics in other parts of the region are as marked by competing resource sovereignties as they are in Bolivia. Other countries in the region also appear to be experiencing political economic and epistemological limitations similar to, or parallel with, those explored above. In connection with this it is of note that similar processes of slippage from radical proposals for overhauling the state and energy-based economy to more pragmatic and standard policies for energetic development have taken place in other countries of the region (Venezuela, Ecuador, Brazil). Latin American countries, Bolivia included, also commonly appear to be caught in the modernist trope of 'energy production and distribution equals development'. This on the one hand represents a reality of imperfect coverage of infrastructure, even energy poverty, but it also indicates a rather rigid high modernist belief (Scott 1999) in the power of technology to improve society. Energetic societies belonging to the recent leftist wave in the region are finding it difficult not only to express contrasting ideas to Occidental notions of development and modernity, but to act outside the box of capitalist economic relations. This places Gardiola's claims of a globalization on Latin American terms, of a freedom to reinvent and reconstruct, in deep question. Indeed, it places the idea of Latin America as an autonomous region very much in doubt.

A great transformation has been under way in Latin American politics and economics, but it is all too reminiscent of the Polanyian 'Great Transformation' (2001 [1941]) that occurred and continues to occur in Europe and other parts of the capitalist market society. Social analysts from Polanyi to Schumpeter (1950) have highlighted the parallel productive creativity and destructive nature of capitalist relations – an ability to draw new value out of land, labour and capital through the invention of market logics based on 'fictitious commodities'. Although a new socialist tone has been adopted in the politics of some countries in Latin America over the last decade, there is no sign of any serious effort to end the destructive revaluing and dispossession of society and nature initiated by the European Industrial Revolution. Although, through a dual movement, new fictions continuously derive new values for land, labour and capital, there remains little direct opportunity for people throughout Latin America to participate in the transformation and definition of these values. In recognizing the basic nexus between economy and the formation of social relationships and the imagination, Graeber (2005) has extended Marxian and Maussian ideas into each other, to argue that the ultimate stakes of politics and social order are the struggle, not to appropriate value, but to take part in the establishment of what value is. Despite improvements including the extension of rights and reduction of poverty, this power to establish the value of land, labour and capital (and with it natural and energy resources), while getting closer, is still distant and abstract in all the countries of the region – and in some more than others. Indeed, because of material and epistemological limitations the attribution of value is not even contained within the region, but through financialization set by stock exchanges, corporations and institutions distant from Latin American shores.

In Latin America significant regional economic and political institutions have been formed to encourage trade, economic integration and political cooperation, and to speed up infrastructural and energy development. A highly relevant example of this, given its direct connection to the TIPNIS road-building project and a series of other socio-environmental conflicts in the region, is the Initiative for the Integration of the Regional Infrastructure of South America (IIRSA), which when launched in 2000 was supported by twelve countries as well as the CAF (the Andean Development Fund), Fonplata (the River Plate Basin Financial Development Fund) and the IDB (the Inter-American Development Bank). The IIRSA initiative seeks to construct the basis of a common integrated communication and transport infrastructure throughout the region with a view to greater trade and cooperation. The IIRSA agreement and the regional constellations that backed it speak of the clear desire of Latin American nations to constitute a confident regional autonomy and identity.

Equally noticeable perhaps is the creation of CELAC (Community of Latin American and Caribbean States) in 2011, with the aspiration to not only deepen international cooperation, but to 'create a new regional financial architecture; more rational use of energy with improved energy access for countries that lack adequate means; enhancement of transportation infrastructure that can permit geographical integration; the definitive eradication of poverty and the provision of health care; water and sanitation projects; and more comprehensive guarantees for human rights of migrants through greater interstate cooperation'.[38] Recognizing not only the membership of the usual suspects of Venezuela and Cuba, but the cross-ideological membership of Mexico, Colombia and Chile, Daniel Ortega called the regional initiative a 'death sentence' for the US Monroe Doctrine.[39] It is important to note, however, that the ink was barely dry on the entity's founding document before the secretary general of the Organization of American States (OAS), a still-active institutional product of the Cold War, downplayed the potential role of CELAC. He called it a 'regional consultative body' that could enhance the coordination of its member states but could never adequately replace the OAS in hemispheric affairs. It is also evident through a closer look at recent regional political economic agreements that there are severe differences of ideological orientation, developmental orientation, energetic reliance and aspiration. Here we see not only the Cuban-style revolutionary internationalist ALBA (Bolivarian Alliance of the Peoples of Our America) pitched against US-friendly free trade blocks such as NAFTA (North American Free Trade Agreement), CAFTA (Central American Free Trade Agreement), CAN (Andean Community of Nations) or Alianza del Pacífico (Mexico, Colombia, Peru and Chile), but changing political backing from member nations in these agreements for non-renewable or renewable resources.

The changing energetic positions reflect new geological discoveries and changing estimates of reserves, but also the influence of domestic political dynamics and decisions such as those witnessed in Bolivia. Indeed, we see competing expressions of resource sovereignty, whereby national state and civil society interests (including the private sector) in energy and development and competing technologies are pitched against regional interests. Perhaps even more salient than the different ideological orientations is recognition of inter-elite competition taking place within and between new regional organizations. With the image of Bolivar commonly placed (as a picture or statue) in the background of many of the meetings of these regional partners, we are reminded of earlier divisions and resource conflicts and of ill-fated attempts to use the Latin American Wars of Independence to create unity in the project of Gran Colombia. As the conclusion of this volume will more strongly propose, economic and political change in Latin America retains the character of a 'civil peace', in which new settlements improve conditions for the upper tiers of society at the same time as continuing the exclusion of the poor. The exceptionalism of neoliberalism is sustained (Ongh 2006).

Conclusions

What I have attempted to make visible in this chapter is not only the wilful naivety of Gardiola-Rivera's claims regarding Latin Americans' freedom to think and act as they please, but the falsity of wider claims to Latin American regional autonomy and the identification of a distinct development path and ethos. While we must accept that great changes have occurred in the region's politics and economy over the last decade, history and globalization have tied Latin America to occidental thought and capitalist material relations. As I have attempted to demonstrate above, the expanding reliance on resource and energy extraction, and resulting socio-environmental conflicts of which the TIPNIS case is illustrative, indicates the manner in which the region's political economy and political ecology are tied to both international and local structures, ideas and dynamics. Indeed, Latin America's possibilities and limitations are revealed by both political economic and epistemological considerations, or, as I have argued in this chapter, what might be called a socio-economics of resource sovereignty. Here local cosmo-political positions, varying levels of claims to sovereignty and territory, and attempts to create social and economic constellations – all connected to energy resources – come into both creative and destructive tension with one another. Here new movements and political agendas are born, but contradictions, co-option, clientelism and failures are also experienced.

It would therefore be fair to conclude that Latin America, while changing, has not been transformed entirely. Indeed, to some extent recognition of the epistemological and political economic limitations described above and of the contradictory forces they generate provides us with a sobering vision not only of the region's recent history of development, but of the mythical nature of Latin America as a region. I am far from the first to recognize this. Many decolonialist writers clearly emphasize the constructed nature of Latin America as a region (Mignolo 2005; Moraña et al. 2008). However, whereas Gardiola-Rivera (2010) attempts to recount the impacts of colonialism and the enlightenment on Latin American history, it is notable that his account of origins attempts to recover a more autochthonous reality, or, as he writes, *the dream of indians*. Here, the broken walls of proto-civilizations and Incan, Aztec and Mayan imperial expansions are skipped over, and the assumption made that before European conquest indigenous peoples had a common sense of themselves, of where they lived, and a project to 'realise the harmony between man and the world' (ibid.: 27). This glance at the past, as with his assumptions about the present, contrasts with more careful considerations. As Mignolo writes, '*America* [...] was never as a continent waiting to be discovered. Rather, *America* as we know it was an invention forged in the process of European colonial history and the consolidation and expansion of the Western world view and institutions' (2005: 2). While indigenous imaginations were clearly present and active in interpreting the world, there was no recognition of separation or territorial delimitation in an Occidental sense. Interestingly, while many post-colonial and decolonialist writers recognize the colonial matrix of power that produced Latin America, and the linkages between this and regional intellectuals' attempts to confront and rework historical experience in the region into a homogeneous Latin American identity, some nonetheless look uncritically at indigenous concepts such as *el buen vivir* as though they were delinked from this history and represent a genuinely regional alternative to Western developmentalist thought (ibid.; Escobar 2010; Mignolo and Escobar 2013). There seems to be a short circuit between analytical ability and political posture. At the same time the blatant material and epistemological complications of Latin America's resource- and energy-dependent past and present highlighted above make this positioning seem deeply flawed.

A good life (*buen vivir* or otherwise) based on harmony and balance between society and nature remains an important ideal, and given the growing recognition of an Anthropocene it perhaps represents a necessary political and economic platform. However, considering current energy politics in what we know as Latin America, there is still no evidence of the substantive practice of a cosmo-political platform, indigenous or foreign, that revalues land, labour and capital in a way that entirely disrupts pre-existing relations and ideas – that heals the Cartesian divide. Left as it is, this might appear a somewhat depressing conclusion. However, before ending this chapter, it is worth emphasizing one last set of observations. Perhaps it does not matter that Latin America is a construct, that there is no cosmo-political platform that is delinked and entirely disrupts pre-existing relations and ideas. The fact that Latin America is changing, but not transformed, indicates that there is enough room within material and epistemological limitations to generate subalternity and to draw on the collision or hybridity of ideas to create positive movements.

Instead of a story of Latin Americanization, a historically and anthropologically grounded story of Latin America's place in the world, while not so heroic, also has energy. Indeed, it is the region's energy politics, needs and limitations which help provide new double meaning to Quijano's (2000) 'socialisation of power'. Importantly for the conclusion to this chapter, Coronil emphasizes that while unpredictable, Occidentalism is undoubtedly creative. Fanon, like Marx, drew on the poetry of the future to imagine a world in which the dead may bury the dead so that the living may be freed from the nightmare of the past. Coronil (1996: 81) writes in concluding his paper on Occidentalism: 'As the future flashes up to a child in the form of a disenchanted, inhospitable, and depopulated world, the safety of those who follow us comes to depend as well on the poetry of the present.' The wild interplay between resource sovereignties, of pluralism and integration, promises no harmony. However, political energies and energetic politics in Latin America are inspiring and continue to encourage regional actors and distant observers to think and act anew.

Notes

1 www.oecd.org/dev/americas/latinamericaabrightfuture.htm.

2 business.blogs.cnn.com/2012/03/14/millennial-david-lloyd-why-the-future-is-latin/.

3 www.manifesttidsskrift.no/mindre-ulikhet-i-latin-amerika/.

4 mrzine.monthlyreview.org/2007/marcano030107.html.

5 In 1960, Latin America accounted for 6 per cent of the world economy (Tissot 2012). In 2009, that share had increased to 7.2 per cent, and outperformed world GDP per capita growth by at least 1.5 per cent. While foreign investments have been falling in developing countries, FDI inflows in Latin America rose 40 per cent between 2009 and 2010.

6 Since 2007, the prices for metals and crude oil have on average been three times as high as at the beginning of the century (UNEP 2013).

7 According to the Economic Commission for Latin America and the Caribbean (ECLAC), from 1990 to 2009 the weight of extractive sector exports increased from 31 to 53 per cent in Peru; from 38.5 to 52.1 per cent in Colombia; and from 12.5 to 38.2 per cent in Brazil. China, which is investing heavily in Latin America and is now the third-largest foreign investor in the region, places almost all of its investments (90 per cent) in natural resources.

8 In Peru, sixty-four hydrocarbon blocks now cover more than 70 per cent of the country's Amazonian territory. In Ecuador, two-thirds of the Amazon is zoned for hydrocarbon expansion. In Bolivia, 55 per cent of the national territory is zoned for potential hydrocarbon exploitation assigned to the state-owned company for exploration. See ella.practicalaction.org/sites/default/files/120928_ECO%20ExtIndCon-Man_%20GUIDE.pdf.

9 Fossil fuels are not equally distrib-

uted in Latin America, in that oil and gas reserves are concentrated in Venezuela while most coal reserves are located in Colombia and Brazil.

10 Brazil accounts for 96 per cent of the region's ethanol production, but Colombia, Paraguay, Mexico and Guatemala also have growing ethanol industries.

11 Between 2002 and 2013, poverty in the region was reduced from 43 to 27 per cent, and inequality also significantly lowered (1 per cent per year).

12 www.economist.com/node/16964135.

13 See www.brookings.edu/research/opinions/2013/11/07-latin-america-growth-rate-talvi-munyo.

14 cedla.org/es/content/2567.

15 An unusual event in a country known for political militancy, but relative to other countries in the region there have been few incidences of political violence in recent times.

16 See www.isj.org.uk/?id=780.

17 nacla.org/blog/2013/11/19/fuel-politics-latin-america-where-begin.

18 www.bolpress.com/art.php?Cod=2012050713.

19 FOBOMADE, TIERRA, LIDEMA, Rainforest Foundation, IWGIA and others.

20 e360.yale.edu/feature/in_bolivia_a_battle_over_a_highway_and_a_way_of_life/2566/.

21 zcomm.org/zblogs/bolivia-dilemmas-turmoil-transformation-and-solidarity-by-kevin-young/.

22 www.bbc.co.uk/news/world-latin-america-15144719.

23 nacla.org/blog/2011/8/26/bolivia-tipnis-marchers-face-accusations-and-negotiations.

24 Between 2005 and 2011, Bolivia's poverty rate declined by 16 per cent (from 61 to 45 per cent), and the extreme poverty rate by 45 per cent (ECLAC 2013).

25 www.aljazeera.com/indepth/opinion/2011/09/2011913141540508756.html.

26 www.worldpoliticsreview.com/articles/print/12763.

27 www.economist.com/node/17851429.

28 nacla.org/blog/2013/2/15/economic-growth-more-equality-learning-bolivia.

29 www.energie-ist-entwicklung.de/download/01_CP_Bolivia_2010.pdf.

30 www.lse.ac.uk/IDEAS/publications/reports/pdf/SU005/kaup.pdf.

31 Iron ore in the country's east is disputed by China, Venezuela and India. Minerals in the high Andes are targeted by France, Japan, Canada, Australia and the United States. Land in the east is already under significant control – via soy marketing and export – by Monsanto, ADM and South American soy capital. In a period in which global prices of crude oil were high the soy and sugar lands in Bolivia were imagined as a new opportunity for biodiesel production. The immense natural gas reserves along the Andean foothills were also coveted by Brazil and Argentina, with Russian, British, American and Spanish capital. All with their own dreams of export to Europe or the United States.

32 Although still at insufficient levels to meet estimated growing demand (the *gasolinazo* price hike was a disastrous effort to increase them) there has been a tripling of the financing going to these sectors of the economy. According to the director of YPFB, Carlos Villegas, this investment has enabled them to meet their contract for the delivery of 7 million cubic metres of gas per day to Argentina. As well as an expansion of oil and gas exploration, the government has also moved ahead with plans to construct two gas to liquid separation plants in Rio Grande and Gran, the construction of two hydroelectric schemes

in Entre Rios, and the construction of thermoelectric plants in Carrasco and Porto Suarez.

33 www.nytimes.com/2009/02/02/world/americas/02iht lithium.4.198777 51.html?pagewanted=all&_r=0.

34 www.lithiummine.com/lithium-mining-and-environmental-impact.

35 This does not mean that different communities agree on extraction, or the form in which this take place and the politics surrounding it. See the recent debate on the new Bolivian Mining Law.

36 Tobacco, sugar, castaña, soya, coca, oil and gas.

37 www.theguardian.com/global-development/2012/nov/05/arturo-escobar-post-development-thinker.

38 www.peacebuilding.no/var/ezflow_site/storage/original/application/7f15782848b5f9b2ac1c03138ac9093e.pdf.

39 is.muni.cz/el/1423/jaro2013/MVZ806/um/from_alba_to_celac.pdf.

References

Achtenberg, E. (2011) 'Bolivia: exploiting the TIPNIS conflict', *North American Congress on Latin America – Rebel Currents*, October, nacla.org/blog/2011/10/7/bolivia-exploiting-tipnis-conflict.

Acosta, A. (2009) *La Maldición de la Abundancia*, Quito: Abya Yala.

Barth, F. (ed.) (1969) *Ethnic Groups and Boundaries: The Social Organization of Culture Difference*, Boston, MA: Little, Brown.

Bebbington, A. (2009a) 'The new extraction: rewriting the political ecology of the Andes?', *NACLA Report on the Americas*, 42(5): 12–24.

— (2009b) 'Contesting environmental transformation: political ecologies and environmentalisms in Latin America and the Caribbean', *Latin American Research Review*, 44(3): 177–86.

Bebbington, A. and D. Bebbington (2011) 'An Andean avatar: post-neoliberal and neoliberal strategies for securing the unobtainable', *New Political Economy*, 15(4): 131–45.

Castree, N. (2002) 'False antitheses? Marxism, nature and actor-networks', *Antipode*, 34(1): 111–46, doi: 10.111/1467-8330.00228.

CEPAL (2002) *Economic Survey of Latin America and the Caribbean 2001*.

— (2012) *Latin American Outlook: Transforming the State for Development*, www.oecd.org/bookshop?9789264121706.

Choquehuanca, D. (2010) *25 Postulados para entender el Vivir Bien*, Unpublished ms, La Paz, www.rebelion.org/noticia.php?id=100068.

Coronil, F. (1996) 'Beyond Occidentalism: toward non-imperial geohistorical categories', *Cultural Anthropology*, 11: 51–87.

— (2002) *El Estado mágico. Naturaleza, dinero y modernidad en Venezuela*, Consejo de Desarrollo Científico y Humanístico de la Universidad Central de Venezuela, Caracas: Nueva Sociedad.

De la Cadena, M. (2010) 'Indigenous cosmopolitics in the Andes: conceptual reflections beyond "politics"', *Cultural Anthropology*, 25(2): 334–70.

De la Cadena, M. and O. Starn (2007) 'Introduction', in M. de la Cadena and O. Starn (eds), *Indigenous Experience Today*, Oxford: Berg, pp. 1–30.

De Sousa Santos, B. (1987) 'Law: a map of misreading. Toward a postmodern conception of law', *Journal of Law and Society*, 14(3): 279–302.

ECLAC (2013) *Economic Survey of Latin America and the Caribbean 2013*, www.eclac.org/cgi-bin/getprod.asp?xml=/publicaciones/xml/3/50483/P50483.xml&xsl=/publicaciones/ficha-i.xsl&base=/publicaciones/top_publicaciones-i.xsl.

Escobar, A. (1999) 'After nature: steps to

an anti-essentialist political ecology', *Current Anthropology*, 40(1).
— (2005) 'El "postdesarrollo" como concepto y práctica social', in D. Mato (ed.), *Políticas de economía, ambiente y sociedad en tiempos de globalización*, Caracas: Facultad de Ciencias Económicas y Sociales, Universidad Central de Venezuela, pp. 17–31.
— (2010) 'Latin America at the crossroads: alternative modernizations, post-neoliberalism or post-development', *Cultural Studies*, 24(1): 1–65.
Fabricant, N. and B. Gustafson (eds) (2011) *Remapping Bolivia: Resources, Territory, and Indigeneity in a Plurinational State*, Santa Fe, NM: School for Advanced Research Press.
Fanon, F. (1967) *Black Skins, White Masks*, New York: Grove Press.
Fatheuer, T. (2011) *Buen Vivir. A brief Introduction to Latin America's new concepts for the good life and rights to nature*, Serie de Publicaciones sobre Ecología, vol. 17, Berlin: Heinrich Boll Foundation.
Garcia-Linera, A. (2012) *Geopolítica de la Amazonia: poder hacendal-patrimonial y acumulación capitalista*, La Paz: Vice-presidencia de la República de Bolivia, www.alames.org/documentos/amazoniaAGL.pdf.
Gardiola-Rivera, O. (2010) *What if Latin America Ruled the World? How the South will take the North into the 22nd century*, London/Berlin/New York/Sydney: Bloomsbury.
Gotkowitz, L. (2008) *A Revolution for Our Rights: Indigenous Struggles for Land and Justice in Bolivia, 1880–1952*, Durham, NC: Duke University Press.
Graeber, D. (2001) *Towards an Anthropological Theory of Value: The False Coin of Our Own Dreams*, London: Palgrave Macmillan.
— (2005) 'Value as the importance of action', *The Commoner*, 10 (Spring/Summer), www.commoner.org.uk/the_commoner_10.pdf.
Grosfoguel, R. (2011) 'Decolonising post-colonial studies and paradigms of political economy: transmodernity, decolonial thinking and global coloniality', *Transmodernity. Journal of Peripheral Cultural Production of the Luso-Hispanic World*, www.dialogoglobal.com/barcelona/texts/grosfoguel/Grosfoguel-Decolonizing-Pol-Econ-and-Postcolonial.pdf.
Gudynas, E. (2010) 'Sentidos, opciones y ámbitos de las transiciones al postextractivismo', in M. Lang and D. Mokrani (eds), *Más allá del desarrollo*, Quito: Fundación Rosa Luxemburgo y Abya Yala.
— (2011a) 'Buen Vivir: today's tomorrow', *Development*, 54(4): 441–7, www.gudynas.com/publicaciones/GudynasBuenVivirTomorrowDevelopment11.pdf.
— (2011b) 'Caminos para las transiciones post-extractivistas', in A. Alayza and E. Gudynas (eds), *Transiciones. Post extractivismo y alternativas al extractivismo en el Peru*, www.redge.org.pe/sites/default/files/GudynasCaminos-PostExtractivismoPeru11.pdf.
Haarstad, H. (2012) *New Political Spaces in Latin American Natural Resource Governance*, Santa Fe, NM: School of American Research Press.
Hale, C. (2006) *Más que un Indio: Racial Ambivalence and the Paradox of Neoliberal Multiculturalism in Guatemala*, Santa Fe, NM: School of American Research Press.
Hall, S. (1996) 'Who needs identity?', in P. du Gay, J. Evans and P. Redman (eds), *Identity: A Reader*, Thousand Oaks, CA: Sage, pp. 15–30.
Heath, J. and A. Potter (2006) *Rebel Sell: How the Counter-culture became the Consumer Culture*, North Mankato, MN: Capstone Publishing.

Kaup, B. (2010) 'A neoliberal nationalization? The constraints on natural gas led development in Bolivia', *Latin American Perspectives*, 37: 123.

— (2013) *Market Justice: Political Economic Struggle in Bolivia*, Cambridge: Cambridge University Press.

Kohl, B. and R. Bresnahan (2010) 'Bolivia under Morales: national agenda, regional challenges and the struggle for hegemony', *Latin American Perspectives*, 37(5): 5–20.

Latour, B. (1993) *We Have Never Been Modern*, Cambridge, MA: Harvard University Press.

Leira, T. (2014) *Brasil – Kjempen våkner*, Oslo: Aschehoug.

Levitsky, S. and K. Roberts (eds) (2011) *The Resurgence of the Latin American Left*, Baltimore, MD: Johns Hopkins University Press.

Lustig, N., L. L. Calva and E. Ortiz-Juarez (2013) 'Deconstructing the decline in inequality in Latin America', Tulane Economic Working Paper Series no. 1314.

Madrid, R. (2011) 'Bolivia: origins and policies of the Movimiento al Socialismo', in S. Levitsky and K. Roberts (eds), *The Resurgence of the Latin American Left*, Baltimore, MD: Johns Hopkins University Press.

Marx, K. (1981) *Capital*, vol. 3, New York: Vintage.

Maxwell, C. and E. Hershberg (2010) *Latin America's Left Turns: Politics, policies and trajectories of change*, Boulder, CO: Lynne Rienner.

McMichael, P. (1990) 'Incorporating comparison within a world-historical perspective: an alternative comparative method', *American Sociological Review*, 55(3): 385–97.

McNeish, J.-A. (2013a) 'Extraction, protest and indigeneity in Bolivia: the TIPNIS effect', *Latin American and Caribbean Ethnic Studies*, 8(2): 221–42.

— (2013b) 'An accumulated rage: legal pluralism and gender justice in Bolivia', in R. Seider and J.-A. McNeish (eds), *Gender Justice and Legal Pluralities: Latin American and African Perspectives*, London and New York: Routledge.

McNeish, J.-A. and O. Logan (2012) *Flammable Societies: Studies on the Socio-Economics of Oil and Gas*, London: Pluto Press.

Mignolo, W. (2005) *The Idea of Latin America*, Malden and Oxford: Blackwell.

— (2011) *The Darker Side of Western Modernity: Global Futures, Decolonial Options*, Durham, NC: Duke University Press.

Mignolo, W. and A. Escobar (2013) *Globalization and the Decolonial Option*, London and New York: Routledge.

Mitchell, T. (2013) *Carbon Democracy: Political Power in the Age of Oil*, London: Verso.

Moraña, M., E. Dussel and C. Jáuregui (eds) (2008) *Coloniality at Large: Latin America and the Postcolonial Debate*, Durham, NC, and London: Duke University Press.

Nash, J. (1993 [1979]) *We Eat the Mines and the Mines Eat Us: Dependency and Exploitation in Bolivian Tin Mines*, New York: Columbia University Press.

Norwegian Ministry of Foreign Affairs (2013) *Evaluation of Oil for Development Programme in Bolivia*, NORAD Report 7/2012, Oslo.

Ongh, A. (2006) *Neoliberalism as Exception: Mutations in Citizenship and Sovereignty*, Durham, NC: Duke University Press.

Orta, A. (2004) *Catechizing Culture: Missionaries, Aymara and the 'New' Evangelization*, New York: Columbia University Press.

Polanyi, K. (2001 [1941]) *The Great Transformation: The Political and*

Economic Origins of Our Time, New York: Beacon Press.

Quijano, A. (1992) 'Colonialidad y modernidad/racionalidad', in H. Bonilla (ed.), *Los conquistados: 1492 y la población indígena de América*, Bogotá: Tercer Mundo/ FLACSO.

— (1998) 'Amercanity as a concept, or the Americas in the modern world-system', *International Sociological Association*, 1: 549–56.

— (2000) 'Coloniality of power. Ethnocentrism and Latin America', *NEPLANTLA*, 1(3): 533–80, www.unc. edu/~aescobar/wan/wanquijano.pdf.

Ranciere, J. (1999) *Disagreement: Politics and Philosophy*, Minneapolis: University of Minnesota Press.

Rappaport, J. (2008) 'Beyond participant observation: collaborative ethnography as theoretical innovation', *Collaborative Anthropologies*, 1: 1–31.

Reid, M. (2010) 'So near and yet so far: a Special Report on Latin America', *Economist*, September, pp. 1–14.

— (2014) *Brazil: The Troubled Rise of a Global Power*, New Haven, CT: Yale University Press.

República del Estado Pluri-Nacional de Bolivia (2009) *Constitución del Estado Pluri-Nacional de Bolivia*, La Paz: Vice-presidencia de Bolivia, www. embajadadebolivia.com.ar/m_publicaciones/libros-pdf/04ncpe_cepd.pdf.

Schumpeter, J. (1950) *Capitalism, Socialism and Democracy*, New York: Harper and Row.

Scott, J. (1999) *Seeing Like a State: How Certain Schemes to Improve the Human Condition Have Failed*, New Haven, CT: Yale University Press.

Soares, S. (2013) 'Can changes in structural heterogeneity explain the fall in earnings inequality? An analysis for Brazil', Unpublished paper, IPEA, Brazil.

Stenger, I. (2005) 'The cosmopolitical proposal', in B. Latour and P. Weibel (eds), *Making Things Public: Atmospheres of Democracy*, Cambridge, MA: MIT Press, pp. 994–1004.

Talvi, E. and I. Munyo (2013) 'Latin America macro-economic outlook: a global perspective. Are the golden years for Latin America over?', Brookings-CERES Macro-economic Report, Washington, DC: Brookings Institution.

Taussig, M. (1983) *The Devil and Commodity Fetishism in South America*, Chapel Hill: University of North Carolina Press.

Tissot, R. (2012) 'Latin America's energy future', Energy Policy Group Working Paper, Inter-American Dialogue, www.thedialogue.org/Publication Files/Tissotpaperweb.pdf.

Tortosa, J. M. (2001) *El juego global. Maldessarrollo y pobreza en el capitalismo global*, Barcelona: Icaria.

UNDP (2004) *Human Development Report: Cultural Liberty in Today's Diverse World*, New York: UNDP.

UNEP (2013) *Recent Trends in Material Flows and Resource Productivity in Latin America*, Nairobi: United Nations Environment Programme.

Walsh, C. (2010) 'Development as Buen Vivir: institutional arrangements and (de)colonial entanglements', *Development*, 53(1): 15–21.

Webber, J. R. (2011) *From Rebellion to Reform in Bolivia: Class Struggle, Indigenous Liberation, and the Politics of Evo Morales*, Chicago, IL: Haymarket.

Weyland, K., R. Madrid and W. Hunter (2010) *Leftist Governments in Latin America: Successes and Shortcomings*, Cambridge: Cambridge University Press.

11 | FROM THE KING'S PEACE TO TRANSITION SOCIETY

Owen Logan and John-Andrew McNeish

Introduction

Previous chapters have examined the social friction generated by energy politics. Many other writers have also explored this, often spurred on by the need to tackle the risks of climate change. However, there is a tendency for the interpretation of the political phenomenon we have called resource sovereignty to become analytically one-dimensional and ideologically rose-tinted. Above all else, what this volume highlights are struggles waged over the socio-political coherency of modern Latin American nation-states, and the possibilities that energy resources provide to achieve or threaten this goal. Our case studies show how flows of energy and money are instruments of statecraft, and so the disturbance of these flows becomes a means not merely to contest power, but to contest the terms of sovereignty itself.

The real complexity of state coherency is the central thread we see running through our case studies, and in this concluding chapter we revisit that theme and set it once again in a global context. We can see how energy resources have played a key role in stabilizing socio-political conflicts in Latin American countries. However, the contradictions of exploitation of energy ostensibly claimed for the benefit of the many, but resulted in the generation of wealth for the few. This is sometimes called the *new extractivism* (Petras and Veltmayer 2014), and it continues to hinder further progress. Despite all the announcements made about transition to non-carbon energy and economic diversification, the sort of 'transition society' we referred to in the Introduction still appears out of reach.

Reflecting back on individual chapters, it is important to stress here that we do not see evidence of the operation of the so-called resource curse. Nor do we see the fulfilment of idealistic claims of a creative economy or technological innovation. Instead we see a more ambiguous and complex picture: *energy resources can promote as well as limit sustainable and substantive development, and it is the socio-economic nature of resource politics which explains these twists and turns.* We also observe that the resources, profits and impacts of the extractivist economy, and the diverse possibilities they generate in the areas of knowledge and technology, remain unequally distributed. This means that the prospects for a progressive Latin American regionalism – and particularly one that would embrace social justice and environmental issues more fully – is anything but secure.

In this concluding chapter we think it important to re-examine this political insecurity in the context of world history. There remains a considerable divergence of aims within and between the countries and peoples of Latin America. Indeed, despite the intellectual hype of a harmony promised by 'living well', or 'the good life', this modus vivendi still appears hard to establish within and between states in the region. As previous chapters have shown, while a progressive tone has been ushered into the centre of public debate, the political left in Latin America continues to have everything to struggle for in terms of reducing structural and institutional imbalances. Notwithstanding the achievements of resource-related politics in stimulating economic development and ameliorating inequality, particularly among countries belonging to the 'pink tide', a recognition of the links between the extractive economy and persisting social, economic and environmental violence makes it clear that the region's overwhelming reliance on energy resource revenues needs to be overcome.

A curse is by definition an inescapable condition for its victims. Having rejected the simplistic idea of a 'resource curse', a fuller and more open debate about energy's role, not as the cause of but as a possible catalyst for political and economic change, is needed. Rather than an abstract entity distant from people, resource sovereignties (while unpredictable and without progressive guarantee) indicate the tangible and therefore political and institutional susceptibility of energy policy. Overcoming dependency on a resource extractive industrial complex will require greater recognition of and confrontation with the persisting patrimonial, and frequently *exceptional*, nature of democratic governance in the region – a form of governance fostered by capitalist-driven competition over natural resources. Lastly, faced with the prospect of increasing rates of socio-ecological conflict, if not ecological collapse, good use must be made of Latin America's global history as an alternative 'resource' for progressive thought and experimentation.

Neoliberalism is far from over and energy-driven politics are on the rise, but current political dynamics in Latin America also provide energetic grounds for a movement away from a present defined by a partial democracy or what might be called a King's Peace, to a future of socio-ecological transition we all need. In the remaining pages we think it important to sketch some final observations about this *energy-fuelled* democratic movement. This description reinforces the insights provided by previous chapters which have shown the contemporary interplay of Latin American resource sovereignties. It also brings us to a critical consideration of their scope and the possible direction of popular sovereignty claims in the future.

From dirty wars to civil peace

By all accounts 20 June 1973 was a crucial day for Argentina. Left-wing Peronists gathered in their thousands to welcome Juan Domingo Perón (1895–1974) back to Argentina after eighteen years in exile. From Perón's platform camouflaged snipers from the Peronist right wing opened fire on the leftists, killing at least thirteen and wounding 365. June 20th entered the history books as the day of the Ezeiza massacre. It is also reported to be one of the first acts of an elite death squad formed at the highest levels of the government and later named the Argentine Anticommunist Alliance. The AAA went on to play a key role in the orchestration of Argentina's 'Dirty War', an infamous and indeterminate period of state terrorism that claimed up to thirty thousand victims and lasted until 1983. The remarkable cruelty associated with this sort of violent repression was not confined to Argentina. With Operation Condor, a secret joint-operations agreement between military and security forces at the time, violent repression of the political left and possible sympathizers took place to differing degrees across the South American continent. From Washington to Santiago and Buenos Aires it was often argued that the catalogue of arbitrary imprisonment, torture and extra-judicial killing was an unfortunate necessity if democracy was to be defended from 'the communist threat'.

With the emergence of guerrilla revolutionaries in many countries of the region in the decade preceding these events, the perception of a real threat was lent credence. However, a more viable interpretation of what occurred in Argentina is that it represented the failure of a populist corporatism that was already heavily tainted by authoritarianism. Following a period in which the power of organized labour had grown and conditions for workers had improved, there was an elite backlash against this movement from below. Here was a replay, albeit an extremely violent replay, of previous cycles of violent repression waged by Latin American generals and political elites on their population in periods following the expansion of political rights and the introduction of social reforms (e.g. Brazil 1945, Guatemala 1954, Colombia 1953, Bolivia 1970 and Argentina 1966). It is notable that in each case the US government either acknowledged, or actively intervened to support, these bloody crackdowns on free association.

While political ideology is without doubt a central factor in driving the repression that spilled out from Argentina to the rest of Latin America and the 'dirty wars' of Central America, it is clear that they also have a strongly material side. While the logics of the defence of fatherland (*patria*), church and home were part of the right-wing populist diatribe, it was clear that the military stepped in when elites perceived a threat to status and property. *Coups d'état* coincided with attempts at land and education reform. It is not only land but the natural resources within which have proved sensitive.

Energy resources and infrastructure have been, and remain, a critical focus of wider political economic conflicts in the region. The case studies included in this volume show how deeply embedded these conflicts are socially and historically. The previous chapters reveal in rich detail the reasons why so many points on maps of socio-economic conflict can be found along the length of the eastern flanks and valleys of the Andes.[1] Indeed, land and territory for narcotics production remain features of the enduring fifty-five-year-old civil war in Colombia, while struggles by warring parties (guerrilla, paramilitary, cartels and military) for the control of mineral and hydrocarbon resources are a central factor in the perpetuation of armed conflict in the course of the last decade (Hernandez et al. 2011; Chomsky et al. 2007). Also worth noting is that the Dirty War in Argentina was reflected in the class make-up of the Malvinas/Falklands war. It was mostly working-class soldiers who were sent by the generals to fight for the islands. And in recent years it has been been made public that the islands are surrounded by rich reserves of offshore petroleum resources.[2]

Somewhat counter-intuitively, the frequent clashes and political violence in Latin America demonstrate that the region's democratic development, albeit restricted, is far from static. The interpretation of liberalism made by Latin American *criollo* elites provided arguments for emancipation from colonial rule, but also delimited citizenship rights and confined the republican state's duties to the needs of a literate and landowning few. Given the self-serving nature of liberal political discourse, it was inevitable that these limits would be tested and political boundaries crossed by others seeking full citizenship and economic sovereignty. Despite elite efforts to virtually erase them from history, indigenous peoples, peasants, plantation workers and a growing urban mestizo labour force have incrementally pushed for the expansion of the state and formal understandings of its political and legal coverage.

In the twentieth century the political spaces for marginalized classes became ever wider, and with time picked up their own forms of political assistance from a changing context of geopolitics. With the proponents of neoliberalism condemning anything more than sentimental expressions of nationalism and the parallel expansion of rights discourses throughout the 1980s and 1990s, the acts of political repression seen under the military juntas of Argentina and other Latin American countries were no longer glossed over by the international community. In this sense the democratic transitions seen throughout the region in the course of the 1980s and 1990s were inevitable. New threats were of course identified as a way to further excuse authoritarian political actions and economic discipline. However, with the fall of the Soviet Union in 1991 these would no longer be so easy to portray as a fundamental threat to economic 'freedom'. Nor would they provide for the sort of right-wing nationalist politics that favoured local capitalists.

Whereas Islamic fundamentalism became a useful counter-insurgency discourse elsewhere in the world, it did not have quite the same traction in Latin America. In recent years the focus on drugs cartels has sometimes looked like the preferred means of demonizing any political policies and economic activities undesirable in the eyes of US-friendly governments. However, in the main no singular enemy was easy to identify. From the point of view of the ruling classes a new civil peace had to be reasserted after the fall of communism, but it was not going to be possible to achieve it through such extreme and brutal means as had hitherto been deployed in Latin America. Under the guise of neoliberalism, and the international banner of security and development (McNeish and Sande Lie 2010), a new basis for both democratic order and occasional *exceptional* actions was made possible (including periods of militarized 'states of emergency'), albeit imperfectly so. The features of this new order are touched on in the economic developments, extractive processes and violence studied in many of the chapters in this volume. They are made particularly evident in Braathen's chapter dealing with Brazil's passive revolution.

Patrimonial capitalism and energetic states

In his book *Capital in the 21st Century*, the centre-left French economist Thomas Piketty (2014) charts the social state which developed in western European countries as a result of the social democratic struggles inside nations and the wars between them. This painful historical process stabilized European government revenues at 45–50 per cent of total national income. In the twentieth century, tax revenues for emerging countries in regions such as Latin America were never higher than 15–20 per cent of national income, and in many cases have declined in the twenty-first century (ibid.: 490–514). At this rate, after paying for the military, the police and legal systems, very little was left for education and welfare. Piketty argues that the neoliberal assault on progressive taxation internationally means that returns on capital outstrip growth, and unless this relationship is reversed the global political economy of this century is likely to increasingly resemble that of the nineteenth century. He calls this patrimonial capitalism. Indeed, he warns that if it is not halted we may expect the world economy in 2050 to be ordered along neo-feudal lines.

The exploitation of energy resources, however, boosts growth and reduces the worrying imbalance between increasing capital returns and growth. In Latin American countries energy resource revenues have supported welfare schemes, and while figures for poverty have dropped in general there clearly remains a lot to be done to further curtail the reproduction of inequality, diversify economies and popularize the ethics of progressive taxation. It is thanks to abundant resources that the ruling classes and economic elites in Latin America have enjoyed relative social equilibrium at comparatively low personal cost – in other words, 'a King's Peace'. This can only go so far in terms of delivering social justice and all-round economic development.

The production and flow of energy can be seen as part of the social and political integration necessary for the social formation of modern states and regional power blocks. However, the caveat to this view of energy dependency is an accompanying potential for social disintegration and political collapse. On the first count we are not alone in seeing energy as a functional basis of integration and centralization. As the editors of the anthropological volume *Cultures of Energy* have written:

> [...] we build our social relationships and cultural understandings to coalesce around the continued flow of energy of familiar qualities in expected quantities. Ensuring access to continued supplies of energy and other resources is one of the central functions of centralized political systems. Shortages of energy – blackouts and queues for gasoline – quickly become political problems and often have political antecedents. However, as people encounter and experience it, the flow of energy in a place tends to be part of the taken-for-grantedness of unspoken, ordinary social life. (Strauss et al. 2013: 12)

Seen through this lens, the production and consumption of energy and the capital-intensive infrastructure required for energy flows appear to be an important function of political systems. The smooth operation of energy systems lends governments and states a degree of legitimacy. In this sense, notwithstanding the failures and the damages of top-down state plans, the same flows and infrastructure remain hegemonic and homogenizing elements that need to be considered when we think about the authority of the state. Even where energy infrastructure is patchy, states still have the unique potential to achieve an egalitarian distribution of energy, and when energy systems are in private hands, serious market failures turn into crises of legitimacy for governments. It is therefore not unusual that even in pro-market environments citizens expect state action to regularize the market.

Looking at the development of modern political hegemony through energy infrastructures and state power – and as we attempt to draw lines between empirical findings and theoretical understanding from our context-specific studies – there are other equally important factors to consider. Principal among them is the relationship between the generation of energy and the generation of political power.

Elias to the liberal peace

In his highly influential work *The Civilizing Process*, first published in 1939, the historical sociologist Norbert Elias – a Jewish refugee from

Nazi Germany – examines the development of vertical interdependent relationships between social groups.

Elias regards the centralization of power in relatively positive terms and he gives implicit credence to a Hobbesian view of state power. Like Thomas Hobbes (1588–1679), who argued in the *Leviathan* (2003 [1651]) that a strong central authority was needed to avoid major civil conflict, Elias sees the modern state's monopoly of violence as the relatively benign alternative to the intra-elite violence of the medieval world. Elias also points out, however, that while toleration for internal violence may have decreased thanks to the modern state, the toleration for externally directed violence only grew in scale as states became more confident of their 'civilization' and moral authority. The same rationale can of course be internalized and used against enemies within, enemies whose civilization values are also deemed to be politically and morally inferior compared to the governing elites'. For Elias it is 'only when these tensions between and within states have been mastered that we can expect to become more truly civilized'. Thus he concludes his book by forecasting the development of 'a worldwide monopoly of force [...] for the pacification of the earth' (Elias 1994: 446).

Writing about the 'conundrum' of the United Nations Security Council, dominated by just five nations out of 191, Paul Kennedy, a historian of the UN, writes that 'everyone agrees the present structure is flawed; but a consensus on how to fix it remains out of reach' (Kennedy 2006: 76). Given the failures and apparent partiality within the UN system, the difficulties in meeting Elias' neo-Hobbesian ambitions are all too apparent. However, Elias still succeeds in showing us how nations and national identities gradually came about in the wake of the formation of states, and not the other way around. From this perspective, the primary motivation in the development of states is to mediate and delimit intra-elite feuding and establish elite authority through the control of peace and war. Elias' now classic study focused on the history of continental Europe. However, in South America, where, for example, the unbridled avarice of the European elites clashed with the Inca empire, we can also see faltering attempts to limit violence and increase mutual understandings. This is evident in historic treaties and petitions all the way back to the treatise on good governance drawn up by Felipe Guaman Poma de Ayala (c. 1535–1616) intended for King Philip III of Spain.

It is worth noting that Elias witnessed the Nazi conquest of the state's monopoly on violence in Germany, an experience which no doubt influenced his view that old social fractures are only covered over by the veneer of 'civilization'. This mirrors Hobbes' earlier reactions to the English Civil War. Essentially, for Elias, civilization is a series of social mediations and restraints among competing elites and ruling classes whose capacity for brutality remains an implicit but constant threat. This is an important argument to bear in mind when considering the contemporary development of progressive politics in Latin America. In Elias' purview there is always the danger that these fractures open up and engulf society in moments when the centre can no longer hold or is simply seen as too weak. In his historical account Elias sees the weakness that is always immanent at the centre as a centrifugal force because rulers always attempt to extend their authority and in so doing cannot help but devolve power outwards (Elias 1994: 195ff.). This is what occurred during the European process of feudalization, and it may also be sensed in the politics of modern regionalism.

In organizations like CELAC, mentioned in previous chapters, we can see elite attempts to forge a regional modus vivendi. In essence Elias' historical sociology of what is still called 'civilization' is really the study of the spread of these interdependent relations, which – when recognized as such within societies – promote cultural reflexivity, foresight and self-constraint. For the European bourgeoisie these capacities also developed as part of the culture of finance and trade. The greatest challenge to the bourgeoisie in the nineteenth and twentieth centuries was the recognition among the working classes of their own chains of interdependence. Thus very different class visions of social, political and cultural agency emerged. While the ruling classes and the bourgeoisie expanded market relations and monetary exchanges over an expanding territory, the organized working class, born in the same process, sought to regulate and limit competition. Socialists drew on this organic experience of cooperation in different countries to formulate internationalist political and economic theory. This can still be differentiated from capitalist regionalism.

The historical insights provided by Elias on state formation have a bearing on how we may see the politics of energy resources operating within and between states. It was not for nothing that V. I. Lenin (1870–1924) coined the slogan 'Communism = Soviet power + the electrification of the whole country' in 1921.[3] Clearly states and their political systems are not static. Even states that are well established can disintegrate with profoundly tragic consequences. Moreover, all states are in ongoing processes of formation and as such carry few guarantees, even when they are also the subject of policies geared to regional integration. Integration at the international level is rarely conducted on an egalitarian basis because the internal politics of nations are not fully articulated at such levels. It is the supra-political class and elites who meet and engage in the necessary political and commercial horse-trading. These dealings often involve agreements about energy, and we have seen that post-communist states as well as colonially founded states still stitched together by the imperial imagination are becoming more or less socially coherent entities as a result of energy politics. In many cases the states' monopoly of violence is directed against articulations of resource sovereignty within their boundaries, and when governments strongly articulate resource sovereignty – for example in Venezuela – there is the not unreasonable fear of the overthrow of the state itself. So energy systems are implicated in what Elias thinks 'civilization' is all about, namely threats of violence on the one hand, and acknowledgements of interdependency and the need for stability on the other. It is on this sociological basis that we may realistically consider the emergence of 'energetic states', i.e. those states where energy resources play an increasing role in upholding a civil peace.

Yet in considering the increasing complexity of energy politics it is clearly important to properly address the Hobbesian worldview that has long been at the root of the right-wing argument that war is the natural state of human affairs. Radical and left-wing discussions of energy politics often tacitly give way to this view more than some on the left may care to admit. There are linkages between the ideas of Hobbes and Elias and liberal ideas of what constitutes 'peace', and these need to be carefully examined if the notion that war is the natural state of affairs is not to become a self-fulfilling prophecy.

In the aftermath of the Cold War there developed a general conviction in the international community about how to address post-conflict situations, i.e. the transformation of war-torn societies through political and economic liberalization. Influenced by the immediate post-Cold War euphoria, captured in Fukuyama's *The End of History and the Last Man* (1992), the roots of this conviction were deeper, stretching back to the Enlightenment. By the mid-1990s this had coalesced into a package of policy interventions under the banner of 'liberal peace-building' that were tested and further developed in the new post-conflict contexts. However, by the late 1990s the presumed success of this package started to be called into doubt.[4] From Bosnia, Rwanda, East Timor and Iraq to Afghanistan, rather than fostering peace, the interventions led to a resurgence of political violence. It also became clear that while claiming to bring 'freedom and democracy' the international community's efforts, led by the USA, also countenanced the use of highly illiberal and undemocratic practices (from rendition to drone attacks) to produce liberal democratic societies and institutions. It became evident too that the stimulation of liberal institutions and values came second to Western capital's interests in securing profits and access to new economic markets and energy resources in particular.

Importantly, while Latin America remained separated from the processes of military occupation and humanitarian interventions of Western peace-building, encouraged by political and new trade agreements, remaining neoliberal governments in the region have enthusiastically adopted liberal peace-building's strong state emphasis as inspiration, if not model, for responding to economic and security challenges. Indeed, the peace processes in Guatemala and El Salvador form the background of earlier experimentation with liberal peace-building. In present-day Colombia, Peru, Guatemala, El Salvador and Mexico a democratic hard hand is seen as necessary to deal with security threats and guarantee defence of the national economy. For critics, however, the operation of this *mano dura* (hard hand) policy has not led to increased peace and prosperity, but more the situation seen in other contexts of the liberal peace, i.e. states of unending war (Duffield 2007).

Political power and resource power

Although it is rarely discussed, how one gets away from the Hobbesian standpoint is an important issue, especially in light of the negative outcomes of the 'liberal peace' outlined above. Indeed, any move beyond this reworking of the old-fashioned King's Peace and partial democracy in Latin America depends on it. In this respect, the work of the Harvard sociologist Talcott Parsons (1902–73) is instructive. Although the Frankfurt School sociologist and philosopher Jurgen Habermas has pursued a very different critical approach from

Parsons, Habermas' theories of political power draw on Parsons' work. Both offer ways of thinking about the nuances of changes in power structures. Their insights can help take us on in time towards what might be signs of realistic alternatives to the modernist positioning of current energetic states.

Compared to Elias, Parsons offers us a positively anti-Hobbesian worldview. Where Elias seems to see a constant threat of political breakdown, Parsons stresses functional integration and a dialectics of power that create stability. Indeed, this is the mainstay of his structural functionalism. From this perspective it would be no surprise that energy can be an incendiary factor as well as a stabilizing influence in Latin American politics. Parsons was a lifelong defender of liberal democracy. Yet unlike some of the more circular accounts of 'goal-orientated behaviour' in functionalist theories, Parsons' (1963) essay 'On the concept of political power' tackles the profoundly dysfunctional elements in systems underpinned by the principle of quid pro quo. In his essay he connects political power and money power in a way that would interest today's anti-capitalists on both left and right. According to Parsons, the circulation or the flow of political power may be broken in rather the same way that fractional reserve banking increases a bank's wealth and power by lending more money than it holds in real terms. Using this example, Parsons insists that power is a diffused open system and not a zero-sum game, i.e. there is not a fixed quantity to be shared out or fought over. Importantly in this view, the increased power of one group does not necessarily come at the expense of another. This view has considerable saliency for the way we see power being contested in Latin America. Historically, increasing or decreasing the pace of energy production from finite resources has conferred power on corporations, states and even paramilitary groups, but not necessarily at the expense of one another.[5] In the case of Latin America, the political strength of the left has grown as a result of resource sovereignty, but from our theoretical and empirical perspective at this point, there is no reason to see the political right occupying a weaker position.

Parsons argues against the Hobbesian tradition of thought, which he reminds us treats power independently of the medium of its operation. Where Elias focuses on the monopoly of violence, Parsons treats it as a medium of power among others. Obviously this treatment can be applied to the control of energy resources. The relationship between money and energy has increased in complexity as a result of the petrodollar settlement, which meant that oil and other major commodities were traded in US dollars. Since the US abandonment of the gold standard under President Nixon this settlement helped support massive US federal deficits and protected the US dollar from the effects of speculation because everyone trading in the dollar has an interest in maintaining its value. Unsurprisingly, the system of dollar hegemony agreed with OPEC is reported to have delighted the historical sensibility of one famous US government adviser so much that he exclaimed: 'we've run rings around the British Empire'.[6] Indeed, the US dollar is still not indexed to a hard value base. This has allowed successive US governments to print dollars almost at will, safe in the knowledge that they will be used in trade and not redeemed. Yet without the decoupling of the world reserve currency from gold reserves, it is difficult to imagine the sense of urgency which the discourses of creativity and knowledge have now acquired in advanced capitalist economies. As we pointed out in our introductory chapter, these discourses are the polar opposite of resource sovereignty.

If power is not a zero-sum game, how then are we to account for apparent losses of power and influence? For Parsons the multiple means through which power is created and articulated means that politics is subject to inflationary and deflationary spirals. Parsons uses McCarthyism as an example of a deflationary spiral which resulted from the over-commitment on the part of the United States after 1945 to becoming the world's dominant military power. With US power *inflated* through the military-industrial complex, McCarthyism sought to regulate, or *deflate*, the rest of the political field in civil society by putting pressure on US citizens to liquidate any commitments that might stand in the way of loyalty to one's country. We need not see political phenomena, such as McCarthyism, having only one root cause, however. Indeed, Parsons' theory of political power serves to underline the way that expressions of power interact dialectically whether in synthesis or in opposition to one another, and also as mediations of one another. In the end the excesses of McCarthyism threatened to deflate the whole power system in the United States. Therefore the repressive movement had to be brought to heel for the US military project to go on. Similarly we can see that the politics of resource sovereignty interact directly and indirectly with other means of accruing power within and beyond the states. Resource sovereignty can be a lever for the redistribution of knowledge and skills, but more often it appears to deflate that political task.[7]

Parsons' theory of human actions was revised and adapted by Jurgen Habermas in the latter's theory of communicative action. Like Parsons, Habermas (1984, 1987) nuances the means of power, but unlike Parsons, Habermas stresses the capitalist system's barriers to integration and proper communication as a result of the social stratification of classes. For Habermas, power is rarely 'hard' and therefore it can seldom be quantified and counted like money, debt or votes, etc. Power is also articulated at the level of communication, influence and culture. Where Parsons may be accused of naturalizing these power relations for the sake of a liberal-friendly theory of systemic and functional equilibrium, Habermas shows how soft forms of power demand qualitative and comparative modes of analysis (Habermas 1987: 179ff.).

There are concrete material relations that underpin both modes of power that are worth foregrounding here. In light of the dialectical processes effecting political power described respectively by Parsons and Habermas, we can return to the Nicaraguan president, Daniel Ortega, and the US guru of cultural regeneration, Richard Florida, both of whom we cited in our introductory chapter. Ortega's political speech, which for us evoked the conversion of energy into political power, centres on territorial resources. On the other hand Florida's claims about post-industrial development having surpassed a dependency on labour and natural resources were based on the knowledge, creativity and *mobility* of the so-called creative class (Florida 2005: 49). Between these two very different but deeply linked expressions of power, there is more to consider than mere political rhetoric. What is at stake here is the competitive structuring of the global market. For example, Manuel Castells (Castells and Ince 2003), a theorist of techno-social change, calculates that a year of strenuous and intelligent labour on the part of farmers in the global South 'cannot compete, in value creation' with one hour of labour on the part of a cutting-edge software programmer in the global North (ibid.: 31). It is not 'the free market' which deflates the value of the former and inflates the value of the latter; rather it is the ever-present political and conceptual management of the market. There is of course nothing new in this insight as it recalls Marx's analysis of the ways 'capitalist production develops technology only by sapping the original sources of all wealth – the soil and the labourer'.

A writer who has drawn on Marx's writings about technology as a means of exploitation is the political ecologist Alf Hornborg. Hornborg addresses the risks of climate change as a symptom of the whole economy, globally, and therefore the inequitable management and distribution of natural resources is his major concern. Hornborg (2013) argues in *Global Ecology and Unequal Exchange* that the development of European economic ideas was calibrated by technological exploitation driven by capitalism. If Parsons shows us that power is not a zero-sum game, Hornborg reminds us that we live in a zero-sum world of finite resources that are competitively devalued by capitalism in the same way the system competitively devalues labour. He argues that our subservience to capitalist ideology surrounding technology (even as we face the risks of a major environmental crisis) is the key to understanding why there is an emphasis on *resilience* rather than revolution in contemporary politics and social science.

For Hornborg one of the main problems in this regard is that theories of value have been overwhelmingly normative on all sides and therefore became excessively scholastic across the ideological spectrum. He argues that theories of value ought to be more ethnographic and descriptive of 'the valuations that people actually make themselves, not on what theorists claim to be an objective source of value' (ibid.: 105). Hornborg's argument deserves to be taken seriously, not least because many commentators on the left of the political spectrum have tended to portray the friction generated by resource sovereignty and environmental activism as a combined and potentially revolutionary threat to the global economic status quo.[8] However, Elias, Parsons and Habermas can show us how some of the inflated political rhetoric surrounding energy politics can be connected to a deflation of political action in other important spheres. Hornborg also deserves attention because he is right that the evaluation and self-evaluation of labour are socially subjective matters. Indeed, the adoption of the labour theory of value by workers ought to be understood not so much in relation to the scholastic traditions Hornborg targets, but as part of the collective acquisition of power. This could not have been achieved without workers' reflexivity concerning the economic structuring of their exploitation.

Machine fetishization

Expanding on the Marxist tradition, Hornborg sees the spread of capitalist discourses of value stemming from 'machine fetishism' (ibid.: 27ff.). He locates this fetishism at the heart of what he further describes as an unwarranted faith in technology as the means to solve the contradictory social and environmental relations created by industrial capitalism. According to Hornborg, technological change is not a solution to the social and environmental problems created by capitalism. He argues that technology, geared as it is to the capitalist-led model of growth, is a central aspect of our mounting environmental problems. This is because the dynamics of exploitation are embedded not only in technological change but also in the notion that structural economic problems can be solved by technology. Conversely, as Marx showed, technological advances were made because of the capitalist requirement to exploit labour and nature more completely. Rather than liberating people from labour, technology drives them further into the processes of capitalist production. This movement has only intensified, not diminished, with every new invention. Indeed, growing numbers of the world's poor are increasingly dependent on cheap goods produced by large-scale capitalist production, the profitability of which is protected by technological advances.

Castells' software designer can certainly afford to purchase 'eco goods' and would also count as a member of the social group that Florida sees as delivering swathes of humanity from direct dependency on labour and natural resources. Even if this 'creative class' was not a highly rhetorical category conjured up by someone with an interest in neoliberal regeneration policies, a rising environmentally conscious class of consumers would not be in a position to change the structural dynamics of capitalist production and consumption. This problem is well understood by Hornborg. Indeed, we may regard Florida's conception of a mobile creative class as a justification of the global redistribution of enrichment and immiseration that is afforded by technology.

So given the weak viability of current models of sustainable development and the inequities of the combined and uneven development that Hornborg addresses, what are his policy proposals? Hornborg is acutely aware of the power of money and also that Luddite-style responses to 'machine fetishism' have led only to political dead ends. He therefore suggests we turn our attention to 'money fetishism'.

He proposes that states create complementary currency systems that would offer tax incentives to support local trade. He illustrates his argument by referring to the pre-colonial exchange values among the Tiv people in Nigeria and contemporary local exchange trading systems (LETS) internationally (ibid.: 102–18). In essence the aim of Hornborg's proposals for reforming monetary exchanges is to limit the international trade and competition seen to damage social relations and the environment.

As we have already indicated, technology has not delivered more free time. Rather it has been used by capitalism to intensify labour and expand a 'social factory' based on stick-and-carrot ethics. Thus capitalism alters moral life worlds. However, from this point of view, capitalist economy is not so different from the Tiv socio-economics Hornborg cites. The secret of capitalist economism is that it seeks to insert itself into every aspect of life, thereby authoring social values. It is precisely this level of integration which is foreshadowed in the integration of pre-colonial Tiv economy and society. For example, in Tiv socio-economics women were naturalized as status symbols for men.

Arguably, political ecology à la Hornborg deflates the socialist project intended to denaturalize the economy for the sake of freedom and equality. Despite drawing on anti-capitalist theory, Hornborg's political ecology appears to underestimate the true scope of capitalist economism. As has been made visible in this volume, Latin American governments on the political left and right sing the international song of taking climate change seriously,[9] but simultaneously argue the economic need to expand all forms of energy and fossil fuel production. The depth of this value system was demonstrated more broadly by government responses to the global financial crisis that resulted in a massive upward distribution of wealth to the rich and austerity for the rest. One lesson from this is that there is a need to reform and rein in the central institutions of finance capitalism. On the other hand the idea of developing a bottom-up economy through local currencies may only be a plausible form of protectionism in an age when economic protectionism is virtually outlawed.

It may well be asked why a radical strand of political ecology which critiques the way revolution is superseded by theories of resilience leaves aside the most obvious structural issues from a moral economy point of view. The answer must be that notwithstanding the need to counter technocratic thought with practical reasoning (something Hornborg evidently understands) there is a tendency in political ecology to want to change human behaviour through a stick-and-carrot approach to socio-economics. The stick is the risk that unless we address climate change there will be a massive eco-systemic breakdown. However, in this sort of scenario equality does not appear to be a public or moral good which deserves to be pursued in its own right; if equality is not entirely sidelined it appears merely as an adjunct to a positive environmental end. Thus its meaning is devalued. It is also a political and academic vanity which ignores the deeply anti-egalitarian argument that war is the natural state of affairs and therefore conflicts over energy and natural resources, now and in the future, may be quite normal and not connected to eco-systemic breakdown. Instead of challenging the right-wing discourse many analysts now regard a major 'socio-ecological collapse' as 'a possibility for renewal', Hornborg points out. 'It is quite possible', he writes in tacit endorsement, 'that nothing short of a system breakdown will force us to change our habits' (ibid.: 145). Here Hornborg's political ecology gives way to the Hobbesian worldview described above, which seems to naturalize disaster while undermining the human capacity to will a better world order into being.

In reality what we are faced with is not an environmental crisis per se. Not only will the planet live on without our species (in the worst-case scenario), but the causes of environmental problems, and therefore the solutions, will continue to be disputed. There are, of course, many images, films and reports that dramatically envisage environmental catastrophe but there are repressive aspects to this wider discourse that are being used to drive people away from the politics of social and environmental justice. As the oft-quoted Marxist cultural critic Frederic Jameson writes: 'Someone once said it is easier to imagine the end of the world than to imagine the end of capitalism. We can now revise that and witness the attempt to imagine capitalism by way of imagining the end of the world' (Jameson 2003: 76).

It may be argued, of course, that the end of capitalism is not actually necessary or even desirable from an ecological point of view. However, it seems increasingly impossible to envisage a transition to a more environmentally friendly economic system if business is allowed to go on as usual. The results are already obvious in many countries – for example, in the expansion of short-haul flights and road transportation compared to the running down and high cost of rail transport if it exists at all. Some environmentalists appear more concerned with futurological thinking and await a dramatic consciousness-altering ecological crisis. But if they are right and this crisis befalls humanity it will be experienced first and foremost at the level of politics. In the face of the combined problems of energy and the environment, policies for energy localization and sustainability, and even the nationalization of energy utilities, are no longer the preserve of the left. We can see, for example, that the far right is effectively colonizing leftist policies, making them part of extreme cultural nationalist projects. Therefore, if we are to take the dynamics of global capitalism into account, we must consider resource sovereignty as a site of contest over the meaning of popular sovereignty and not simply an expression of it. Here we recognize an echo of our introductory emphasis on Graeber's insight that the ultimate stake of politics or social order is the struggle, not to appropriate value, but to establish what value is (Graeber 2005).

Ethno-nationalism in Latin America hinges on the high demographics of people of mixed race and mestizo origins making the historic social and political trend towards pluralism. Indeed, it should be evident from the political processes, clashes and movements covered in this volume that if a King's Peace exists, the King's orders are far from unquestioned. However, a warning needs to be underlined. The influence of the far right on democratic political discourse everywhere is a pressing issue that calls for further research. So far the rise of the far right internationally has been measured mainly in quantitative terms, not in the sort of qualitative terms distinguished by Habermas through his critical engagement with Parsons. If we follow this line of analysis we can see that the issues of patrimonial capitalism are certainly not confined to the problems of agreeing about progressive taxation or the parallel rise of rentier energy states. The left-leaning economist Piketty (2014: 487) highlights that the widening gap between growth and accumulation is connected to deeper value-creating systems concerning education, meritocracy and the institutional reproduction of inequality – all with serious implications for economic development. The liberal willingness to ameliorate inequality through various conservative models of welfare but *not* to hinder the reproduction of inequality by instituting strictly egalitarian education systems is a key area where liberal and far-right beliefs coincide.

Piketty is sensitive to the way the level of mutual respect among the class adversaries in the political economy may well hinge on the shared experience of education or the fracturing of education. What Piketty misses is the way that such processes also relate to value-creating systems rooted in nature and controls over environmental resources and markets. Here another element of liberal patrimonial capitalism is the creation of governance institutions and a common market for goods and services to supposedly protect and conserve the environment. But, it must be asked, for whom? Other logics, not driven by GDP-measured growth, and therefore other more sustainable relationships to nature, are ignored. As if in a mood of denial, the more society institutionalizes the reproduction of social and environmental inequality – from private schooling to life in gated and almost wholly privatized communities – the more likely it is that the winners in the system will be convinced of their own inborn merit and become alienated even from their kin. This is the major threat to the social coherency of states, and on its own resource sovereignty does little to counter the trend.

Conclusion: contested powers

Our understanding of politics and state formation in *Contested Powers* is drawn from empirical qualitative analysis, but also from an ideologically diverse range of theorists. Whatever explanatory strength theory may have comes from critical comparison and synthesis, not ideological dogmatism. And of course, in a neo-Kantian view, this process of thesis, antithesis and synthesis is important because theoretical issues inform and shape empirical studies. As we have shown in this chapter, there is a danger of inadvertently strengthening the Hobbesian worldview by ignoring it. We take two important anti-Hobbesian points from Parsons; first, that the various means by which power is accumulated ought to be differentiated and that they need to be analysed relationally. Secondly, that when applied, our theoretical inquiry shows that power in Latin America is certainly not a zero-sum game. Chapters in this book make evident that sections of the organized working class as well as indigenous, rural and urban social movements that have been severely repressed for decades have in many cases recovered power by pursuing the politics of energy and natural resources. Nevertheless, this recovery has not taken place at the expense of the long-established elites who (*pace* Elias) have not completely lost control over state violence, although the willingness to deploy it quite as brutally or overtly as in the past seems to have diminished. But this cannot be taken to mean that the elites are in any sense losing power.

The underlying balance of power is not as clear in many nations as it would appear in popular reports of the Latin American 'left turn'. Latin America is not transformed, but under transformation. Easier to discern are new forms of governmentality that are coming into being on the basis of the national and international politics of energy resources. This political field is characterized by an uneven and rather hierarchical politics of regional integration, which hangs together with strategies for the foundation of Latin American interdependencies reliant on agreements related to the exploitation of particular energy resources. What this competition for resource sovereignty writ large, or in other words the contest over important levers of power, will really amount to depends on the extent to which regional integration can take place from below, and not just at the level of a political class which has integrated the left within its ranks and in many ways remade itself. This does not remove the discursive issue of resilience rightly identified by Hornborg as a barrier to more revolutionary-minded reforms. Clearly, integration from below can be managed from above in order to deflate its potential radicalism. There is then a serious danger that the politics of resource sovereignty will continue to be shaped according to the requirements of regional capitalist development. The idea of Latin America may be further shaped according to the logic of resilience, rather than a true socio-economic revolution or a serious attempt to tackle patrimonial capitalism.

Notes

1 www.ejolt.org.

2 www.economist.com/blogs/americasview/2014/02/oil-and-gas-falklands.

3 This slogan has its origins in Lenin's speech 'Our foreign and domestic position and Party tasks', given in November 1920. See: www.marxists.org/archive/lenin/works/1920/nov/21.htm, accessed June 2014.

4 www.lse.ac.uk/alumni/lseconnect/pdf/winter2009/liberalpeace.pdf.

5 In recent years Nigeria has provided examples of paramilitary groups interrupting oil flows and gaining political leverage by doing so.

6 Herman Kahn (1922–83), quoted in Naylor (1987: 26).

7 The relative low priority given to educational reform in Venezuela exemplifies an internal deflationary tendency associated with resource sovereignty. However, Venezuela's attempts to acquire technical knowledge from other countries, such as China, also suggests the potential to use resources to provide leverage externally. These two

movements, internally and externally, are related because the class and racial composition of those Venezuelans who are likely to benefit from international links will be shaped by the reproduction of long-standing educational inequalities in the country.

8 See 'Naomi Klein: How science is telling us all to revolt', www.newstatesman.com/2013/10/science-says-revolt.

9 Including the Morales government's organization of the World Peoples' Conference on Climate Change in 2010 and introduction of the Law of Mother Earth and Integral Development in 2012.

References

Castells, M. and M. Ince (2003) *Conversations with Manuel Castells*, Cambridge: Polity Press.

Chomsky, A., G. Leech and S. Striffler (2007) *The People behind Colombian Coal*, Casa Editorial Pisando Callos.

Cleaver, H. (2000 [1979]) *Reading Capital Politically*, Leeds and Edinburgh: Anti/Theses and AK Press.

Duffield, M. (2007) *Development, Security and Unending War*, London: Polity Press.

Elias, N. (1994) *The Civilizing Process*, revised edn, Oxford: Blackwell.

Florida, R. (2005) *Cities and the Creative Class*, London: Routledge.

Fukayama, F. (1992) *The End of History and the Last Man*, New York: Simon and Schuster.

Graeber, D. (2005) 'Value as the importance of action', *The Commoner*, 10 (Spring/Summer), www.commoner.org.uk/the_commoner_10.pdf.

Habermas, J. (1984) *The Theory of Communicative Action*, vol 1: *Reason and the Rationalization of Society*, Cambridge: Polity Press.

— (1987) *The Theory of Communicative Action*, vol. 2: *Lifeworld and System: A Critique of Functionalist Reason*, Cambridge: Polity Press.

Hernandez, L., H. Prieto and F. Garcia (2011) *Petróleo y Conflicto: en el Gobierno de la Seguridad Democratica 2002–2010*, Universidad Nacional de Colombia.

Hobbes, T. (2003 [1651]) *Leviathan*, Renaissance Books.

Hornborg, A. (2013) *Global Ecology and Unequal Exchange – fetishism in a zero-sum world*, London: Routledge.

Jameson, F. (2003) 'Future city', *New Left Review*, May/June.

Kennedy, M., B. Lietaer and J. Rogers (2012) *People Money, the Promise of Regional Currencies*, Axminster: Triarchy Press.

Kennedy, P. (2006) *The Parliament of Man – the United Nations and the Quest for World Government*, London: Allen Lane/Penguin.

McNeish, J. and J. Sande Lie (2010) *Security and Development*, Critical Interventions in Anthropology, Forum for Social Analysis, vol. 11.

Naylor, R. T. (1987) *Hot Money and the Politics of Debt*, London: Unwin.

Parsons, T. (1963) 'On the concept of political power', in R. Bendix and S. M. Lipset (eds), *Class Status and Power – Social Stratification in Comparative Perspective*, London: Routledge & Kegan Paul, pp. 240–65.

Petras, J. and H. Veltamayer (2014) *The New Extractivism: A Post-Neoliberal Development Model or Imperialism of the Twenty-First Century?*, London: Zed Books.

Piketty, T. (2014) *Capital in the 21st Century*, Cambridge, MA: Belknap/Harvard University Press.

Scott, J. (1999) *Seeing Like a State – How Certain Schemes to Improve the Human Condition Have Failed*, New Haven, CT: Yale University Press.

Strauss, S., S. Rupp and T. Love (2013) *Cultures of Energy: Power, Practices, Technologies*, Walnut Creek: CA: Left Coast Press.

ABOUT THE AUTHORS

Edith Barrera is a research professor in the Institute of International Relations at the Universidad del Mar (México, Oaxaca). Barrera has authored various articles and co-authored the following books: *Environmental Governance* (University of Veracruz, Plaza y Valdés, 2013), and *Traditional Wisdom and Ethics and Environment* (Ed. Academía Española, 2011). For the past six years, Edith Barrera has conducted research on renewable energy in the Isthmus of Tehuantepec, Oaxaca and its impact in indigenous communities.

Dominic Boyer is professor of anthropology at Rice University and founding director of the Center for Energy and Environmental Research in the Human Sciences (CENHS, culturesofenergy.org), the first research centre in the world designed to promote research on the energy/environment nexus in the arts, humanities and social sciences. He is an editor of the journal *Cultural Anthropology* (2015–18) and also edits the *Expertise: Cultures and Technologies of Knowledge* book series for Cornell University Press. His most recent book is *The Life Informatic: Newsmaking in the Digital Era* (Cornell University Press, 2013) and with Imre Szeman he is preparing *Energy Humanities: A Reader* for Johns Hopkins University Press.

Einar Braathen is a senior researcher in international studies at the Norwegian Institute for Urban and Regional Research (NIBR). He was the leader of a work package 'Politics and Policies to Address Urban Inequality' in the EU-financed project *Urban Chances – City Growth and the Sustainability Challenge* (2010–14). From 2015 to 2018 he is heading the project *Insurgent Citizenship in Brazil: the Role of Mega Sports Events*, funded by the Research Council of Norway. Braathen has authored and co-authored a series of articles and books including *Poverty and Politics in Middle Income Countries* (Zed Books, 2015) (see: http://www.nibr.no/staff/70/Braathen).

María Victoria Canino is a sociologist at the Central University of Venezuela (UCV), assistant director of the Centre for the Study of

Science, and director of the Laboratory for Political Ecology at the Venezuelan Institute for Scientific Study (IVIC). She is the author of over fifty publications and has supervised many students at graduate and postgraduate level. She currently coordinates the state-financed research projects 'Poverty, Environment and Climate Change' and 'Socio-environmental Conflictivity and Community Participation: Punta Cardón'. She teaches courses in social studies at postgraduate level at IVIC and in research methodology at the Department of Sociology at UCV.

Cymene Howe is associate professor of anthropology and a core faculty member in the Center for Energy and Environmental Research in the Human Sciences at Rice University. Her research interests are broadly concerned with rights, ethics, knowledge, materials and political movements in Latin America. She is the author of *Intimate Activism: The Struggle for Sexual Rights in Postrevolutionary Nicaragua* (Duke, 2013). Her forthcoming book with Cornell University Press, *Ecologics*, analyses renewable energy transitions in Oaxaca, Mexico and queries the overlapping conversations between feminist and queer theory, new materialisms, multispecies ethnography, ontologies and imaginaries of the future in the Anthropocene.

Virgilio Reyes is a senior researcher and current director of the Faculty of Latin American Social Sciences (FLACSO) in Guatemala. He holds a PhD in rural sociology from Wageningen University in Holland and a masters in social anthropology from the Universidad San Carlos and Oslo Universities. Until recently Reyes was the research director for the section on population, environment and development at FLACSO. He has a long publication and research history including multiple reports, articles and book chapters in regional and national publications.

Iselin Åsedotter Strønen is a researcher at Chr. Michelsen Institute (CMI) in Bergen, Norway. She gained her PhD in social anthropology from the University of Bergen in 2014. Her publications include 'Development from below and oil money from above. Popular organization in contemporary Venezuela', published in McNeish and Logan (eds), *Flammable Societies: Studies on the Socio-Economics of Oil and Gas* (Pluto Press, 2012) and 'A feminist revolution? Women, women's struggle and activism in Chavez' Venezuela', published in the *Norwegian Journal of Anthropology* (2013). Strønen is currently project coordinator and researcher for the CMI-based research project *Everyday Maneuvers: Military–Civilian Relations in Latin-America and the Middle East*.

Fernanda Wanderley is professor at the Postgrado en Ciencias del Desarrollo de la Universidad Mayor de San Andrés (CIDES-UMSA). Wanderley has authored and co-authored a series of articles and books including ¿*Qué pasó con el proceso de cambio en Bolivia? Objetivos acertados, medios equivocados y resultados transtrocados* (Plural, 2013) and 'The economy of gas and the politics of social inclusion in Bolivia', in McNeish and Logan (eds), *Flammable Societies: Studies on the Socio-Economics of Oil and Gas* (Pluto Press, 2012). She is currently the director of the Doctorate Program on Development in CIDES-UMSA and carrying out research on solidarity, economy, employment and gender relations (www.fernandawanderley. blogspot. com).

Cecilie Vindal Ødegaard is associate professor at the University of Bergen, Department of Social Anthropology. Ødegaard is author of the monograph *Mobility, Markets and Indigenous Socialities: Contemporary Migration in the Peruvian Andes* (Ashgate, 2010), and has also published a series of articles in journals such as *Journal of the Royal Anthropological Institute*, *Ethnos Journal of Anthropology*, *Forum for Development Studies* and *Journal of Ethnobiology and Ethnomedicine*, among others. Her research interests include questions of indigeneities, landscape and animism in the Andes, as well as a focus on work, gender, urbanization, informal economies and the state.

INDEX

Abramsky, Kolya, 22, 23
Abreu, José Antonio, 217, 219, 224, 225, 228, 230, 232, 243, 244, 247
Agamben, Giorgio, 19, 161
agarradas strategy (Guatemala), 46
Agnew, John, 41, 60
Alaniz, Julio César, 66
ALBA *see* Bolivarian Alliance of the Peoples of Our America (ALBA)
Alba Eólica energy company, 71
Alba Generación energy company, 71
Albanisa energy company, 71
Alcorta, Lourdes, 148
Alemán, Arnoldo, 68
Alianza del Pacífico, 282
Álvaro Obregón (Mexico): blockade of, 92, 94, 95, 103
Amatique Bay (Guatemala), 45
Andean Amazonian Capitalism, 269
Andean Community of Nations (CAN), 282
Andean Development Fund (CAF), 242, 281
anthropology, 20-2 *passim*, 168–70 *passim*, 190, 278; of energy, 20, 22
anti-globalization movements, 104
Arequipa (Peru), 143, 147, 154, 161, 163; protests against gas prices, 146
Argentina, 9, 293; Dirty War, 293, 294
Argentine Anticommunist Alliance (AAA), 293
Arias, Manuel, 122
Arica (Bolivia), 15
Armas, Castillo, 44
Arts Councils (UK), 244
Arzac, Jonathan Davis, 94
Asamblea de los Pueblos Indígenas del Tehuantepec en Defensa de la Tierra y el Territorio, 105, 106

Bagua (Peru), 2009 uprising in, 145, 146

Baker, Geoff, 226, 229
Balibar, Étienne, 231
Banco del Sur, 16
Barcelona Centre for International Affairs (CIDOB), 264, 271
bare life, 19, 161
Barth, Fredrik, 278
Basic Petroleum International, 49
Basic Resources oil company, 49
Batahola Sur (Nicaragua), 77, 78, 79
Bebbington, Anthony, 261
Belo Monte dam (Brazil), 13, 14
Berger, Oscar, 56
binnizá community (Mexico), 107
biodiversity, 48
biofuels, 14
biopolitics, 21
Blom Hansen, Thomas, 8
Bolaños, Enrique, 68, 72, 83, 84
Bolívar, Simón, 282
Bolivarian Alliance of the Peoples of Our America (ALBA), 15–16, 282
Bolivarian Revolution (Venezuela), 117, 128, 134, 231
Bolivia, 9–11 *passim*, 15, 163, 169, 258–82 *passim*; 2009 constitution, 175, 176, 264; anti-smuggling measures, 162; blackouts in, 274; development in, 268; economy, 173, 180, 188, 189; energy demand, 273, 274; Energy Development Plan (2009), 176; fishing industry, 210; foreign investment in, 274; fuel subsidy, 144, 178, 181, 182, 273; gas production, 180; hydrocarbon policy, 175, 177, 272; indigenous communities, 275, 276; investment in energy sector, 273; lithium mining in, 275; Ministry of Hydrocarbons and Energy, 181; National Development Plan (2006), 176; nationalization of energy

INDEX | 317

resources, 175, 177, 184, 185, 188, 189, 272; 'neoliberal nationalization' in, 267; public policy in, 172; removal of fuel subsidy, 144, 169, 178, 179, 183, 184, 186; transport in, 181, 182, 187
Bolsa Familia programme (Brazil), 258
Bor, Domingo Sánchez, 229
Borchert, Gustavo, 226, 237
Bourdieu, Pierre, 27, 223, 231; *Distinction*, 230
Boyer, Dominic, 21, 142
Brazil, 9, 10, 16, 200, 258; and renewable energy, 13; auctioning of oil fields, 195, 196, 213; environmental movement, 208, 209, 212; nationalization of oil resources (1939), 201; oil industry, 201, 208; oil royalties, 205, 206, 212; *pré-sal* oil fields, 16, 195, 196, 201–4 *passim*, 213; social movements, 196; trade union movement, 202, 203, 211
bribery, 101, 151, 161, 162
bricolage, intellectual, 27, 28
British Gas company, 267
Brookings Institution, 259
Buci-Glucksmann, Christine, 197
buen vivir, 261, 262, 277, 278, 284, 292

Cabral, Sérgio, 206, 207, 213
CAF *see* Andean Development Fund (CAF) *and* Development Bank of Latin America (CAF)
Calderón, Felipe, 99
Camisea gas project, 145
Campos basin oil spill (2011) (Brazil), 209
CAN *see* Andean Community of Nations (CAN)
Canelas, Iván, 186
Cantarell oil field (Mexico), 98
capitalism, 217, 218, 280, 305, 306, 307; end of, 308
carbon capture and storage (CSS), 208
Cardoso, Fernando Henrique, 198, 200, 202, 205, 206
Castells, Manuel, 304, 306
Castree, Noel, 262
Catia TV (Venezuela), 241
CELAC *see* Community of Latin American and Caribbean States (CELAC)
Central American Free Trade Agreement (CAFTA), 282
Central American Institute for Fiscal Studies (ICEFI), 41
Central Unica dos Trabalhadores (CUT) (Brazil), 202
Central University of Venezuela (UCV), 129
centralization, 24
Chacón, Cecilia, 271
Chalfin, Brenda, 142, 163
Chamorro, Violeta, 68, 70, 72
Chapman, Andrew, 92, 94, 95, 110
Chávez, Hugo, 7, 117, 128, 134, 227, 238, 245, 258
Chavismo, 238, 243
Chevron oil company, 11, 209
Chile, 15, 258, 282; and renewable energy, 13
Chirino, Orlando, 245
CIDOB *see* Barcelona Centre for International Affairs (CIDOB)
civilization, 298, 299, 300
class, 299
Cleaver, Frances, 26–9 *passim*
climate change, 305, 307, 308
Coalition of Workers, Peasants, and Students of the Isthmus (COCEI) (Mexico), 97, 104, 105
Collier, Paul, 6
Colom, Álvaro, 52
Colombia, 9, 236, 241, 282, 301; civil war, 294
Committee for the Defence of Health and Life (CODESVI) (Venezuela), 124, 129, 133
commodity constructs, 142, 143, 163
commons: tragedy of the commons, 25
community media projects, 242
Community of Latin American and Caribbean States (CELAC), 245, 281, 282, 299
CONAMAQ *see* National Council of Ayllus and Markas of Qullasuyu (CONAMAQ) (Bolivia)
CONAP *see* National Council of

318 | INDEX

Protected Areas (CONAP) (Guatemala)
Copalar dam (Nicaragua), 87
Coronil, Fernando, 272, 279, 285; *The Magical State*, 9
cosmo-politics, 276, 277
Coutinho, Carlos Nelson, 196, 197, 198, 200
creative class, 304, 306
crime, 236, 237, 247
Cuba, 16, 282
culture, 230, 244; cultural policy, 218
Cupples, Julie, 75

dams, 13
De la Cadena, Marisol, 276
De Soto, Hernando, 150
democracy, 7, 301, 302
Desaguadero (Bolivia/Peru), 152, 153
development, 16, 117, 255, 258–61 *passim*, 278, 280, 295
Development Bank of Latin America (CAF), 235
developmentalism, 269, 278, 284
Douglas, Mary, 27
Dudamel, Gustavo, 217, 225, 245

Economic Commission for Latin America and the Caribbean (ECLAC), 257
Ecuador, 9, 11, 261, 278
education, 222; musical, 224
Education for All Development Index (EDI), 220
'El Gran Sueño de Bolívar' refinery (Nicaragua), 71
El Perú heritage site (Guatemala), 48
El Salvador, 301
El Sistema *see* Sistema youth orchestras
El Vergelito (Guatemala), 52
electricity: as a right, 88; theft of, 80, 81
Elias, Norbert, 298, 299, 300, 302, 305; *The Civilising Process*, 297
ENCOVI *see* National Living Conditions Survey (ENCOVI) (Guatemala)
energetic societies, 255, 280
energetic states, 28, 300
energo-politics, 21
energy, 304; consumption, 297; cost of, 15, 75, 76; cultures of energy, 297; distribution, 14; energy conflicts, 18; energy infrastructure, 297; export of, 157; production, 297; renewable, 13, 14, 97, 98, 100, 111, 257
Engels, Friedrich, 305
Espinar (Peru), 148
Esquipulas Agreements (Guatemala), 45
ethanol, 13
ethics of illegality, 158
European Union (EU), 245
excessive presidentialism, 226, 245
exclusion, 18, 19, 218
extractivism, 60, 135; *see also* neo-extractivism
Ezeiza massacre (Argentina), 293

Fabricant, Nicole, 275
Falklands war, 294
Fanon, Frantz, 279, 285
FEMSA drinks company, 100
Fernandes, Sujatha, 241
Florida, Richard, 5, 304, 306
Fonplata *see* River Plate Basin Financial Development Fund (Fonplata)
Foucault, Michel, 19
Friedman-Rudovsky, Jean, 268
FSLN *see* Sandinista National Liberation Front (FSLN) (Nicaragua)
Fujimori, Alberto, 145, 150
Fukuyama, Francis: *The End of History and the Last Man*, 300
Fund for the Economic Development of the Nation Law (FONPETROL) (Guatemala), 54
FYDEP *see* Investment and Development Company of the Petén (FYDEP) (Guatemala)

Gabrielli, Sérgio, 203
Garcia, Alan, 147
García, Anibal, 55
García, Romeo Lucas, 49
García Hernández, Rodolfo, 56, 58
Garcia-Linera, Álvaro, 269
Gardiola-Rivera, Oscar, 258, 280, 283; *What if Latin America Ruled the World?*, 254

gasolinazo, 10, 179, 183–9 *passim*, 267
Gates, Leslie, 227
Gazprom gas company, 267
General Directorate of Cultural Heritage (Guatemala), 48
getulismo, 205
Giddens, Anthony, 27
Gledhill, John, 9, 10, 148
Global Environment Facility, 29
Gonzales, Eduardo, 148
González, Humberto, 245, 246
good faith economy, 223, 224, 246
governance, 292; environmental, 25, 27, 29, institutional, 25
government organized non-governmental organizations (GONGOs), 246
governmentality, 19
Graeber, David, 28, 281, 309
Gramsci, Antonio, 196, 197, 200
Green Battalions (Guatemala), 59
Greenpeace, 208, 212
Grupo Preneal energy company, 100
Guatemala, 11, 40, 43, 44, 53, 61, 301; Agrarian Reform Law (Decree 900), 44; Chamber of Commerce, 51; clientelism, in, 53; consitution of, 43, 45; contract 2-85 (1985), 49, 51, 55, 56, 58, 59, 61; economy, 42; environmental protection in, 45, 47; Hydrocarbons Law (1983), 55; indigenous population, 41; military dictatorship, 43; oil industry, 42, 44, 60; oil royalties, 53; Oil Tax Fund Law (2007), 54; Peace Agreements (1996), 44, 46; Petroleum Code (1955), 44; poverty in, 41
Gudynas, Eduardo, 128, 135, 260, 261, 268
Gustafson, Bret, 267, 275
Guzmán, Jacobo Arbenz, 44

Haarstad, Håvard, 260
Habermas, Jurgen, 301, 302, 304, 305, 309
Hall, Stuart, 276
Harris, Olivia, 160
hegemony, 196
Herrera, Ruth Selma, 83, 85
Hidrogesa hydroelectric plant (Nicaragua), 87
Hispanoil oil company, 49
Hobbes, Thomas, 299, 300, 302, 308, 310; *Leviathan*, 298
Hobsbawm, Eric: *Bandits*, 236
Holloway, Richard, 217, 218
Homo sacer, 19
Hornborg, Alf, 306, 307, 308, 311; *Global Ecology and Unequal Exchange*, 305
Hulme, Mike, 104
Humala, Ollanta, 148
human capital, 242, 246
human rights, 225, 226
Hunt Oil company, 145
hydropower, 13, 257

ICEFI *see* Central American Institute for Fiscal Studies (ICEFI)
ikojts community (Mexico), 107
import-substitution industrialization (ISI), 259
inclusion, 222, 223
inconformes, los see Mareña Renovables, anti-Mareña resistance movement
indigenous communities, 12, 276, 277, 278; indigenous imaginations, 284; International Labour Organization convention 169 on indigenous rights, 12, 276
INE *see* National Statistics Institute (INE) (Guatemala)
inequality, 18, 19, 309
INIA *see* National Institute for Agricultural Research (INIA) (Venezuela)
Initiative for the Integration of the Regional Infrastructure of South America (IIRSA), 281
INSOPESCA *see* Socialist Institute of Fishing and Aquaculture (INSOPESCA) (Venezuela)
institutionalism: critical institutionalism, 26, 27, 28, 29, 97, 117; new institutionalism, 25, 26
Inter-American Development Bank (IDB), 70, 73, 96, 99, 102, 234, 281
International Energy Agency (IEA), 257, 259

International Monetary Fund (IMF), 70, 73, 267
Investment and Development Company of the Petén (FYDEP) (Guatemala), 47
Izabal (Guatemala), 45

Jackson, Jesse, 247
Jameson, Frederic, 308
Jones, Owen: *Chavs – the Demonisation of the Working Class*, 231
Juarez, Benito, 97
Jungle Brigades (Guatemala), 59
Junta de Vecinos de Punta Cardón *see* Punta Cardón Neighbourhood Association (Venezuela)

Kaup, Brent, 266, 267
Kennedy, Nigel, 228,
Kennedy, Paul, 298
King, Martin Luther, 247
King's Peace, 296, 301, 309

La Botija (Venezuela), 121, 122
La Libertad refinery (Guatemala), 50
La Mancha journal, 243
Labour Party (UK), 218
labour theory of value, 217, 305
Laguna del Tigre National Park (Guatemala), 40, 46–52 *passim*, 55, 61; drug trafficking in, 52; forest fires in, 53; Master Plan 2007–2011, 48, 58; militarization of, 59
Latin America, 9, 10, 254, 256, 259, 260, 262, 272, 280–5 *passim*, 291–6 *passim*, 301, 302, 307, 311; and renewable energy, 13; demand for energy, 256, 257, 259; democracy in, 294, 295; energy prices in, 143; environmental struggles in, 270; ethno-nationalism in, 309; far right in, 309; hydrocarbons sector, 256; inequality in, 18, 19; left-wing mobilization, 10, 258, 259; liberalism in, 294; oil dependency, 255; political violence in, 294; transportation in, 257
Latour, Bruno, 17, 276
Lenin, Vladimir Ilyich, 299

Lévi-Strauss, Claude, 27
Levitsky, Steven, 266
liberalism, 295, 301, 302
Libra oil field (Brazil), 213
living well *see buen vivir*
Logan, Owen, 141; *Flammable Societies*, 17, 23
Lorenzana, Waldemar, 52
Lote 88 (Peru), 145
Lula da Silva, Luiz Inácio, 195, 197–206 *passim*, 211, 213
lulismo, 197, 199, 200, 205, 212

machine fetishism, 306, 307
Macquarie Group bank, 100
Madrid, Raúl, 266, 267
Maduro, Nicolás, 117, 128, 243
Malcolm X, 247
Malvinas war *see* Falklands war
mano dura policy, 301
Maraven oil company, 123
March, James, 25
Mareña Renovables, 92–6 *passim*, 100, 101, 104, 111, 112; anti-Mareña resistance movement, 103–9 *passim*, 112; effect on fishing, 102, 109, 110; occupation of communal land, 102; use of bribes, 101; use of violence, 101
Marx, Karl, 218, 279, 285, 304, 305, 306; *The 18th Brumaire of Louis Bonaparte*, 199
material culture, 16
Maya Biosphere Reserve (MBR) (Guatemala), 40, 45, 46, 47, 48–9, 50, 51, 55
McCarthyism, 303
McGuigan, Claire, 76
McMichael, Philip, 17
McNeish, John-Andrew, 141; *Flammable Societies*, 17, 23
media: community-run, 241
Medina, Franca, 122
Medinaceli, Mauricio, 181
mega-projects, 111
Mendez, Charles, 238, 239, 241
Meneses, Oscar Sotillo, 243
Mensalão scandal (Brazil), 198, 203
Mercado Nuevo (Peru), 143

INDEX | 321

Mercado Oriental (Nicaragua), 75
meritocracy, 231, 232, 242; see also Sistema youth orchestras, as meritocracy
Mexico, 9, 10, 97, 282, 301; and renewable energy, 96, 99; *autoabastecimiento* policy, 99, 100; indigenous movements, 107; oil industry, 98
Mignolo, Walter, 284
Mitchell, Timothy, 21, 142, 255; *Carbon Democracy*, 7
Mitsubishi Corporation, 100
modernity, 16, 17
Mokrani, Leila, 181
money fetishism, 307
Monroe Doctrine, 282
Montero, Alfred, 200
Morales, Evo, 162, 183–9 *passim*, 258, 263, 265, 266, 269, 270, 273, 274
Mother Earth, 268; see also pachamama
Movement for Socialism (MAS) (Bolivia), 173, 184, 185, 186, 188, 266, 269
Movimiento Comunal (Nicaragua), 86
Movimiento de Renovación Sandinista (Nicaragua), 69
music, 244; classical, 228

National Consumer Defence Network (RNDC) (Nicaragua), 70, 77, 83, 84, 85, 86
National Council of Ayllus and Markas of Qullasuyu (CONAMAQ) (Bolivia), 264
National Council of Protected Areas (CONAP) (Guatemala), 45, 47, 48, 50, 51, 55, 58, 59; Monitoring and Evaluation Centre (CEMEC), 51
National Institute for Agricultural Research (INIA) (Venezuela), 129, 130, 132
National Living Conditions Survey (ENCOVI) (Guatemala), 41
National Petroleum Institute (Guatemala), 44
National Statistics Institute (INE) (Guatemala), 75, 77, 78
nationalism: resource nationalism, 259, 262, 269

Negarestani, Reza, 99
neo-extractivism, 260, 261, 291; see also extractivism
neoliberalism, 295, 296; green neoliberalism, 112, 113
Nicaragua, 14, 16, 87; and renewable energy, 69, 71; blackouts in, 75; Chamber of Commerce, 75; cost of energy, 75, 76; electricity distribution in, 69, 70, 74; electricity sector, 70, 73; Energy Stability Law (2005), 76; Ministry of Energy and Mines, 66, 78; national sovereignty in, 87; non-governmental organizations in, 82, 86; popular protest in, 83; privatization of utilities, 66, 73, 74, 75, 84; structural adjustment programme, 70; support from Venezuela, 85
non-governmental organizations (NGOs), 246; NGOism, 246
North American Free Trade Agreement (NAFTA), 282
Nueva Esperanza (Guatemala), 53

O petróleo tem que ser nosso campaign, 201–4 *passim*
Occidentalism, 279, 280, 284, 285
October Revolution (1944–54) (Guatemala), 43
oil dependency, 7, 142, 173, 174, 292
Oil Fund (Guatemala), 53
oil industry: investment in, 259, 260
Olivas, David, 66
Olsen, Johan, 25
Operation Condor (Argentina), 293
Organization for Economic Co-operation and Development (OECD), 257
Organization of American States (OAS), 265, 282
Orientalism, 279
Organization of the Petroleum Exporting Countries (OPEC), 260
Ortega, Daniel, 68, 69, 71, 73, 82, 83, 84, 89, 282, 304
Oslender, Ulrich, 41, 61
Ostrom, Elinor, 25

pachamama, 159, 160; *see also* Mother Earth
Paraguaná Refining Complex (Venezuela), 118, 126, 135
Parsons, Talcott, 301–5 *passim*, 309, 310; 'On the concept of political power', 302
Partido de la Revolución Democrática (PRD) (Mexico), 102
Partido dos Trabalhadores (PT) *see* Workers' Party (PT) (Brazil)
Partido Revolucionario Institucional (PRI) (Mexico), 101, 102
passive revolution, 197, 200, 211, 213
peace, 300, 301
Peace Brigades International (PBI), 50, 51
peak oil, 98
Pemex oil company, 10, 98
Perenco oil company, 40, 49–55 *passim*, 58, 59, 61
Pérez, Carlos Andrés, 233
Perón, Juan Domingo, 293
Peru, 10, 12, 145, 163, 301; border with Bolivia, 151; cost of energy, 146, 147; free trade zones in, 150; kinship in, 155, 158, 159, 160; Trade Promotion Agreement with USA, 145, 163; violence against protestors, 148, 149
Petare (Venezuela), 238, 241
Petare TV, 238, 239, 240, 242
Petén (Guatemala), 45, 46, 48
Petro-Andina oil company, 264
Petrobras oil company, 10, 202, 203, 204, 210, 211; founding of, 201; oil leakages by, 209
Petroleos de Venezuela SA (PDVSA) oil company, 10, 71, 116, 123, 124, 127, 129, 132–5 *passim*
PETRONIC oil company, 71
PetroPeru oil company, 145
PGGM pension fund, 100
Philip III, king, 298
Pieterse, Jan Nederveen, 17
Pietri, Arturo Uslar, 232
Piketty, Thomas, 309, 310; *Capital in the 21st Century*, 296
Pinelo, José Enrique, 186
pluralism: legal pluralism, 12

pluri-verse, 276, 279
Pluspetrol oil company, 145
policy, public, 170, 171, 172
Polyani, Karl, 280
Poma de Ayala, Felipe Guaman, 298
poverty, 219
power, 302, 303, 304; 'popular power' discourse, 128; socialization of, 285;
pre-salt oil deposits *see* Brazil, *pré-sal* oil fields
prior consent: principle of free and prior informed consent (FPIC), 12
Propetén conservation group, 51
Puno (Peru), 154, 161, 163
Punta Cardón (Venezuela), 116–26 *passim*, 134, 135, 136; brothel in, 121; community organizations in, 133; environmental impacts of refinery, 121, 124, 125, 127; fishing industry, 118, 123, 126, 129, 130, 131, 132; health problems in, 124, 125, 127; oil production in, 116, 123; public services in, 126
Punta Cardón Neighbourhood Association (Venezuela), 124

Quijano, Aníbal, 285
Quintana, Juan Ramón, 187

racism: class racism, 231, 237
Radio Bemba (Venezuela), 241
Ramsar Convention (1971), 46
Rappaport, Joanne, 278
renewable energy *see* energy, renewable
Repsol oil company, 145, 264, 267
resilience, 305
resource curse, 5, 6, 24, 27, 29, 141, 292
Rio de Janeiro (Brazil), 205, 206, 207, 211, 213
Rio+20 summit (2012), 22
River Plate Basin Financial Development Fund (Fonplata), 281
RNDC *see* National Consumer Defence Network (RNDC) (Nicaragua)
Roberts, Kenneth, 266
Robinson, William, 272
Roitman, Janet, 158
Romero, Ricardo, 238

Rousseff, Dilma, 195, 197, 201, 207, 213
Rydin, Yvonne, 25

San Andrés (Guatemala), 46
San Dionisio del Mar (Mexico), 94, 101, 107
Sánchez de Lozada, Gonzalo, 185
Sandinista National Liberation Front (FSLN) (Nicaragua), 67, 68, 69, 82, 83, 84
Schiller, Naomi, 241
Schumpeter, Joseph, 280
Scotland, 217, 222, 245; 2014 independence referendum, 223, oil revenues, 223
Scott, James, 26, 28; *Seeing Like a State*, 24, 297
Scouting movement, 244
Sen, Amartya, 225
Shell oil company, 120, 121, 123
Simón Bolívar University (USB) (Venezuela), 129
Simón Bolívar Youth Symphony Orchestra, 245
SINDIPETRO oil workers' union, 202
Singer, André Vitor, 197, 199, 200
Sistema youth orchestras, 217, 219–31 *passim*, 235, 237, 239, 241, 243, 244, 246; as meritocracy, 226, 228, 229; cost–benefit analysis of, 234; effects on crime, 224, 225, 229, 230, 232; *El Sistema* documentary, 230
smuggling, 10, 15, 141, 142, 150–9 *passim*, 162, 164, 179, 183, 187; counter measures, 162; role of women, 150, 153
socialism, 280
Socialist Institute of Fishing and Aquaculture (INSOPESCA) (Venezuela), 129–32 *passim*
sociology, 168, 169, 170, 190
Solón, Pablo, 271
sovereignty, 142, 143, 158, 163; resource sovereignty, 4, 5, 8, 23, 27, 40, 88, 112, 211, 222, 223, 263, 270, 277, 279, 280, 282, 292, 302, 303, 304, 309
Starn, Orin, 276
stateness, 16
states of exception, 19

Stenger, Isabelle, 276
Stepputat, Finn, 8
Strauss, Sarah, 21
sumaq qamana see buen vivir

Taiwan: investment in Nicaragua, 71
Tarrow, Sidney, 196
technocracy, 24, 87, 117, 167
technology, 306, 307
Tehuantepec peninsula (Mexico), 14
Teresa Carreño theatre (Venezuela), 227
Territorio Indígena y Parque Nacional Isiboro Secure (TIPNIS) (Bolivia), 263, 264; protests in, 10, 11, 263–71 *passim*, 275, 276, 277, 279
Texaco oil company, 49
Therborn, Göran, 197
Tiv community (Nigeria), 307
Todd, Cecilia, 224, 227
Total, 267
Tsing, Anne, 18

Union de Comunidades Indígenas de la Zona Norte del Istmo (UCIZONI) (Mexico), 108 Unión Fenosa energy company, 70, 73, 74, 79, 80, 82, 84, 87; and over-charging, 76, 77, 78; complaints procedure, 77, 78; protests against, 72, 83
United Kingdom (UK), 231; education in, 221
United Nations Children's Fund (UNICEF), 41
United Nations Clean Development Mechanism, 99
United Nations Environmental Programme, 29
United Nations Security Council, 298
United States Agency for International Development (USAID), 271
United States of America (USA), 15, 294, 303; blockade of Nicaragua, 68; identity politics, 247; relations with Guatemala, 44
University of San Carlos (Guatemala), 47
US dollar, 303

Vaides, Federico Ponce, 43

value, 17
Vargas, Getúlio, 199, 201, 204–5, 211
Véliz, Sergio, 58, 59
Venezuela, 7, 9, 10, 11, 15, 116, 117, 128, 217, 219–20, 222, 226, 245, 246, 258, 282; and renewable energy, 13; community media in, 241; corruption in, 238; crime, 230, 232, 238, 241; education, 221, 234; fishing industry, 128, 129; foreign policy, 15; investment in Nicaragua, 71, 85; oil industry in, 123; Plan de la Patria 2013–2019, 135
Víctores, Humberto Mejía, 49
violence, 19, 20, 236, 237; political, 301; state violence, 19, 298, 300
Virgin of Copacabana, 159
Voltaire, 228

Waka heritage site (Guatemala) *see* El Perú heritage site (Guatemala)
Walsh, Catherine, 278
Washington Consensus, 185
Webber, Jeffery, 269
Weber, Max, 19
Wikileaks, 271
Wilhite, Harold Langford, 21
Wilson, William, 247; *The Truly Disadvantaged*, 238
wind power, 14
Wisnik, José Miguel, 246; *O Som e o Sentido*, 244
Workers' Party (PT) (Brazil), 198, 202, 203, 211
World Bank, 70; Multilateral Investment Guarantee Agency, 84
World Commission on Dams (WCD), 13
World People's Conference on Climate Change (2010), 268

Xan oil well (Guatemala), 40, 49, 50, 55
Xtrata mining company, 148

Yacimientos Petrolíferos Fiscales Bolivianos (YPFB), 176, 184, 273
Young, Michael, 232; *The Rise of the Meritocracy*, 231

Zapatista movement (Mexico), 105, 106, 112
zones of awkward engagement, 18